Hollywood Chemistry

When Science Met Entertainment

ACS SYMPOSIUM SERIES **1139**

Hollywood Chemistry

When Science Met Entertainment

Donna J. Nelson, Editor
University of Oklahoma
Norman, Oklahoma

Kevin R. Grazier, Editor
Planetary Scientist/Writer
Sylmar, California

Jaime Paglia, Editor
Writer/Producer
Venice, California

Sidney Perkowitz, Editor
Emory University
Atlanta, Georgia

American Chemical Society, Washington, DC

Distributed in print by Oxford University Press

Library of Congress Cataloging-in-Publication Data

Hollywood chemistry : when science met entertainment / Donna J. Nelson, editor, University of Oklahoma, Norman, Oklahoma Kevin R. Grazier, editor, planetary scientist/writer, Sylmar, California, Jaime Paglia, editor, writer/producer, Venice, California, Sidney Perkowitz, editor, Emory University, Atlanta, Georgia.
pages cm. -- (ACS symposium series ; 1139)
Includes bibliographical references and index.
ISBN 978-0-8412-2824-5 (alk. paper)
1. Science in popular culture. I. Nelson, Donna J., editor of compilation. II. Grazier, Kevin R. (Robert), 1961- editor of compilation. III. Paglia, Jaime, editor of compilation. IV. Perkowitz, Sidney, editor of compilation.
Q172.5.P65H65 2013
791.43′66--dc23
2013027668

The paper used in this publication meets the minimum requirements of American National Standard for Information Sciences—Permanence of Paper for Printed Library Materials, ANSI Z39.48n1984.

Distributed in print by Oxford University Press

PRINTED IN THE UNITED STATES OF AMERICA

Foreword

The ACS Symposium Series was first published in 1974 to provide a mechanism for publishing symposia quickly in book form. The purpose of the series is to publish timely, comprehensive books developed from the ACS sponsored symposia based on current scientific research. Occasionally, books are developed from symposia sponsored by other organizations when the topic is of keen interest to the chemistry audience.

Before agreeing to publish a book, the proposed table of contents is reviewed for appropriate and comprehensive coverage and for interest to the audience. Some papers may be excluded to better focus the book; others may be added to provide comprehensiveness. When appropriate, overview or introductory chapters are added. Drafts of chapters are peer-reviewed prior to final acceptance or rejection, and manuscripts are prepared in camera-ready format.

As a rule, only original research papers and original review papers are included in the volumes. Verbatim reproductions of previous published papers are not accepted.

ACS Books Department

Contents

Science Reaches the Public

Indexes

Foreword

How I Learned To Stop Worrying and Love the Story

I used to be kind of a jerk.

This is not an easy admission to make, but science is about—above all else—honest and open assessment, even of ourselves. Especially if it's painful. Those truths we hold most dear are the ones we must examine most critically. I am a scientist, so I have to follow this path no matter where it leads.

My jerkdom wasn't by nature, nurture, or by choice. Not immediately, at least. It grew over time, and in the interest of that honesty and openness, I'll readily admit I was to blame, but I had help. Hollywood was my enabler.

It had always been there for me. *Ghost Host* on Friday nights, and *Creature Feature* on Saturday nights: Weekends were made for science fiction movie marathons on local TV.

Oh, I watched them all. From every black and white, rubber-suited reanimated dinosaur terrorizing Tokyo to the color Hammer horror flicks with budgets so small you could hardly see them on my family's 18-inch TV set. I was addicted to the television. I dreamed of doing what the protagonists in those flicks did: fly a starship across the galaxy, meeting aliens, fighting those with nefarious intent, exploring strange, new worlds. I couldn't get enough, and I loved those movies unconditionally.

At least, at first. Because I was a budding scientist, it was not a blind love. I began to see the flaws in those movies, and over time was all too willing to point them out to friends, or anybody else who would listen. "That's so fakey!" I would cry out loud when a spaceship on strings sputtered past, or a hobbling actor in a slimy alien costume stalked Our Heroes.

Eventually, my verbal exclamations evolved into written ones. That was at the same time I became a professional scientist, researching the real Universe for a living. I never got to find another planet or meet aliens, but it was my hope that maybe my work made it a little easier for others to do so.

Even if the real Universe didn't allow me that privilege, that didn't stop me from plunging in to the fictional ones crafted by others. I still loved watching movies, but now I went to the local theater with friends and we would dissect the films afterward, sometimes going over their scientific trespasses late into the night. We'd feed off of each other's comments, laughing and escalating the snark until we got downright vicious.

Yet we'd always go back to the theater.

Eventually, the Internet came along. Not long after creating my first website (back in the Dark Ages of 1998), I decided it would be fun to continue this grand tradition of critiquing the science of cinema, and I dove in with both glee and fervor. No movie was safe: from *Armageddon* to *Austin Powers*.

I was right; it *was* fun. It was surprisingly easy to deconstruct Hollywood accuracy, or lack thereof. Any mistake was fair game; a flubbed line with bad math was just as likely for me to mock as a plot device upon which the entire movie turned. Blowing up a giant asteroid? *Pshaw*. Saying "million" instead of "billion"? *Please*. Shadows moving the wrong way at sunset? *Let me sharpen my poison keyboard.*

Movie after movie came and went, and I watched each in the darkened theater, off to the side, hunched over my notepad with my pen clicked and ready, and—literally—a flexible red-filtered flashlight (an old astronomy trick to keep the eyes dark-adapted) wrapped around my neck like a scarf to illuminate my writing in case the scene I was destroying was too dark for me to see my own words.

Within hours an update to my site was ready for others to read. I tore apart movie after movie, savaging them for any scientific slight I perceived. Astronomy was my specialty, but any field of science was fair game.

Then, one day, I had an epiphany. Well, actually, the epiphany was forced upon me. I was at a science meeting, a gathering of thousands of astronomers to present and share research. Taking a break between sessions, I was touring the rather expansive exhibit hall, looking over posters of research in progress, chatting with old friends and colleagues as we came across one another, and generally enjoying myself. One section of the hall was set up for professional telescope manufacturers, and there was some fairly large and sophisticated equipment on display. I stopped at one in particular to admire, and the gentleman who worked for the manufacturer came over to talk to me. We talked about the telescope for a while, and he told me he helped design quite a bit of the electronics that helped steer and point the lens.

He got a gleam in his eye, and he asked me, "Did you ever see the made-for-TV movie *Asteroid*?"

I should have thought this through before responding. Sadly, though, (as you may recall) I was a jerk.

I told him I *had* seen it, and I lit into it. I told him how the science was *awful*, and how the way they depicted asteroids was laughable. Nodding toward the telescope, I told him how badly the movie portrayed the observatory used by astronomers to track the killer rock. They even observed the asteroid during the daytime!

I laughed, expecting him to join in.

He didn't. He had borne my diatribe in silence, waiting for me to finish. When my laughter died off, he spoke up.

"Did you get a good look at that telescope in the movie?" he asked me. Still not realizing what I had walked into, I nodded. "Sure," I replied. "It was actually a fairly accurate depiction of what a 'scope in an observatory looked like. Really, it was one of the few accurate things in the whole flick."

"I helped build it," he told me. "The studio called me and asked me to work with them on that part of the set. I put one together that was pretty much the way it would be in a real observatory, but after the director looked at it, he decided he wanted it to look more complicated, to make it seem more like what the audience expects for a piece of scientific equipment." At this point he fixed me with his stare, leaning in just a bit closer to me. "So I added a bunch of electronic boxes and wires and other stuff that had no real purpose at all. They were just there for show, to make it more exciting."

He paused pointedly, then continued. "You thought it was real, didn't you? But it wasn't. If you were fooled by that, and even thought it looked good, then why do you care if there are other little mistakes in a movie?"

What he said cut right through me, and there was no way I could avoid the truth: *He was right.*

It was a sea change, a pivotal moment for me. If I had seen our conversation in a movie, ironically, I would've scoffed at such hackneyed writing. But it was real, as was the absolute certainty that I had been a colossal ass.

I had been watching movies for all the wrong reasons. I was doing it solely for the purpose of reviewing and eviscerating them, and *no longer for the purpose of simply enjoying them.* Amazingly, it was as if a weight was lifted from my life and, in that moment, it all changed. I accepted that while the science is important in science fiction, the story *must* come first. Don't get me wrong: I'd prefer the science be accurate. In fact, I **strongly** believe that a writer who knows the science (or has access to it through a science consultant) will find plot developments he or she may not have considered otherwise.

Science can, and should, lead the story where it needs to go. In the end, though, it is the telling of the story that must win out, even if there is a scientific stumble or two along the way. After all, look where I was standing: at a meeting of astronomy professionals, and I was there because I was one of them. And I was one of them in large part due to the inspiration all of those movies, good and bad, that I had once so gleefully torn apart.

See? I *was* a jerk, but emphasis on the *was*. I had faced my own flaw, and had found redemption. My own story has character development, as any good story should.

I still review movies when the opportunity arises, and I still point out the flaws. Now though, when I do so, I make sure it's in the context of helping others understand the actual science behind the moment, showing them the joy and wonder of how reality really is. I used to do it to take the movie down a peg; now I do it to allow that movie to inspire others the way I was inspired. As I continue to be.

How can I not? The introduction of bad science after good led to my own character development. What might good science after bad do for yours?

Phil Plait
Blogger for Slate Magazine

Preface

In the 1989 romantic comedy *When Harry Met Sally*, a chemical reaction between two initially immiscible people eventually results in a wonderful Hollywood equilibrium, but only after several misfires. Over the decades, the interactions between science and Hollywood have proven every bit as unpredictable and volatile as the journey taken by Harry and Sally. Sure, there have been fleeting encounters, but on those occasions the two never produced a strong reaction, blending more like oil and water: sometimes parting on unfriendly terms, sometimes parting as friends, sometimes simply parting. The relationship never crystallized into something solid until very recently.

Hollywood and science have found each other again, and seem to have formed the strongest bond to date. The increasing use of science consultants in science fiction and science-themed productions, from comedies like *The Big Bang Theory* to dramas like *Breaking Bad*, and the creation by the National Academy of Sciences of the Science and Entertainment Exchange—an organization founded to connect scientists and Hollywood productions—suggests a new level of dedication to science and scientific fidelity that can only benefit both sides of the equation.

What finally catalyzed this reaction? This eclectic collection of essays examines the connections between Hollywood and science, with a primary focus on the current state of the relationship. It features contributions from screenwriters, producers, directors, scientists, science advisors, science writers, even a music composer and a dramaturg. The formats of the chapters contained herein are equally eclectic: some take the form of academic journal articles, some are written as less formal interviews and some are narratives. Similar to a romantic comedy, the tones of the offerings range from the purely serious to the comedic.

In the first act of this production, "Science Reaches the Screen," contributors explore the various approaches that different television series and movies employ to incorporate accurate science into their productions. In other instances, authors explore the more fundamental aspects of the science—like sound, music and light—that enable audience appreciation of television and film.

For the second act, entitled "Science Reaches the Public," authors explore the effects that science in television and film has on the viewing public. Some authors explain the science, both explicit and implied, that can be found within various Hollywood productions—and explore those moments where, as happened between Harry and Sally, Hollywood and science fail to click instead of meshing. Other authors examine the influence that Hollywood science has on the science community, public policy and the legal system. Still others describe pedagogical applications of television and movie science to education—as well as Hollywood's role in motivating future generations of scientists and engineers.

Like Harry and Sally, it took science and Hollywood a few tries to discover their mutual chemistry. Only time will tell if the bond is ephemeral, or one of substance. Based on the enclosed collection of essays, it would seem that this mercurial couple has a solid future indeed.

Donna J. Nelson
Department of Chemistry
University of Oklahoma
Norman, Oklahoma 73019

Kevin R. Grazier
Science Advisor
SyFy Channel and Turner Network Television
Sylmar, California 91342

Jaime Paglia
Co-Creator/Writer/Executive Producer of Syfy Channel's Eureka
Venice, California 90291

Sidney Perkowitz
Department of Physics
Emory University
Atlanta, Georgia 30322

Science Reaches the Screen

Chapter 1

"Creation": When Art and Science Collide

Natalia Reagan and Jon Amiel*
***E-mail: jon@jonamiel.com**

Science and art are oftentimes presented as two distinct fields with very little overlap. But the anatomist and the artist, scientist and storyteller, have far more in common than meets the eye. Upon further investigation, it is clear that science and art cannot exist without the other; they are co-constitutive. Scientists must think creatively in order to form theories (the Earth is round, it circles the Sun) that often appear outlandish not only to laymen but even to their own colleagues. The theory of evolution, presented by Charles Darwin in "*On The Origin of Species*," required creativity that was then supported by the evidence-revealed empirical analysis. Essentially, it took Darwin's right brain to creatively form the question and his left brain to try and refute the question using the scientific method. Similarly, art requires the brain to function in ways that would not be possible without science. For instance, writing music is highly reliant on mathematics, sculpting the human figure requires a basic understanding of our anatomy and physiology, and the technology used in cinema would be not be possible without the leaps made by scientists and engineers during the last century. In this chapter, writer/director Jon Amiel discusses his film "Creation" with anthropologist Natalia Reagan. Their aim—to explore just how deeply science and art are intertwined within the process and the product of film-making.

Jon Amiel's Preamble

What follows this introduction is the (mercifully!) condensed account of several conversations I had with Natalia Reagan over a six-month segment of this year, from spring through autumn.

Initially conceived of as a straightforward interview in which she'd ask and I'd talk about my film of Charles Darwin's life, *Creation*, we found instead that the movie became the spindle of a pin-wheeling conversation about the convergences of art and science, fact and fiction, scientific academia and the Academy of Motion Pictures Arts and Sciences.

Our backgrounds could hardly be more different. Natalia is a burgeoning force of nature who's currently breaking down the barriers between scientific enquiry and populist entertainment. She's that rarest of poly-hyphenates, an anthropologist/primatologist/actor/comedienne/writer. I myself am a more humdrum species—the common or garden feature film director. I am however one who's been lucky enough to have a career spanning thirty or so years of theater, film and television from the *Singing Detective*, through *Copycat*, *Sommersby*, *The Core*, *Creation*, and *Entrapment*.

What set us on this collision course was our own set of convergences—she's a scientist with a passionate involvement in the expressive arts, I'm an artist with a lifelong love of science. Our conversations revealed to us connections that spanned gender, generation and background and gave us both deeper insights into the shared DNA of the creative and the scientific mind.

If this account gives you a fraction of the pleasure it gave us, we shall consider our time well spent.

Natalia Reagan's Preamble

The Art of Creating

pro·cre·ate *verb* \-ˌāt\
transitive verb
: to beget or bring forth (offspring) : propagate

Science and art may appear to be two disparate fields that share few connections to one another, however this could not be further from the truth. Science and art, on the contrary, are two entangled fields that are integral to one another. The film *Creation* is an excellent example of this hypothesis, that science and art are co-constituted. In the film, Charles Darwin wrestles with his internal discourse about whether he should publish his groundbreaking scientific findings at the risk of outraging those with strict religious convictions—including his wife. *Creation* is a beautiful piece of art that richly depicts the unraveling of a man before he published his prolific *On the Origin of Species*.

Similarly, the production of a film is much like a living and breathing organism—its biology is not dissimilar to that of a siphonophore. Like the Portuguese Man of War, a film is a colonial organism made up of many zooid components, writers, cinematographers, cast and crew. Unlike the Man of War however, a film must have a "brain"—one central prism through which the potentially divergent creative energies of this multi-cellular organism are focused. For the profoundly beautiful feature film *Creation,* director Jon Amiel was that prism. Amiel puts it slightly differently: "I view myself as the conductor. Not just in the musical sense—though I do certainly conduct the orchestra that's often playing someone else's composition—but also in the electrical sense. I serve as the lightning rod to conduct the collective energies of hundreds of people into a single channeling purpose."

If God breathed life into the clay that was Adam, it was Amiel who breathed life into the man who turned the scientific community on its ear. As Charles Darwin took great care to present his findings, Amiel was equally as mindful—and even reluctant to—direct a film about arguably the most influential man in the past two-hundred-years.

I had the honor of sitting down with Amiel—director, writer, and fiercely brilliant interlocutor—whose great love of science and respect for the art of creation is clear in his words and his work.

Science and Art Collide...Over Brunch

Natalia Reagan: Just as Darwin was influenced by his peers and predecessors like Lamarck and Linnaeus, who influenced you most as you were growing up?

Jon Amiel: Everybody has someone outside of their parents that is transformative in their lives. He wasn't exactly Lamarck or Linnaeus but he was my mad uncle David. Uncle David was a fitness instructor in the army and a pugilist in life's great boxing ring. He swore like a trooper, had an unquenchable thirst for knowledge, and would encourage all forms of bad behavior in his nephew. He was an autodidact who loved fact and despised fiction. When other kids were getting comic books and (later) thumbed copies of *Playboy* from their Bad Uncles—mine would thrust copies of *Scientific American* and *New Scientist* at me. I'd feverishly wade through densely incomprehensible articles in the hope of grasping a few key concepts that I could wield in a discussion with him. Paradoxically, my Uncle David not only gave me an abiding affection for cantankerous people who need to engage through intellectual combat—he also fostered in me a fascination for all things scientific.

NR: Uncle David sounds like a man after my own heart. Isn't it funny that some of the most cantankerous people in our lives so often have soft underbellies beneath those curmudgeonly carapaces?

JA: Indeed. And what about your upbringing led you to being this unique poly-hyphenate? Isn't your mother a professor?

NR: Yes, and actually my parents are a great example of art and science colliding! My mother is a professor with a PhD in psychology, my father is an art director who is also a very talented musician and artist. They were a great case of opposites attracting! It was a fun household to grow up in—lots of musical instruments and books. They both instilled in me a love of scientific inquiry and the arts—I feel fortunate to have had that in my life. National Geographic, the late great Omni magazine and Architectural Digest were magazines I would thumb through as a kid. Because of Nat Geo, really, I became quite fascinated with primates—nonhuman primates to be exact. Well, Nat Geo and recurring King Kong nightmares!

JA: Really? Well that's a terrifying mix.

NR: Yes it was! But even though King Kong was the antagonist in my dreams I still was drawn to him and gorillas. It's interesting how such a film—a piece of art—created a burning love of primates and science.

JA: Yes, it's fascinating how many scientists acknowledge a movie as the germinating factor in their interest in their field. In the work I've done for the Science & Entertainment Exchange, I've met a number of eminent scientists who have talked about films as diverse as *Star Trek, Fantastic Voyage, Contact* and even *The Core*—my attempt to bring a little science to a geophysical disaster movie—as inspirations.

NR: So, I'm curious—given that *Creation* deals with Darwin's conflict between his scientific insights and his religious beliefs—were you raised in a religious household?

JA: Hardly! My parents were lifelong members of the Communist Party. I was raised a devout atheist and have so continued to be. To be honest with you, I've never been deeply exercised by whether God exists. It's always felt to me axiomatic that man created God in his own image. Unlike Darwin, I've led a life relatively untroubled by doubts on that topic.

NR: So if not his struggles with the idea of an omnipotent, omniscient deity, what then was it that drew you to Darwin's story?

JA: Initially—nothing! When my friend John Collee called me from Australia he told me he'd been sent Randal Keynes' book on Darwin for adaptation to film and asked if I wanted to work with him on it. I said: "Absolutely not!"

NR: Because you didn't want to do a biopic about the most prolific scientific mind of the 19th century?

JA: Aaah the dread B-word! Exactly! There's a reason why most so-called biopics are poor movies. A great life doth not a great movie make. Chronology is not plot. Plot is a Newton's Cradle of consequences, one event caroming onto the other and setting it in motion. A life is usually a series of events loosely ordered by chronology—what I call an "*and then, and then*" narrative. Take away the epic spectacle of *Gandhi* for instance and you have a series of incidents loosely yoked together by history and glued together by a great performance. Not a great film.

Another problem with the biopic is the reverence trap. Reverence is the worst attitude with which to embark on an examination of someone else's life. Respect them, sure. Love them, absolutely. But never revere them! Biographical drama at its best is the unflinching examination not only of the facts but the emotional shadow lines of somebody's life. The greater the life, the more beloved the Great Man or Woman, the more difficult the biographer's act becomes.

If the subject has any living descendants who are likely to be offended by a warts-and-all account, the job becomes still more difficult. If you're treating an intellectual giant like Darwin, you're certain to stir a hornet's nest of fervent ideologues. And God forbid you should tackle a religious figure like Mohammed or Jesus! Martin Scorsese's *Last Temptation of Christ* was savaged by the zealots; yet his *Raging Bull* totally transcends the biopic genre for two reasons—firstly, he finds the dramatic spine of the story—the relationship of the two brothers—secondly, he's completely unencumbered by reverence!

And that was only the first reason I said no to my friend John Collee! The most important was that I had no real emotional connection to Darwin. To me, he was as remotely forbidding as one of those statues on Mount Rushmore. The scale of those characters is so massively dwarfing of us mere mortals it's impossible to imagine any emotional connection with them.

NR: They certainly create a buffer…some distance…

JA: Yes. And then there was the age thing. It seemed to me that Darwin was *old.* He must surely *always* have been old. The only image I had of him was the one most of us have—it's the one which graces the back of our British ten-pound note: a prematurely aged man with a forbidding Victorian beard and haunted hollow eyes. The eyebrows seemed to have grown over the eyes, throwing them into even deeper protective shadow, like bushes concealing the mouth of a cave…Imagine trying to get an actor to emote from behind all that latex and prosthetic hair… Impossible!

NR: All right, so I'll ask the obvious question! What changed your mind?

JA: Collee persuaded med to at least read the first 100 pages of the book—it was called *Annie's Box* and Randal Keynes had an inside track on his subject—he's a great, great grandson of the Great Man. After a lot of material contextualizing the intellectual landscape of Darwin's times, I came to the core segment of the book. It describes through letters, journal entries and contemporary accounts the gradual and unbearably sad death of Darwin's favorite daughter, Annie. She was ten.

Fortunately the Victorians were relentless chroniclers, journal-keepers and letter-writers. After his death, all of the Darwin archive was zealously preserved by Emma and Darwin's second favorite daughter, Ettie. Randal Keynes was going through an old chest of drawers that had belonged to his great aunt when he discovered a small wooden box hidden away in the back of one of the drawers. It turned out to be Annie's little reliquary of special treasures—a small wooden time-capsule of memories put away by her grieving parents after her death.

Among the treasures was a piece of paper in Darwin's unmistakable handwriting, minutely listing the medications, the dates given and their effects. It became suddenly obvious to Randal that Darwin had been much more intimately involved with his daughter than had been previously imagined. Increasingly, he began to explore the effect Annie's death had on Darwin's creative journey towards writing *Origin*...

Reading the letters that Emma and Charles wrote to each other during that time was like listening to the voices of any contemporary parent going through the unthinkable horror of losing a child. They reached across the intervening void of a hundred and fifty years and utterly gripped my heart. Suddenly I found that a picture was emerging, of a person completely different from the one I had imagined. Someone to whom I would not only relate but could connect to with a force that quite surprised me. Someone, astonishingly, I even came to love.

NR: It sounds like this book- this piece of art, if you will—made you fall in love with this relatable character who just happens to be father of evolutionary science. It's amazing how these figures of history seems so foreign until you imagine them as the doting and goofy father and loving husband. And seeing their flaws, whether they are self-doubt, greed, or fear, also allows these historical figures to seem relatable. Darwin, in particular, came to life in the film through the portrayal of his comfortable and playful relationship with his children. I feel like your own life may have influenced the focus on Darwin's role of a father in *Creation*. As an anthropologist I enjoyed this aspect since reproductive success, whether we are aware of it or not, is the goal of individuals. Speaking of reproductive success, you told me earlier you had four kids. So you started to connect to him first as a father?

JA: Yes. Paul Zak has made a very interesting study (funded by DARPA slightly worryingly) of the neuro-chemistry of storytelling...

NR: His TED talk focused on the discovery of the moral molecule, right? And he discussed how women were nicer than men—he attributed it to all that extra oxytocin women naturally have!

JA: Exactly. And testosterone inhibits the production of oxytocin, the caring sharing hormone. It's not our fault we're crass insensitive creatures compared to you gals!! It's our hormones! In his work on storytelling, Zak shows how the story of a dad going shopping with his daughter produces no change in our blood chemistry. Tell that same individual the story of a father who knows his daughter is going to die and fights hard against loving her because the pain of losing her will be too great... The listener shows elevated levels of cortisol (emotional stress and attention) and oxytocin (bonding and empathy). People who came away from that story were much more likely to give generously to charity because they were awash with oxytocin.

NR: So I see where you're going with this. It was the relatability of Keyne's account that released all that cortisol and oxytocin in you and made you much more charitable to poor Chuck Darwin?

JA: My point exactly. And in telling this story on film I'm seeking to generate the same chemical responses in my audience. Darwin was an exceptionally contemporary kind of dad. The opposite of the stern Victorian father, he was intensely playful with his kids, who seemed to be allowed to run around in happy Bohemian disarray. When the kids toddled into his study, Darwin was always delighted for the distraction, happy to peer up from the microscope, sit them on a knee, and involve them in his experiments. At one point he had all his kids chasing bumblebees round the Sandwalk and trying to figure out how they marked their territories.

Annie was his special love, his oldest daughter, the one he shared most with, his hope for his old age. That she, the most robustly energetic and loving of his kids, should be the one to die, was unbearable. So, after all my initial reluctance I found myself completely drawn into this man at this very specific point in his life. Not so much his adolescence, nor so much his old age... But at this moment—this very specific juncture in his life, the years leading up to the publication in 1851 of his masterwork. He's a man in his early forties. He's a doting father who has a dying child, he's married to a woman who's his childhood sweetheart and best friend but who also holds religious views deeply and diametrically opposed to his, he's on the threshold of giving birth to an Idea that disturbs him so profoundly it's literally making him ill… I found myself connecting to him as a father, and as a husband and—in my own poor way—as a creative artist.

That's why Collee and I settled on the title. *Creation* was not only a poke at the Genesis nuts, it spoke to the mysteries of childbirth and parenthood, and it spoke to the travails of the creative artist.

NR: And Randal Keynes' thesis was that in a sense the death of his daughter was a transforming event, it was *really* a fulcrum point. The point where the balance of his belief in God versus his beliefs in a world dictated by laws that had nothing to do with God swung irrevocably.

JA: Exactly. And that fulcrum point gave us the spine of our drama. The father-daughter relationship became the center of our story as the brother relationship became the center of Scorsese's story. It gave us a human and deeply emotional hook on which to hang some of the other central dilemmas of Darwin's life—his relationship with his wife and his relationship with his god. With Annie's death, any attenuated belief that Darwin might have retained in a deity who organizes the world according to some grand benevolent design went out the window. To use the techno-babble of film story structure, Annie's death was the *inciting incident* of a story that culminates with the birth of *Origin*... In other words, I could see the starting point of a film that wasn't a biopic!

Collee and I very quickly decided that we wouldn't do a soup-to-nuts story that started with childhood and ended in death. We wouldn't make a reverential account of a great life. We started out instead to create a portrait of a great mind in a great state of turmoil at a singular point in his life. We decided to make it a subjective account rather than an objective one—an account in which the stories he told to his children, his dreams and nightmares, had as important a role to play as the "facts" of his life. A feature film isn't a medium that lends itself well to the exposition of abstracts—if you want an account of Darwin's ideas, read his book or watch a PBS special. We tried instead to allow his ideas to percolate through the confrontations of characters and through the language of the film's imagery. We strove, in other words, for the art that conceals art, the science that conceals science.

Taking Creative License

NR: So did this influence your decision to write the film using Darwin's daughter, Annie, as his embattled inner monologue? She's so perfectly woven into the story as Darwin toils over his decision to publish his theory. She seems to be both a source of joy and pain for the tormented Darwin. Losing Annie drove a wedge between Darwin and his wife and nearly pushed him over the edge. Annie was his creation and now she was gone…

JA: …which presented us with one of the great challenges of telling this story. In reality, Darwin spent nine years between her death and the publication of *Origin*—mostly dissecting barnacles. Not exactly the stuff of cinematic legend! And yet the events were deeply inter-connected. Freeing ourselves from the tyranny of chronology and allowing the narrative to be emotionally rather than chronologically driven allowed us to get to the core emotional truths. The clinching idea was the notion (possible in a non-biopic account) that Annie's spirit should become his companion, interlocutor and the voice of his creative conscience.

NR: Which of course ruffled so many neo-Darwinist feathers—all those fervent ideologues you were talking about.

JA: Yes. We were in the happy situation of offending two opposing groups of fervent ideologues almost equally! The Creationists were of course deeply offended that we should dare depict one of their great ideological Satans as a warm sympathetic human being. Equally, a number of entrenched Darwinists were outraged that we should dare show one of the greatest scientific minds of all time unhinged to the extent of talking to an empty room.

NR: Every year the Sloane Foundation gives a prize to the outstanding science-oriented film of the year. In my view, yours was surely that film. Yet they chose not to give it to you. Do you think they were part of that offended group of ideologues?

JA: Oh…probably! Perhaps our film was not 'scientific" enough for them or perhaps they considered that we had taken too many liberties with the facts. I don't think we did. Is there factual evidence to prove Darwin talked to his dead daughter? No, of course not. Even if that evidence had existed in a letter or journal entry, it's likely that the zealously protective Ettie would have expunged it from history's record. The biographer's duty is to the known and attestable facts, but the filmmaker's is to responsible and imaginative surmise. And I think we made a number of responsible and imaginative surmises about Darwin—all of them supported by Randal Keynes, by the way.

For instance we know that when Annie died, Darwin wrote a four-page *in memoriam* about her, a deeply affecting account of his wonderful child. He then folded it up, put it in a drawer and never looked at it again. A perfect paradigm of repressed memory! We know Annie was barely talked about in the house, and we know Darwin became physically and psychologically more and more severely afflicted in the years following her death. He essentially had what we would now call a nervous breakdown, suffering from a series of physical ailments, which we're increasingly coming to believe were psychogenic in origin. When his closest friend, Hooker's daughter, died, Darwin wrote a beautiful letter to him saying that not a day passed when he didn't see Annie's little face and hear her voice. Is it therefore so unlikely that Darwin "communed" with his dead daughter? I don't think so. There's also speculation that some of Darwin's symptoms were caused by withdrawal from the large quantities of laudanum (a tincture of opium) that he took to control his symptoms. Read between the lines of Darwin's letters and you'll find much evidence that this great rationalist's mind was capable of acting irrationally.

NR: So in a way your film owes more to *A Beautiful Mind* than to *Gandhi.*

JA: Yes. And a great deal both in terms of style and content to a series I made for the BBC many years ago—Dennis Potter's brilliant *Singing Detective.*

NR: No songs in this one though!

JA: No songs, sadly. But the same stylistic collision of anecdotes and dark dreams, the same sense of a beautiful mind unraveling, even the same character—the shrink who may or may not be a quack but who provides our hero with the key to unlocking the door of an old deep trauma and begin to heal. Just in case anyone missed that point—I had him played by the same actor—the wonderful Bill Patterson.

NR: Talking of anecdotes....The astonishingly affecting account of Jenny the orangutan and the way in which that story connects with the story of the Fuegan children and ideas of children's vulnerability and mortality...Talk about what those meant to you.

JA: From very early on we knew that these were stories that didn't just need to be told but visualized. Visualizing them through the eyes of a child allowed us to pull this sort of fabliau-like quality from the stories, to give them a condensed and extracted narrative and visual quality. Telling the stories in this fashion allowed us to conceal any didactic intent and still get to the heart of some of Darwin's ideas about the ways in which natural selection could work to the brutal detriment of the young and the vulnerable. The sequence in which the fledgling falls out of the nest, dies, decomposes and feeds the ongoing cycle of life, equally served triple duty. It was a visual illustration of the famous meadow bank passage in *Origin,* it was a third piece in the interlocking triptych of stories about the young in jeopardy, and it served to address another element of our story—a sort of unasked question: how must it feel to be a naturalist, accustomed to see the young of many species predated on by others, then to pass these observations through the experiential prism of losing a child?

NR: Yes, people don't go to two-hour movies to be lectured, they want to be entertained. There has got to be a payoff. As well as the more somber elements of his personality, I loved the way you caught Darwin's playfulness, especially between him and Annie. You took a scientific figure who had been placed on a pedestal and who was revered to the point where they've been stripped of human-like qualities and you infused him with humanity.... Which brings me to my next question: what was behind your decision to cast Paul Bettany as Charles Darwin? Casting is often seen as an 'art,' but was there any science to your casting of the film?

JA: Yes, the casting process is indeed multilayered. There is the artistic/aesthetic aspect of casting—you want the actor to somewhat resemble the character being portrayed. But beyond that, there is a scientific component to casting; looks alone are not enough. There also needs to be a degree of complexity and intelligence embodied by the actor. Finding the perfect actor for the part is much like a chemistry experiment. What the actor *looks* like is the physics part, what he/she makes you *feel* is the chemistry. Casting is like titrimetry. You're looking for an actor who can make you feel a certain range of emotions—whether you realize it or not, you're titrating, quantitatively analyzing the concentrations of emotions in those actors. So both of these artistic and scientific aspects were great considerations of mine when casting this film. There's a famous portrait of Darwin by George Richmond. It depicts a Darwin that was close to the age of the man in our movie. He looks just like Paul Bettany. I certainly wanted an actor who could look like Darwin without intervening layers of latex. I also wanted one who could convey his physicality. Darwin was 6'2," very tall for those days, and clearly awkward in his own body. Darwin, like Bettany, had this high domed forehead with an early receding hairline and a quick-to-color sandy English complexion.

As I said, it was also very important to cast an actor of intelligence. I consider myself a good director of actors but there are three things I think it's impossible to direct an actor into: a sense of irony, a sense of humor—and intelligence. You can put a pair of glasses on a thirty year-old TV actor and pretend he's a Nobel Prize-winning nuclear physicist, but the sad fact is you can tell by closely scrutinizing their face and body that they're saying the words but *not* having the thoughts. Real intelligence, of the kind that Paul Bettany has in spades, is a rare commodity, especially amongst the younger male segment of our acting community.

NR: I have no idea what you're talking about. Aren't they precious?

JA: Bless their little six-packed, gym-built hearts! So many of our young leading boys (leading men are oddly in short supply) are much better at *doing* than thinking. They have a thinking look, which essentially involves furrowing their brow, and looking as though they're trying to pass a rather stiff motion. It's this intensely constipated look that's supposed to pass for deep thought. Couldn't have one of those guys playing Charles Darwin!

NR: It's funny. The list of what you need in an actor is the same list I need in a boyfriend.

JA: Right? Well, Good luck.

NR: Thank you. Especially, in this town [Los Angeles]. A lot of people have agendas…

JA: Yes, for a town that is all about collective imaginative endeavor and creative thought, people are curiously dug in about who they are, what they want and what they think.

NR: Being that dug in has got to limit you from really thinking freely. So being intensely focused on a goal can equally hold you back—in both Hollywood and science. Sometimes you have to step back and re-assess the big picture in order to find answers and opportunities. Creativity requires focus, but also needs wiggle room for adaptation and exploration. It's difficult to explore those options when you have blinders on. I wonder if that hinders creation rather than helping it?

JA: I think so. For sure.

NR: This morning you sent me a very touching NPR interview that Maurice Sendak did a few years ago where he talked about how creation can hit a writer out of nowhere and that it can be one's salvation. Sendak then said his latest book "saved him." So maybe there is an evolutionary component in humans that selects for creativity and innovation? And creativity not only ensures survival for our species but also encourages individual growth. Do you ever feel like the creative process saves you from yourself?

JA: Ah, the inexplicable of the nature of creation! I think the creative process not only helps me evolve—it keeps me alive. With apologies to Descartes, I create therefore I am—I think! Probably my greatest fear is the fear of my own mediocrity. I think that keeps me moving forward the way a shark needs to swim. William Wyler's wife described watching her husband make a movie like "Standing on the edge of a pool and watching him swim underwater six feet below you." It's a beautiful simple simile that describes what it's like to be with someone who is immersed in the creative act. I try to be as inclusive as I can to those I love and who love me, but the fact is, it's an intensely excluding business, creating. And thus, intensely lonely.

NR: Does your family go on location with you?

JA: Yes, as much for selfish reasons as any others, I always try and have my family travel with me.

NR: A lot of directors and actors don't want to be encumbered with spouses and children. They want the freedom to immerse themselves in their "roles." Do you think your family affects the way you direct?

JA: When I first started directing I think I created my directing persona (because all directors need a persona) that would work for my family. I realized I wanted my kids to be able to come on set and not see some sort of red-faced screaming monster that they didn't recognize as their dad.

NR: So how would you define your persona?

JA: I suppose its "Stern-but-Kind Father" meets "Mischievous Older Brother." It's one that seems to work well for me and creates the kind of working environment I most want. I'm not a screamer on set. Instead, I affect a sort of jokey relaxation that's entirely faux. I'm not in the least bit relaxed and everybody knows it! And yet everyone gratefully participates in the deception. It's a benevolent charade that invites—almost demands—that everyone else on set enter into it, too. The fact that I'm scared shitless and lying awake at three in the morning in the dark maw of horrible self-doubt is nobody's business but mine. Bertrand Tavernier, the great French director, said (and I'm paraphrasing I'm sure) that when you're a director you should act like you're the host of a party. That's the best advice ever.

NR: Do you find it's gotten easier for you through the years?

JA: Nope. On the contrary. I think if anything it's become slightly harder. Cortisol and norepinephrine and all those other stress hormones seem to have an accumulative effect. It's not uncommon for great stage actors (Lawrence Olivier and Ian Holm among them) to develop stage fright in later life. Not only am I *not* less scared than I used to be, I'd be worried if I were. I think the fear keeps me honest. Fear of failure, the fear of mediocrity that I talked about before… The fear, which keeps me awake at three in the morning, is also I think, an essential part of the imaginative process. It's that process, the obsessive rolling and rolling around of an idea, like a stone in a stream, until it's as smooth and polished as an alluvial pebble, when it has a shape so perfect that you cannot imagine it having any other—that's the thing that happens in those dark hours of the night. Great films have that quality—the sense that each image has been imagined with an almost lapidary precision. The current of that stream is fear and the compulsion that goes with it. I think that I have the résumé of an ambitious man not because

I had burning vocation to direct movies—I did not. I think it's because at most crossroads in my life, faced with two choices, I generally did the thing that scared me more.

NR: Most animals are driven by fear, too. But oftentimes it's defying getting eaten by predators or taken down by the elements. Humans use fear in a myriad of ways—creativity being one of them.

JA: When I look at the worrisome question as to why so many great directors make bad movies later in life, I think it's because they forgot to stay scared. To become a creative artist you have to accept fear as your constant companion.

The Future of Evolution

NR: In *Creation* there is a profound line uttered by Darwin saying that humans select for beauty but nature selects for survival—a true example of humans combining science with an artistic eye. And nowadays we are seeing people selecting the sex of the their child, genetically modifying our food, and there is a constant race to create antibiotics that can combat constantly evolving bacteria (due to our already high use of anti-biotics). Is selecting for a preferred aesthetic such a good idea? Especially since a 'preferred aesthetic' is far from static—our preferences are always changing, trending, and evolving. One day it's hip to be a blonde haired, blue-eyed waif, the next week it's all about the hourglass figure with darker features. On Monday the up-turned 'button' nose may be what people are pining for but by Thursday they've moved on to the Greek variety. It's rather difficult to keep up *and* it begs the question, *should* we keep up or let nature take its course. Take a museum full of various pieces from the masters ranging from Renoir, to Gauguin, to Picasso, to Kandinsky, to Caravaggio—there are so many mediums, styles, and modalities on display. All pieces are striking in their own way. And yet some individuals are drawn to the array of color fancied by Kandinsky while others prefer Caravaggio's use of light. Is the human form any different? Beauty, whether it be in the form of art or science or a melding of the two, is truly in the eye of the beholder. And those opinions are always changing! So what do you think about this idea of humans attempting, if you will, to play God by manipulating science?

JA: We seem to be seeing over and over again that our attempts to play God have been about as mixed as God's attempts to play God. If one looks at the scale of global catastrophes and the scale of man's inhumanity to man (and man's inhumanity to woman in the case of our Republican party) one has to ask, if this is all the product of "intelligent design"—what kind of designer is this God, and of what kind of intelligence is he/she/it possessed?!

So in that respect, I can't say we've done any worse than our so-called Creator. On an individual level, we've created many things of beauty, but our larger attempts at engineering—whether on a bio-molecular level (you talked about anti-biotics) or a grand civic level (damming rivers for hydroelectricity and destroying vast ecosystems) or a grand social level (China's discredited policy of allowing only one child per couple) or on an ecological level (global warming)—over and over again, our attempts to play God with our destinies end up proving catastrophic. By and large, Darwin's natural selection has been replaced by human selection. We probably now select for more species of plant and animal than Nature. We breed livestock for meat yields, crops for pest resistance, dogs for appearance, fruits for shelf life.... Since we now control our environment, environment no longer determines selection. We have no other natural predators but disease and other humans.

NR: I can't help feeling that we've outstayed our welcome at this restaurant, and well exceeded the 5,000 words allotted for our chapter...

JA: And barely scratched the surface of our themes. But I do think we've come a little closer to understanding some of the connective tissue between the creative and scientific mind...

NR: Indeed. The movie director has spoken to the anthropologist and has been understood. Though this may be just a conversation begun, I think we've come a good step closer to determining that science and art are truly co-constitutive.

JA: Co-constitutive and co-mingled. At their very foundation, science and drama will always have these two things in common—they're about great characters, pursuing elusive dreams, creating great stories and illuminating ever-changing truths.

Chapter 2

Science Fictions and Fictional Science: A Brief Tour of Science in the *Star Trek* Universe

Andre Bormanis*
***E-mail: abormanis@adelphia.net**

Incorporating a level of scientific credibility into a television series set in the far future, with a long, established history and legions of dedicated fans, presented a unique set of challenges. The writers and producers of the various incarnations of *Star Trek* made every effort to build their fictional universe on a foundation of solid scientific thinking, with the assumption that many new technologies will be invented in the coming centuries, many new phenomena will be discovered in deep space, and other intelligent species will be encountered as humans explore the Milky Way galaxy. The need to convince the audience to willingly suspend its disbelief was the primary goal in production design, story development, and dialogue. This article examines the evolution of the technical dimensions of the Star Trek universe, and the role played by its science consultants in shaping that universe.

The telephone rings. The voice on the other end of the line asks, "Is there any way to use the transporter to get the fetus of a pregnant woman into another woman's body?"

This was not the first time someone from the *Star Trek* production office had called to ask me a question I had never considered before. My initial reaction was that the notion of "beaming" a fetus from one womb into another was utterly ridiculous, absurd on its face, completely impossible. Of course, that isn't what the person who called me wanted to hear. That person was a writer, and he had a deadline. To have any hope of meeting it, he needed a solution to a problem in his story, a story that in a few days would become a script that would be filmed for a television show that was produced on a very tight schedule. I kept my initial reaction to myself, and kicked my sci-fi imagination into high gear.

"Well, I think there are a few possibilities here," I said, feverishly trying to devise some remotely plausible-sounding ideas. "But I'm pretty sure we're going to have to do more than simply transport the fetus. Let me check with a couple of medical experts and get right back to you." And so another typical day in the life of the *Star Trek* science consultant had begun.

The original *Star Trek* premiered in the United States in the fall of 1966. Getting it on television was a minor miracle. Gene Roddenberry, the man who created the show, had written a pilot two years earlier. It was filmed, and while it was a very impressive hour of television, it was ultimately rejected by the American network (NBC) that ordered it. They thought the plot was "too cerebral" to attract a network audience. But the good people at NBC were sufficiently intrigued by the concept of *Star Trek* that they made the unprecedented decision to order a second pilot (with a few changes in cast and a new story). That pilot was accepted, and as everybody knows, the show eventually became astronomically popular.

Roddenberry insisted from the beginning that *Star Trek* would be aimed at an adult audience. He had no interest in producing another *Captain Video* or *Buck Rogers*. And he knew that in order to attract an adult audience, the starship he launched on its five-year mission would have to look and feel like something that could someday be built—maybe not for hundreds of years, but someday.

The "willing suspension of disbelief" is a phrase writers use to describe the nonjudgmental state of mind they need to lull their audience into as they tell their stories. For one hour a week, Roddenberry needed his audience to accept the reality of the (out-of-this) world he had created. A starship with a crew of hundreds, capable of traveling interstellar distances in mere days, was bound to strike most adult viewers in the 1960's—several years before the first Moon landing—as highly implausible. How do you make something as far-fetched as that seem real?

Roddenberry called in a team of consultants. He met with local scientists and engineers from the Jet Propulsion Laboratory (JPL) in Pasadena and the RAND Corporation in Santa Monica, picking their brains and pushing their imaginations to sketch out the technologies that would be required to build a starship.

What would propel the ship through space, for example? Roddenberry was told that the chemical rockets being developed to take men to the Moon would never be powerful enough to carry humans to the stars. So, he asked, what is the most powerful source of energy known to man? Antimatter. Could it be made in sufficient quantity to power a starship? Possibly, but of course no one then (or now) knows how. But let's assume, hundreds of years in the future, that problem has been solved…

The starship *Enterprise* was modeled in many respects on a contemporary naval vessel. As Roddenberry and his consultants pondered other aspects of the ship, the limitations and, frankly, primitiveness of current technologies became more and more apparent. This was particularly so in the medical section, or sickbay. In a circa-1960s era hospital, doctors stuck a thermometer into a patient to get his temperature, put an inflatable cuff on his arm to measure his blood pressure, stuck him with needles to draw blood that often took days or weeks to test for abnormalities—all of this struck the would-be starship designers as embarrassingly crude. In an age of computers and automation and space travel, hospitals seemed distinctly low-tech. So Roddenberry and his team asked themselves how space technology might influence medicine in the future. It seemed plausible that vital signs could someday be taken by remote sensing devices, and that the power of computers could be harnessed to perform tests more rapidly, analyzing tissue scans in mere seconds. Most importantly, from the point of view of a television writer, making the captain wait days or weeks for the ship's surgeon to diagnose some alien pathogen that was killing off the crew was generally not good storytelling. Thus was born the biobed, the hypospray, and the medical tricorder.

A few weeks after *Star Trek* premiered, Roddenberry began to receive letters from various medical technology firms asking him how he knew they were working on such things. He didn't, of course. He and his advisors simply asked logical questions about where medical technology was lacking and how it might evolve given ongoing advances in electronics and other fields.

And so, piece-by-piece, Roddenberry assembled his starship. In addition to the warp engines and sickbay, it included a bridge, crew quarters, recreation rooms, "turbolift" elevators, a shuttlecraft bay, an engineering section, and the iconic transporter room. I suspect that if human beings ever build real starships, they will not even remotely resemble the good ship *Enterprise*. But if that day ever comes, I also believe that Roddenberry's approach will still be considered visionary for its time.

Providing technical and scientific notes to the writers of a show like *Star Trek*, which takes place hundreds of years in the future, is quite different from working on a present-day medical drama or police procedural (or even a science-fiction series that takes place in today's world or the near future). I didn't have to work within the constraints of current medical practice or forensic science or engineering. My objective was to stay true to what we know today about the laws of physics, the properties of stars and planets, and the nature and evolution of life, but to stretch my imagination (and that of the audience) into a future universe where starships, alien life forms, and exotic space phenomena were integral elements of the stories being told.

Star Trek also has a well-established technical language, words and phrases that have been coined on the show over the course of decades. Much of this language has found its way into popular culture: "Set phasers on stun," "Hailing frequencies open," "Beam me up Scotty," and so on. As Jeri Taylor, one of the producers of *Star Trek: The Next Generation* and *Star Trek: Voyager* once told me, writing *Star Trek* is like writing period literature. In order to do it successfully, you have to be fluent in the language of the age—in this case, the Space Age (fortunately, I didn't have to learn Klingon).

One thing that was never allowed in a *Star Trek* script is scientific exposition. Stopping the action to have a character explain wave/particle duality or some other arcane technical subject was strictly forbidden. Characters could talk about a scientific phenomenon or technical problem, but not deliver a lecture on it. As every writer learns in Screenwriting 101, exposition is the mortal enemy of drama. If my suggestions for dialogue ever veered in the direction of scientific speechifying, the producers quickly let me know about it.

When real scientists or engineers get together to talk shop, they tend to speak in a kind of scientific shorthand. So, whenever I was asked to provide technical language for dialogue, I constructed it in a way that sounded like a couple of scientists in a hallway at Fermilab having a conversation about their work. The audience might not be able to follow every detail, but the characters had to sound like real scientists who know what they're talking about. Their exchange had to have a scientific ring to it. If an actor playing an astrophysicist can deliver a line about the impending collision of a pair of neutron stars in a way that sounds convincing, the audience will be able to suspend its disbelief and stay involved in the story. And, hopefully, any astrophysicists in the audience will be satisfied that the dialogue is credible.

Some *Star Trek* dialogue (okay, maybe more than some) consists of what the writers fondly called technobabble. But even purely fictional terms have to be used consistently from episode to episode. If we've defined the physical properties of the nadion particle in episode 32, it must have the same properties if it comes up again in episode 132. We kept extensive notes on all of our technobabble to try to stay consistent. Even the slightest deviation from an established definition would be caught by somebody in the audience. Fearing the wrath of our fan-base kept me on my toes.

An important lesson I learned at the beginning of my tenure as science consultant was the challenge this kind of dialogue posed for our actors. On *Star Trek: Deep Space Nine*, actress Terry Ferrell played Lieutenant Dax, a science officer. Since she was the only scientist in the crew, the bulk of technobabble was often carried by her. Terry is a tall, striking, beautiful woman. The first time I met her was at a "wrap party," the fling the studio throws at the close of each season to celebrate the end of production. I saw her standing at a table by herself, and shyly introduced myself as someone who worked on the show. She smiled warmly and took my hand:

"So what is it you do?" she asked.

"Uh, I'm the guy who puts the technobabble into your dialogue," I replied.

Her expression quickly changed from friendly to horrified. She grabbed me by the lapels, literally lifted me off the floor, and began screaming expletives at me. After the color had drained from my face, she started chuckling. She was just ribbing me. Terry has a great (though sometimes wicked) sense of humor.

A few moments later she told me something I never forgot. When her character was doing a scene by herself, she said, just talking one-on-one to our voice-activated computers, she didn't mind if every word out of her mouth was some kind of technical verbiage she didn't understand. But if she was in a scene with other actors—where she had to be sure to hit her marks, stay aware of what the other characters were doing, and find her best dramatic choices—it was extremely challenging to try to remember a string of words that ultimately meant nothing to her. What she told me made perfect sense, and I felt a little embarrassed that it hadn't already occurred to me. But after that conversation, I was sure to be conscious of the fact that whatever technical language I suggested for a line or a scene wouldn't be too much of a burden for the actor reciting it. And Terry never swore at me again.

Even the most obscure details of particle physics can lead to trouble. In an episode of *Star Trek: Deep Space Nine*, a shady alien brings a device aboard the space station that can locally change the laws of probability. He uses it to win big at Quark's Bar, the station's gambling parlor. Quark, the owner, suspects that the gambler is cheating, but he can't prove how. The writer of the episode asked me if Science Officer Dax could find some highly unlikely distribution of quantum variables on the station that would prove someone was messing with the basic laws of physics. I thought for a moment and suggested an improbable alignment of neutrino spin states. The writer thanked me and wrote his script.

A couple of years later, physicist Lawrence Krauss wrote a highly successful book called *The Physics of Star Trek*. He went into great detail deconstructing the "future science" of transporters and warp drives, explaining that it would be essentially impossible to build such technologies even in the 23rd century (at least in terms of the way we depicted them on the show). Dr. Krauss also noted some of the science errors that popped up in specific episodes. He was particularly annoyed by the neutrino spin reference on *Deep Space Nine*. He noted that unlike every other subatomic particle, neutrinos have just one spin state, so there was no way to get an improbable alignment in one state or another—they're all, always, in the same state. I had either forgotten this from my college education in physics or never learned it. My embarrassment at this subatomic snafu is somewhat tempered by the more recent discovery that neutrinos have a slight mass and can oscillate between different flavors, yielding two possible spin states (take that, Dr. Krauss!). Science marches on.

Some mistakes are harder to catch. In an episode of *Star Trek: The Next Generation*, Lieutenant Worf, a big, strapping Klingon, was badly injured in an accident. He was rushed to Sickbay, where the doctor called for 75 cc's of the (fictional) drug Inaprovaline. She grabbed the hypospray from the nurse and injected Worf directly in his neck, stabilizing him.

The episode was filmed before my tenure as science consultant began, but even if I had worked on that script I never would've caught the errors in this scene. They came to my attention through a letter from a fan with a Ph.D. in pharmacology. Her letter started, as most letters from fans who are scientists do, with a few lines about how much she enjoyed our show, and how the original *Star Trek* helped motivate her interest in science. She then pointed out that drug dosages are rarely given in cc's—typically they're given in milligrams, even if the compound is dissolved in a few cc's of saline or some other liquid. Secondly, 75 cc's of anything is a lot of fluid to pump into somebody's neck, even if that someone is a Klingon. Presumably the stuff is going into his carotid artery and then straight to his head. And thirdly, the little vial at the end of the hypospray that carries the medicine is *too small* to hold 75 cc's! After I read that letter, I made sure to tell the writers to always express drug dosages in milligrams. And I always tried to imagine how a given scene would look on the screen, and whether there was anything incongruous in what the audience would be seeing that might take them out of the story.

Perhaps the most problematic (in terms of feasibility) technology on *Star Trek* is the transporter system. When Gene Roddenberry created the original series, he needed a way to get his characters from the ship to the surface of a planet as quickly as possible. In a one-hour TV show, the writer only has 45 or 50 minutes to tell a story. Landing a massive starship on a new planet every week not only consumes valuable story time, it's also an extremely expensive special effects sequence. Roddenberry's solution was the transporter. The crew step into the transporter chamber, Scotty works the controls, and seconds later they appear at the desired location.

The transporter was one of the most effective story-telling devices on *Star Trek*, but as a technological device it misses the mark. In one of the original *Star Trek* episodes, Captain Kirk explained that the transporter is a "matter-energy scrambler" that takes a person apart, atom-by-atom, converts those atoms into a beam of energy, and reassembles the transportee at the desired location. No wonder Doctor McCoy was reluctant to use it.

Objections to the feasibility of the transporter are too numerous to mention here. All of us who worked on the show understood that turning a body into a stream of plasma and shooting it through space would be an even more difficult way to travel than flying coach. A fundamental problem stems from the Heisenberg uncertainty principle. It is impossible to simultaneously know, with great accuracy, both the position and momentum of a subatomic particle. If we use the transporter to break Humpty Dumpty into his constituent atoms, we can never have enough information to put him back together again. Either his atoms will be in the wrong places, or moving in the wrong directions. To let the physicists in the audience know we understood this problem and that Starfleet engineers had solved it, Mike Okuda, one of our graphic artists and technical advisers, coined the term *Heisenberg compensator*. We made reference to this device several times when the transporter was offline or suffering some kind of malfunction. Viewers familiar with quantum mechanics seem to get a kick out of it. And whenever someone asks Mike, "How do the Heisenberg compensators work?" he replies, "Very well, thank you."

There were plenty of times our gallant crews found themselves in situations that couldn't be explained or resolved by present-day science. In an episode of *Star Trek: The Next Generation*, Captain Picard comes to the aid of a world that's suffering from massive quakes and tsunamis. The problem lies in the planet's core. It's cooling rapidly, causing it to contract. Is there any way the powerful starship *Enterprise* can reheat it?

The core of an Earth-like planet is likely to be a ball of iron and nickel several thousand kilometers in diameter. Changing its temperature by even a fraction of a degree is beyond the reach of anything we can imagine today. The solution clearly had to involve a fictional technology. But whenever I had to come up with a fictional fix, I always tried to make it an extension of, or analogy to, something that exists today. In thinking about the molten core of a planet, it occurred to me that it is not dissimilar (thermodynamically speaking) to the core of a nuclear power plant. In a nuclear core, a small, invisible stream of neutrons initiates a chain reaction that ultimately raises its temperature hundreds of degrees. So I suggested that the fictional nadion particles used in the *Enterprise* phaser beams could, in a similar way, trigger a chain reaction in some exotic elements in the core of the alien planet. The energy released by the reaction would reheat the core. Obviously there are many problems with this scenario: even if such a reaction could take place, controlling it would be problematic, and it would probably take centuries if not millennia to increase the temperature of the core appreciably. In all probably it would make the quakes that were wreaking havoc on the planet worse, at least in the short term, not better. But it was an idea that the audience could understand—it was relatable. Most people are aware of the incredible power of a nuclear chain reaction, so they would be less likely to dismiss an explanation based on something similar than on a few lines of pure technobabble.

Science, of course, is a process that continually leads to new knowledge, and sometimes overturns widely accepted theories. This is something I often kept in my mind during my tenure as science consultant. In an episode of *Star Trek: Voyager* from the mid-1990's, Captain Janeway has a line of dialogue in which she mentions the age of the universe. At the time, there was more uncertainty in that number than there is today, so I called my friend Dr. Laura Danly, an astronomer currently at Griffith Observatory in Los Angeles, to ask her what the latest "consensus" number was. She said it was around 13.5 billion years, but then told me I should have Janeway say 48 billion, "just to tweak all the astronomers in the audience."

I thought that was very funny, and was briefly tempted to suggest it to the writer, but in the end I went with the consensus number. When a script called for it, I always tried to use the best available numerical data for things like the number of genes in the human genome, stars in the Milky Way galaxy, and so forth. But of course we keep refining those numbers, and making discoveries that call into question basic assumptions behind our estimates. Near the end of the 1990's, astronomers discovered that several billion years ago the expansion of the universe accelerated. This is truly shocking, and someday it may lead us to revise our estimate of the age of the universe. By Captain Janeway's time, we may have very different ideas about our cosmic origins.

More so than any other fictional television series I can think of, *Star Trek* has played an impressive role in shaping the future. Countless scientists and engineers who grew up with the show in its various incarnations have cited it as a formative influence. *Star Trek* may not be a reliable guide for predicting the future, but it has certainly inspired many of the people who are building the future. The engineer responsible for the original "flip" phone has stated that he was explicitly thinking of Captain Kirk's communicator when he designed it. A cash prize is currently being offered to anyone who can build the equivalent of a medical tricorder. The NASA Glenn research center has explored the potential for using anitmatter for propulsion. The Mexican physicist Miguel Alcubierre has worked out the theoretical foundations for creating a warp drive. Work in metamaterials has led to crude "cloaking screens" and research in quantum mechanics has led to the "teleportation" of photons. Is there any other show in the history of television that has had a comparable impact on science and engineering research?

In retrospect, probably the hardest part of being the *Star Trek* science consultant was keeping ahead of the rapid advances in science and technology that take place in the real world on a seemingly daily basis. The computer and communication technologies of the various *Trek* series seem particularly primitive in light of the Internet and touch-screen telephones. Warp drives, teleportation devices, and fetal transplants remain, of course, in the realm of fiction—at least for now. The great French science fiction writer Jules Verne once said, "What one person can imagine, another can create." *Star Trek* has helped inspire countless young people to pursue careers in science and engineering. What might they create in the coming years that will help set us on the path to the stars? I can only hope that as they conjure the technologies of tomorrow, the bright, optimistic, and deeply humane future that *Star Trek* envisioned will guide their imaginations.

Chapter 3

A Gram of Prevention Is Worth a Kilogram of Cure: Teaching Writers Science

Michael S. Brotherton*

Deparment of Physics and Astronomy, University of Wyoming, Laramie, Wyoming 82071
***E-mail: mbrother@uwyo.edu**

The easiest way to correct a mistake is to prevent the mistake in the first place. I describe a preemptive program, the Launch Pad Astronomy Workshop for Writers, that educates participants in the basics of astronomy and related sciences. The goal is to empower them with the fundamentals, a network, and research capability in order to put more science in front of their audiences and to decrease the error rate in that science. Science within entertainment, which now comes in myriad forms including TV, movies, books, video games, and more, serves as stealth education when it is correct. Furthermore, many scientists self-report that they chose to pursue their careers because of the inspiration of movies and TV shows like *Star Trek*. The program is very well-received by participants and may serve as a model for other fields.

Introduction

Writers don't set out to make mistakes.

Writers strive to create realistic worlds in which people behave like people, and the world behaves like the world, except when exceptions are made to include speculative elements like stargates, sparkly vampires, or the Force. Those speculative elements still have to follow the rules established for them, or some fraction of the audience will find the stories internally inconsistent and ultimately reject them.

The thing is, the world is a complicated place, even the limited world of a novel or a movie. Few people are experts at writing, let alone the myriad details of science, history, foreign cultures, technology, law, medicine, and the human heart. Painting a rich and perfectly plausible world without making any errors is not always a realistic goal, but it is what good writers strive to achieve within the context of their stories and themes. Many are more than willing to invest significant quantities of their time, do their research, and produce stories as interesting and mistake-free as possible.

A few years back I set out to help them achieve that goal within my own area of expertise: astronomy. In 2002, I achieved two milestones by selling my first science fiction novel to a major publisher in the field (*Star Dragon*, 2003, Tor Books) and getting hired as a tenure-track faculty member. Soon thereafter, I started noticing something.

At science fiction conventions, people were a lot more interested in me as a scientist rather than just another science fiction writer. Among my colleagues in astronomy, I sometimes got more questions about science fiction than my research into quasars and black holes. Synergies existed that I could exploit in a unique niche, and I began to look for ways to combine my loves for science and science fiction in productive ways.

One of my efforts involved securing a multi-year educational/public outreach grant from NASA to put on a workshop to teach writers astronomy, in order to increase the quality and quantity of correct science in their work, and to help inspire the next generation of scientists.

I can trace my interest in space and astronomy back to when I was six-years-old and my parents called me into the living room, turned on the TV, and said, "Michael, we think you'll like this." It was an episode of *Star Trek*, the original series, and I did like it. I liked it a lot. From then on science fiction was part of my life, and so was science. I am not unique among scientists, either. The entertainment we consume as kids can and does inspire many of us to pursue future careers.

Writing my own novels and pursuing my own research, I might reach out and inspire some future scientists out there. Teaching other writers, many much more successful than myself and enjoying significantly larger audiences, is a way of reaching a lot more people. In addition to impressionable, bright kids, books, movies, and television also reach adult audiences who are out of school and unlikely to actively study science, but who still vote and should have informed opinions about the complex, technological civilization we live in today.

The Launch Pad Astronomy Workshop for Writers

With that initial NASA funding, and with the collaboration of science education expert Jim Verley, Launch Pad was born. More recently it has continued with National Science Foundation funding under very similar parameters.

It is impossible to teach any science topic in depth in a single week, but few professionals have the time to take off multiple weeks. Keeping this in mind, we do a crash version of an introductory Astro 101 course. There are approximately fifty hours of lecture in a college semester, so a week isn't a totally crazy time period, but giving or receiving fifty hours of lecture in a single week would probably make heads ache if not explode.

We mix up traditional lecture with lab and computer exercises, discussions, nights out with our local telescopes, and issues of science education and communication.

The goal is to provide writers a basic background and enough knowledge to seek out more information when necessary, and to know when it is necessary. Also, a number of authors seem more likely to have the confidence to tackle science-rich concepts after an experience like Launch Pad.

More information may be found at the website for Launch Pad at http://www.launchpadworkshop.org. In order to provide guidelines for anyone considering the creation of a similar workshop, the particulars of our program are described in the subsections below.

Budget and Logistics

The total budget for each annual workshop is only $15,000, and a stripped-down version for less money is possible. We take advantage of inexpensive University of Wyoming housing and free classroom facilities to keep costs down, as well as the assistance of the department office staff to help with logistics.

The exact values vary from year to year, but our expenses bringing in about 15 people usually breaks down like this:

- $6000 for Participants Requesting Travel Assistance
- $3000 for Guest Instructor Stipends and Travel
- $2500 for Dorm Rooms
- $1500 for Van Rentals
- $1000 for Food
- $1000 for Reference Textbooks (to take home)

We solicit applications over the Internet, and find that some portion of attendees do not request travel assistance, stretching the budget. We also cap the travel at about $500 per person.

We have travel days to fly in participants and sandwich between them six days of activities. We usually have lectures in the morning and much of the afternoon, along with afternoon and evening activities. One morning is set aside for a hike, which is a welcome opportunity. Continental breakfast, snacks, and lunches are all provided in the classroom to save time. Participants are on their own for dinners.

All in all, some 50-60 hours are scheduled for the six days of the workshop. While participants are often tired and mentally exhausted by the end of the week, the enthusiasm of the self-selected group typically remains high. We don't waste their time.

Content

So, what exactly happens during the workshop?

We teach a set of tight core lectures starting with size scales and a Douglas Adams quote from *The Hitchhiker's Guide to the Galaxy*: "Space is big. Really big. You just won't believe how vastly, hugely, mind-bogglingly big it is. I mean, you may think it's a long way down the road to the chemist's, but that's just peanuts to space."

The vastness of space is one of the things that writers—let alone scientists—often fail to appreciate. Spaceships really shouldn't be viewed as versions of naval ships with all the associated misconceptions that brings, such as travel times of mere weeks between ports, not without invoking magic tricks like hyperspace at least.

We follow up with some other fundamental concepts about the seasons and phases of the Moon, core concepts that are relevant in areas well outside science fiction.

We cover some basic physics concerning light and gravity that are relevant to multiple areas of astronomy. Essentially everything in the subject depends on the detection and interpretation of light reaching us from space, while gravity is the fundamental force most important in many astronomical situations.

We cover the Solar System, as well as the exploding field of extra-solar planets.

A large number of the lecture hours are devoted to stars: their properties, their births, their lives, and their deaths. Stellar nucleosynthesis creates the heavier elements on the period table and we are in fact made of "star stuff," as Carl Sagan told us. We also get in some of the stranger and more fun aspects of astronomy here, such as supernovas and black holes.

Large enough groups of stars constitute galaxies, and we cover the Milky Way, as well as galaxies more generally. Galaxy clustering leads to large-scale structure and cosmology. Usually the Big Bang is the final topic of the week, with a discussion of dark matter, dark energy, and the creation of the light elements hydrogen and helium.

Beyond the above part that consists of crash-course lectures, stripped to the bone, we build in hands-on exercises, telescope viewing experiences, computer activities, and both formal and informal discussion opportunities. We throw at least one party in which writers and scientists get to mingle, and the writers get to see that scientists aren't all bearded men wearing glasses and white lab coats, but can be tattooed and sport Mohawks.

A visit to our 2.3-meter telescope at the Wyoming Infrared Observatory (WIRO) is a highlight. It's an impressively sized fully operational professional facility that lets the writers see how astronomers actually work when observing. When not observing, they learn that astronomers sit in front of computers in their offices, like much of the world these days.

Several hours are devoted to issues of science education and communication. For instance, education researchers have learned that many people carry around misconceptions that make it difficult to approach certain ideas, and those misconceptions must be knocked down before the correct explanation can take hold.

In astronomy, and all areas of science, misconceptions abound. A lot of people think they know what an asteroid field looks like because of scenes in *The Empire Strikes Back*, but the reality is that if you stood on one space rock in our own asteroid field, you'd be very unlikely to see another. Many viewers of the old movie *Outland* probably believe that humans exposed to vacuum explode, and may have incorrectly learned from *Mission to Mars* that liquids leaking into space won't (instead freezing into solid ice). The list goes on and on.

Many lessons of science communication are incorporated into the regular lectures, too. We have discussions about units to use (Americans are more familiar with English units, but metric units should be used when the scientists are discussing issues amongst themselves), and how they're redefined to be convenient numbers (as light-years and astronomical units avoid large and confusing powers of ten). We discuss how concepts in space are abstract and remote, but that there are ways to relate them to every day experience (e.g., density waves in spiral arms aren't too dissimilar from some aspects of highway traffic).

Finally, we do evangelize somewhat, urging writers to assume some personal responsibility for making careers in STEM (science, technology, engineering, mathematics) fields seem interesting and plausible. Unattractive stereotypes reflect lazy writing and do not reflect the real diversity present among scientists.

Participants

Because of my background as a science fiction novelist, and my original vision of the workshop as science education for science fiction writers, most of the participants have been science fiction writers. I know how to advertise to them, and the field is relatively small and well connected. Most participants have applied due to the good word-of-mouth advertising.

We currently get about 60-70 applications for about 15 slots each year, and have nearly 100 alumni of the program. Despite the original bias toward science fiction writers, strong applicants with different backgrounds have made me rethink the workshop. Now I see the benefits of broader applicants, with the main discriminate being their ability to put astronomy and related science in front of large and diverse audiences. A best-selling author of werewolf novels has the opportunity to teach phases of the Moon to many people who would never pick up science fiction.

Our participants reaching the largest audiences have been involved with major motion pictures and TV series from the big networks. Stephen Gould's novel *Jumper* was a big budget movie release in 2008, while Rob Sawyer's novel *Flashforward* was a TV series on ABC during the 2009-2010 season. These two and other participants may have similar deals in the future.

Participant Josepha Sherman wrote a number of *Star Trek* novelizations, while Jeffrey Carver wrote the novelization of the new *Battlestar Galactica* mini-series. The written form often allows for more explanation of science issues than the original versions, and benefits in a readership derived from TV audiences.

Participant Dani Wolff writes for television programs, including cartoons featuring *Spider-man* and *The Avengers*. Writer/director Robin Christian Peters has made movies such as *Disconnect* (2010) and the forthcoming *My Dog the Space Traveler* (2013).

Best-selling novels also reach large audiences, and participants Scott Sigler, Carrie Vaughn, Julie V. Jones, among others, have landed on bestseller lists in the past and likely the future. In the ensemble, participants have written hundreds of novels and short stories with a total audience measured in the millions, encompassing everything from science fiction to fantasy to young adult and children's books.

We have also had interest from editors of novels and magazines and a handful have attended. While they're less likely to have a direct hand in the initial creation process, they exercise quality control over the stories they select, and their background at Launch Pad may positively influence them to select more science-rich stories.

Traditional venues like movies/TV and books/magazines are not the only media reaching large audiences. We've had several authors who work in the video game industry, notably Marc Laidlaw, whose *Half-Life* games sell in the millions. Another innovative writer is Brain Malow (working with his partner Tara Fredette) who is a science comedian who has appeared on national television (*Craig Kilbourn*) and created a series of educational videos for Time.com.

Our guest instructors come and provide several hours of presentations, complementing the regular astronomy lectures. They also participate in the workshop and while they're less likely to learn a lot of new things, they do seem receptive to the educational message of Launch Pad. Past instructor and planetary scientist Kevin R. Grazier is a science consultant who has worked on *Battlestar Galactica* and *Eureka* for the SyFy channel, among other projects. Award-winning novelist Joe Haldeman has written for movies and his classic novel *The Forever War* is currently under option by Ridley Scott. Award-winning writer and NASA scientist Geoffrey Landis speaks at science fiction conventions. Jerry Oltion and Christian Ready reach out to amateur astronomy groups. *Analog* magazine editor Stanley Schmidt has influenced a generation of science fiction writers. Bad Astronomy blogger Phil Plait reaches large audiences through his blog, non-fiction books, and TV show *Bad Universe*.

Outcomes

We have several measures of effectiveness of the program, although we expect that the intangible effects will be great and amplified over time, although difficult to quantitatively measure. We give participants pre and post tests in astronomy every year, as well as evaluation forms for more general feedback. The evaluations indicated an overall high satisfaction with the workshop, individual presenters, and the overall organization (>85% satisfaction). They also indicated through the survey instrument that they greatly increased their astronomy content knowledge and their ability to recognize misconceptions (this number was 80%). There was a high positive response (>85%) to their newly developed awareness of and interest in the educational implications of their work. The response was 100% positive when asked if the workshop was worthwhile, met their expectations, and whether they would recommend it to others. That measure is remarkable, and speaks to the strengths of the concept, execution, and funding that permitted us to maintain a small, tightly focused program addressing participant interests very directly.

We annually send questionnaires to past participants and maintain an alumni email list to communicate with them. We don't get 100% compliance, but usually about 2/3 chime in with information. We've had approximately 45 short stories and about 15 novels published that have been directly inspired by Launch Pad or benefited from the knowledge generated by the workshop (also a screenplay, not yet sold). Additionally, editor participant John Joseph Adams (of *Lightspeed* Magazine) has published several hard science fiction stories that got their astronomy checked and corrected, including one by participant Vylar Kaftan that ended up on the Nebula award ballot. Similarly, writer/editor Jody Lynn Nye successfully pitched an anthology series that will feature science-rich stories by Launch Pad alumni. The first book is expected to be published in 2013.

More than 50% of those attending the workshop continued to report having a new sense of confidence in their ability to write about science in general and astronomy in particular. One former Launch Pad participant discovered a scientific error on a PBS NOVA webpage and felt compelled to write and seek a correction because NASA just paid to have them become a more effective communicator of science.

There were several astronomy children's books (by Josepha Sherman) that were written and published post-workshop that were more scientifically accurate, and which were more carefully scrutinized because of the author's participation in Launch Pad. Another former participant says that work at Launch Pad has been an essential part of the critique process used when looking at other authors SF writing.

On the educational side, one Launch Pad participant (novelist Samantha Henderson) developed a unit on astronomy for a fourth-grade science class because of newfound confidence at the workshop. A university professor who teaches literature reports having repeatedly utilized information at the Launch Pad workshop to enhance the classroom experience and dispel misconceptions about science. Another university faculty member (Christine Stebbins) reports redesigning portions of a Science Fiction writing class because of the Launch Pad experience. Another incorporated concepts from the Launch Pad that were critical in developing an art class project with a science theme at the high school level.

There was also a report by one Launch Pad participant that they have become a member of the Science Fiction Writers of America's SIGMA SF group, an informal association of SF writers who make themselves available to policy makers who are interested in outside-the-box thinking. Groups who have solicited their input include the Defense Department, the Department of Homeland Security, and an advanced research group in the intelligence field.

Here are some participant comments about the impact of attending the Launch Pad workshop:

"My experience at Launch Pad affects practically everything I write, in part because it affects what I choose to write about. With greater confidence in my astronomical knowledge, I choose to write about astronomical themes more often."

"I am on sabbatical in the fall for the express purpose of working on a novel, one that I expect to be significantly informed by my Launch Pad experience."

"Launch Pad was a big help to me with my book due out next year. The knowledge I gained there also helped me with scientific accuracy for the next several books in the series and I plan to rely heavily on the information (and experts!) I met at UW when it's time to start writing them."

"Launch Pad was an amazing experience."

"...in my novel writing, is a heightened appreciation for the value of scientific accuracy in my stories, or at least intelligent plausibility in situations where I am extrapolating beyond the bounds of what we know today."

"...the experience was both incredibly enjoyable and educational. The lessons I learned there have infiltrated every level of my fiction in production; I step carefully with science in the wake of that week, researching every natural point I can to be sure I get it right."

"I simply cannot laud the workshop enough, and I can only hope that more and more writers and educators will have the chance to partake in what is truly a once-in-a-lifetime experience."

"...the best straight science-science fiction idea I've ever had came directly out of that workshop."

"Launch Pad has influenced my writing and teaching of writing almost every day since we were in Laramie. I cant thank you and NASA enough."

Frankly we have been overwhelmed by the universal praise and powerful recommendations participants have been giving us and the program exceeded our expectations. There are occasional critical comments but we have never had anyone report a negative experience overall. This effort has been hugely rewarding.

A One-Day Version as an Alternative Model

It can even be challenging for would-be participants to take off a week for an intensive workshop. At the invitation of Dragoncon held annually in Atlanta, Georgia, a convention that attracts about 50,000 attendees, we have recently initiated a one-day version of the workshop there.

Any topic can be taught in any length of time, with varying degrees of success. A one-day event is challenging, but was well received in its inaugural year of 2012. In this case all we can hope for to provide background, networking, and inspiration. Time will tell how effect we are compared to the version we hold in Wyoming.

Preventing Armageddon: A Call for Every Field

The movie *Armageddon*, about preventing an asteroid impact, directed by Michael Bay and starring Bruce Wills, has grossed hundreds of millions of dollars. The film averages about one error per minute, many of them scientific in nature. That's a lot of errors, many of which viewers will not realize are errors. We do learn from what we watch, even if we deny it, or advertisers would not spend the billions of dollars that they do.

We should strive to prevent another *Armageddon*, but the truth is that there is a similarly error-ridden movie, or many, for every subject under the sun. Law, history, geology, medicine, etc., all suffer numerous and unnecessary errors in many productions.

I have received communications from scientists in other fields (e.g. biology and neuroscience) who learned about Launch Pad and asked about how it was funded and for copies of my proposals. The National Science Foundation, for instance, likes to have educational components to grant proposals and see broader impacts from their funding. Launch Pad is a positive element by those criteria and helps a proposer stand out from others who describe common activites like teaching students and giving an occasional public talk. Versions of Launch Pad geared to other fields like chemistry may be an easy sell to some review panels and a significant plus to a grant proposal.

While movie productions and TV shows can hire science consultants to correct mistakes in scripts, often a better product may be produced if the writers themselves know enough to avoid mistakes in the first place. There is interest from writers in getting things right, and science is a particularly challenging subject for many, especially if they arre beyond traditional education. Build the program, and they will come.

Writers are well educated about literary sources and can quote Ben Franklin about how an ounce of prevention is worth a pound of cure, but many don't necessarily know that scientists use the metric system or other field-specific units. Scientists can help by telling them that, and a lot more.

Chapter 4

How *The Universe* Is Made

Colin Campbell*
***E-mail: cc@colincampbell.net**

What really goes into making an episode of *The Universe*? How do we pick topics and find experts? As with most journeys, it begins with an idea that launches a series of events, some of which can be controlled and some of which include industrial freezers, footage-eating cameras, and "The Icy Ravine of Death."

When it comes to the most elemental of all of the great mysteries of existence, my guesses are no better than yours. I really have no idea how the universe was made.

I can, however, tell you a great deal (*1*) about how *The Universe* is made (*2*).

For almost a decade, the fine folks at Flight 33 Productions in Sherman Oaks, California, have been producing a popular science show that, for the most part, sticks to straight science. The series has been a consistent top seller on home video. When The History Channel (*3*) was looking to rebrand its "History International" channel as "H2," it ordered up new episodes of *The Universe* to help seal the deal. And why not? As a show, it's reliable. It's solid. And the content? Well, it's universal.

As the writer and director of four episodes (*4*), I've been lucky to work with some top-flight people on both sides of the camera. I'm hoping here to provide some insight into how the show comes together. What really goes into making an episode of *The Universe*? How do we select topics? How do we locate experts? And how does it end up on your TV/on your iPad/being pirated to YouTube by a guy in Malaysia with the screen-name "KittehLuvr32"?

As with most journeys, it begins with an idea.

Three Exciting Days

I'm a freelance writer/director on *The Universe*. From my experience, the three most exciting days in the production of an episode are as follows:

1) The day you confirm you've booked the gig. (Freelance, remember?)
2) The day you find out what your topic is going to be.
3) The day it airs on TV. (Yay! Still a unique thrill.)

As a freelancer, I have very little control over any of these days, exciting as they are. That's especially true when it comes to the selection of the topic that will be at the forefront of my thoughts for the next few months. So, if you're writing and directing an episode, where does your topic come from? (Or, to put it more properly, from where does your topic come (*5*)?)

First, you embark on a journey that includes aspects of both gradual and sudden enlightenment, not unlike those undertaken by Buddhist monks, only with a lot more caves.

<BEGIN SEQ="DREAM">

You start your trip, ironically enough for a "hard" science-based show, by flying in to Roswell, New Mexico. Grab a rental car, and head east and a bit south, to nearby Bottomless Lakes State Park. You're looking for Lea Lake, where scientists still aren't sure how water circulates in and out of what should be a self-contained system.

There is, of course, a simple answer as to where the water goes—it flows through a system of sparkling underground aqueducts (*6*). These ten-foot diameter tubes provide an underground connection between Lea Lake and Carlsbad Caverns National Park, 100 miles to the south. But how to find them?

Park in the lot on the north shore, just off County Road 409, and dive to the bottom of the lake (you brought your SCUBA gear, right?), shooting for a spot approximately 300 meters south/southeast of the lake's geographic center. On the lake floor, locate the small triangle of artificial rocks placed to precisely replicate the relative positions of Deneb, Altair, and Vega (*7*), press down on "Deneb," then swim 75 meters west to a mountain-like underwater rock formation. Scan the underside of the most prominent overhang; you're looking for the now-open hatch to an underwater "moon pool (*8*)." Pop up inside, then get changed back in to civvies for the next leg of your trip, via a one-person submersible, launched through the curving, twisting hydro-pneumatic tubing.

After your underground/underwater journey, you'll be deposited at the bottom of a spectacular cave, far more grandiose and glittering than anything the National Park Service shows for the tourists at the "official" Carlsbad Caverns (*9*). Here, you are off the map, in a cave that doesn't officially exist. Hop out of your pod, and you'll see, in the center of the cave, a shaft of multi-colored light, shining down on an exquisitely set Victorian-era dining table.

Take a seat.

For the next two hours, you are treated to the meal of your life. Tankards of ale. Flagons filled with Kosta Browne Pinot Noir Sonoma Coast 2009 (*10*). A *sommelier* to explain to you the difference between a tankard (a silver or pewter stein) and a flagon (a leather, metal, or ceramic pitcher). Mountains of only the finest meats and cheese are produced and presented to you, shimmering cuts of beef so succulent you can almost smell the Japanese alfalfa fed to the virgin heifers (*11*) as they began their own journey, first, to Wagyu, and then, to your now incredibly satiated stomach.

Finally, as your dessert is cleared away, the tablecloth is scraped clean, and the lighting inside the chamber changes just a bit. You now hear a low throb reverberating through the cavern, a sort of pulsing heartbeat, but of who or what? The planet? That would be silly, of course, but it does add to the overall air of anticipation that suddenly seems to be building.

Now, the *fromagier* (*12*) brings you a small royal blue envelope, sealed with artisanal beeswax, the wax imprinted with a logo familiar to Sagan-ites and space probes - the naked forms of a man and a woman, standing next to each other, the man's right arm bent and raised in a gesture of greeting.

Your hands tremble as you open the envelope.

On the inside, in simple white text on a black index card—a motif that evokes the very universe itself—is your topic:

"The Outer Planets."

You contemplate this for a moment. And as you do, you hear the quiet hiss of a previously hidden elevator door sliding open in the cave wall behind you. The message is clear: It's time for you to go.

You leave a generous tip for the *fromagier* and begin the return to civilization. You get up from the table and head for the elevator. And as you take one last, long look at this cave of wonders, and contemplate all you've seen and all you've experienced, you realize that there is more to see in this world than can ever be seen, more to do than can ever be done (*13*). But you have your assignment. You now know what you must do.

As the elevator doors begin to close, you take another look at the card in your hand, and suddenly, you call out, "Wait! 'The Outer Planets'? Did you want me to include Pluto???"

But the elevator door has closed.

</SEQ>

At least, that's how I daydreamed it would go when I took the gig.

But as it turns out, the topics for our show emerge from a much more mundane process.

Someone tells you what your topic is.

That's it.

Now, there's probably another whole chapter (and perhaps even a corresponding reality show—pitch alert!) to be written concerning the generation of topics for shows such as this. After all, there are only so many different combinations of planets, galaxies, stars, perils, space phenomena, dangers, cataclysms, world-ending calamities, and end of existence scenarios to cover (*14*).

So, at the start of the season, the office staff generates one list of ideas, the network has its list of ideas, the EPs have a third list of ideas. Occasionally, the writer/directors are asked to contribute some ideas (*15*). All of these topics end up on a master list with the EPs, which, after some trimming, then goes to the network. My understanding is the network ends up blessing, say, 12 or so, a number that is then further whittled down by the EPs to the final 10 or so that go into an average season order.

Among those final 10, there may be episodes that contain themes similar to those we've done before. Multiple times. By about Season 7, and this is certainly a concern of all TV shows, not limited to just ours, things are getting a bit dicey on the "original idea" front. Between rejected crossover ideas ("Ice COMET Truckers—huh? Think about it!") and the inevitable infiltration of "whatever is popular at the moment" ("What if ... and stay with me ... but what if Honey Boo Boo went to Neptune..." "BUZZ! NEXT!"), the key to making the show work is to find people who can bring a fresh, new approach to material that may have been previously covered.

It's time to bring in... The EXPERTS™.

Expert Storytellers

The best part of working with the experts we interview for *The Universe* is that "a-ha" moment, when something they say in response to something you say leads to that golden moment of clarity we all seek as storytellers (*16*).

With luck, the first of those moments (and you're hoping for many along the way of producing this episode) will come early in the process. Solving the riddle of how best to bring your topic to exciting television life is the first order of business.

Before production starts, pre-interviews with the experts are critical to developing and fleshing out the bare-bones skeleton that the writer/director has in mind. I may have found what I think is a bunch of cool, fun stuff for the show, but I can guarantee you the experts I'm talking to know so much more. It's not uncommon for me to pitch an idea or topic to a scientist, and hear, "Well, that sounds good, but what's **really** interesting is..." And then we're off in a whole new direction.

Once the writer/director has figured out what's in and what's out of the show, the collaboration with our experts continues. We discuss who is going to be covering what—not every expert covers every topic in the show. We'll also go over possible interview locations. I'm a big fan of trying to tie interview locations in thematically with the episode's topic (why not, right?), so for "Deep Freeze," all of our interviews were either in very cold places (Ice rinks! Industrial ice warehouses!) (*17*) or places we could light to achieve a similar look.

Sometimes, a topic suggests a course of action or research that leads to an inevitable structure. "Jupiter, the Giant Planet" is likely to be a fairly straightforward episode. On the other hand, coming up with both the content and the structure of a show called "Deep Freeze"—well, that might be a little trickier.

Side note: So how did we solve that chilly conundrum? After getting no satisfactory answer to my repeated questions (most of them along the lines of,

"Hey, what exactly is supposed to be in this episode? Anyone? Hello?"), I finally settled on a double-helix-inspired parallel track (*18*). As the episode progressed, we would "move" both chronologically and geographically; from closest to Earth to furthest away in space, and as that was happening, we'd also travel from the Big Bang to the end of time. (Simple, right?) I think it worked... but then again, this structure was buried beneath the snow, so to speak, so it probably wasn't noticed by anyone. Which is how it should be. No one wants to see how hard the duck is paddling underwater; they just want to watch it glide along the surface.

Questions, Anyone?

With the basic idea in mind, and the first inklings of structure starting to take shape, we're ready to start shooting interviews. You'll need a long list of questions, and a group of interesting people (and interesting locations) for this step. So how does someone get to be an expert on The Universe (*19*)?

Producers and associate producers are always on the prowl for fresh voices and new faces. We've seen people on YouTube whom we thought would be a great fit for our show, and they usually work out pretty well. (You never know which random interview or podcast could be the one that gets seen by someone somewhere, and ends up leading to something else. That's Hollywood, baby!) We've also booked interviews with people we've seen on NASA-TV, or (after a phone interview to make sure they can speak) authors of papers or research related to the topic at hand. We even booked someone after a Flight 33 staffer (not working on our show) saw them giving a TED talk. And yes, sometimes geography plays a part—Flight 33 is based in southern California, so if you're at UCLA, USC, CalTech, or JPL, the chances of you ending up in one of our shows are probably slightly higher than those of your colleagues at University of Alaska - Chukchi Campus (*20*).

Speaking of dangerous snowy wastelands...

Unearthly Glows and Sunrise Bears

Let us talk now about a place that shall hereafter be referred to as "The Icy Ravine of Death."

It should go without saying that this is a place that no television writer/director should end up, ever, much less at 3:00 a.m. Much much less in an area named for (presumably carnivorous) bears.

And yet, in our dedication to bring to you, the home viewer, the most authentic, most vibrant, most exotic locations, this is the exact place where I found myself during the production of season one's magnum opus, "The Outer Planets."

Let me explain.

One thing our scientist experts are very good at is suggesting new and unusual locations for shooting. After all, there's only so many times you can interview someone in a generic field, or standing in the woods. So when astronomer Dr. Henry Throop, along with my associate producer and I, were discussing places on Earth that we could possibly shoot that might pass as a plausible simulation of the icy surface of Pluto, my ears perked up when he mentioned he knew of just the right spot. And before you knew it, the trip was set up. It was to be an early morning shoot. A very early morning shoot, Dr. Throop said, because we would try to capture that unearthly glow that sometimes comes over the horizon just before sunrise. "Great," I said. "Where did you say this place was again?"

Turns out, it was the charmingly named "Bear Lake." Named for the **bears** that were often found in the that area. In the middle of Rocky Mountain National Park.

Special bonus: It wasn't going to be a short hike. In order to get to the area Dr. Throop was thinking about, we'd have to hike in, at least a couple miles. (And what's this you say about "bears?")

The Sun Also Rises

Okay, let's back-time this puppy. We have to be set up and ready to shoot the interview by pre-sunrise. We're going to need at least 45 minutes to get the camera and lighting gear set up. But we can't take too much gear, because we'll be hiking through the Rocky Mountains, and we don't have a production assistant with us—it's just Dr. Throop, the camera guy and myself. (This is actually somewhat of a bonus—it can't take THAT long to set the gear up, because we aren't bringing much.) But we will need to budget 90 minutes to walk in. Because, as previously mentioned, it's a bit of a hike. Like, through the woods, on a snow-covered trail through the mountains. Let's assume that it'll take a good 15 to 30 minutes for us to prep for this HIKE WITH CAMERA GEAR THROUGH THE SNOWY BEAR-INFESTED MOUNTAINS IN THE DARK.

Time for the math. Sunrise? It's at about 5:45 a.m.

Pre-Sunrise? 5:15 a.m.

Set up? Gotta start that by 4:30 a.m. Because, you know, instead of "losing the light," we're going to be gaining it. And we've got only one shot.

Hike? Must depart by 3:00 a.m.

Meet up with Dr. Throop? Better schedule that for 2:30 am in the parking lot near the trail.

Which is all well and good, except for the fact that this was the SECOND shoot on our trip. Which meant the day before, we'd be working with another scientist in Denver until 6:00 p.m. or so. At which point, we'd have to break for dinner, then make the two-hour drive to RMNP (*21*).

Those of you who have deduced that this sounds like an insane plan, congratulations. Those of you who are simply nodding your heads, thinking "this sounds like television," you also get a gold star.

So, when the time comes, the camera guy and I wrap in Denver, grab some grub on the road (in Boulder), and make the night trip to Estes Park, CO (where, sadly, the production company missed the opportunity to book us in to the famed Stanley Hotel; just as well, seeing as we'd be at the hotel for all of about 4 hours). We crash out for a couple hours, then wake, shower, and head for the mountains.

“Hiking” on the “Trail”

I’d like to report that “all’s well that ends well.” I’d like to, but I can’t.

Because your humble correspondent is the dumbest person alive.

You see, when they were telling this city boy all about a “hike through the snowy mountains,” all I could think was, “Yeah, but it can’t be THAT snowy! I mean, we’re civilians, not hikers. For goodness’ sake, we’re from TV (*22*)!” In my head, for some reason, I’m picturing a “trail” in the sense of that trail that runs all over Tom Sawyer Island, in the middle of the Rivers of America at Disneyland.

It quickly becomes apparent the Bear Lake trail is not that kind of trail.

For starters, it isn’t paved.

Additionally, your humble correspondent failed to heed the numerous, “Ah, make sure you wear appropriate hiking shoes” warnings issued by my producer, and by Dr. Throop himself.

Yes, I showed up for this snowy mountain hike in a $60 pair of Nike Air Whatevers.

“Hmm, that could be a problem,” said Dr. Throop, who, fortunately, just happened to have brought an EXTRA SET OF SNOWSHOES with him. (Note to self—if you ever find yourself scheduled to do another interview where SNOWSHOES are required, reconsider the booking. Or at the very least, try pitching History/Flight 33 on an episode featuring the most tropical places in the universe instead.)

Strapping on the snowshoes(!), we set off into the dark.

Now, I don’t want to bore you with further details (*23*). Let me just point out the following.

1) We were not eaten by bears.
2) I’m not what you would call a “strong” snowshoer. I believe this remains the only time in the history of *The Universe* that a guest actually had to carry a director out of the filming location.
3) It’s a good thing the hike to the location was conducted in the middle of the night (we used coal miner-style lamps on our heads to light the trail), because when we hiked back out in the sun after the interview, I pretty much freaked out when I saw that just off the trail on both sides of us was a very steep drop to hundreds of feet of “ravine,” at the bottom of which the bears were probably waiting.
4) It would have been a GREAT FANTASTIC ULTRA-WONDERFUL shoot... if only the damn camera hadn’t eaten some of the footage.

Yes, that’s right, after risking life and limb (and bears) to get this wonderful, one-of-a-kind location, we were let down by a failure of non-linear media storage (*24*). Most of the interview (thank goodness) made it. Most of the “beauty” shots did not. Which is why, if you watch the episode, you’ll see plenty of Dr. Throop’s interview, conducted while he was sitting in what does look like a quasi-Plutonian snowy mountainous area. But you will see very few of the establishing shots needed to really sell the idea of why we were there.

This is television: In turn, equal parts fabulous and frustrating. And almost always produced under the threat of being eaten by bears.

Get Me Rewrite!

Assuming you survive all of your shoots, you return to home base with loads of video (drop those off at the office to have transcripts made and to get the footage loaded in to the edit bay) and tons of new information about your topic.

At this point, your TV shows approaches its most "term paper"-like state. You make another pass through the script, adding in bits and pieces of information gleaned through interviews and research (*25*). When you get your transcripts back, you review the sound bites and try to make sure people said what you think they said. Mix and match, bite here, announcer track there, let the story breathe, and voila, you have a script. My first drafts tend to run a tad long; 80 pages or so is not that unusual (*26*). In the two-column format we use for scripts, the standard TV guideline of "a minute a page" holds up remarkably well. So given that the show runs about 44 minutes, there is definitely some editing to do before the script goes off to the bosses for approval. Trim out all the fat, figure out what stories are working the best and focus on those, do another pass to remove any duplicate information, then send a nice, tight 50-pager off (*27*).

As the director of the episode, you also get to help design the CGI graphics needed for your episode. It's one thing for astrophysicist Alex Filippenko or astronomer Pamela Gay to talk about why the Boomerang Nebula is colder than the space surrounding it, but it's another to explain that explanation to CGI artists in such a way that they can bring it to life in a visually exciting (and educational!) way (*28*).

When you get your script back, you apply the notes given by your bosses (and the network), and head in to the edit bay. Let me state here, for the record, that the editors of *The Universe* are as responsible as anyone else for how great the show looks. Again, with multiple seasons under their belts, there's no need to reinvent the wheel; the format of the show is the format of the show. But the editors on our show are true collaborators, often able to tease out exciting segments from moments that might have otherwise been buried in the footage. They also do a fantastic job of trimming up segments that may seem to play okay on paper, but feel long when you actually watch them. A team of associate producers and researchers feed the edit bays with footage, footage, footage (we can never have too much), often from NASA (yay for free footage!) or other acquired sources. We sometimes organize small pick-up shoots for things that we may have missed in the field. (Whoops, forgot to get that shot of the comet research website!)

At this point, we'll also send the script to a few of the experts interviewed in the show for a fact-checking review, just to make sure we didn't botch anything so badly in the editing process that we've accidentally changed the meaning of a key fact or statistic. Sometimes, and this should come as a surprise to no one, experts send back contradictory notes. One thinks you've gotten something wrong one way, and the other thinks you've gotten it wrong another way. Or worse—one expert thinks the other got something wrong. When that happens, it's up to the writer/director to reconcile the two notes—or, as has been known to happen—cut the offending material out entirely. Unless it's a critical point, it may be better to let it die in the edit bay than go on the air "wrong." (By this point in the process, we're usually starting to feel some time pressure, because this show does have to go off and be on TV at some point.)

Once the editor has the show cut to the proper time, our ace post production producer takes charge. Juggling multiple responsibilities, she supervises the final round of notes, makes sure we put a polish on the video, works with a sound mixer to punch up the audio, gets the final version closed-captioned, and then outputs the finished show to tape and ships it off to the A&E Networks (parent of HISTORY and H2) in Stamford, CT (*29*). FedEx, *The Universe* is in your hands...

Expanding with the Seasons

One of the questions we addressed in "Deep Freeze": Will the universe continue to expand indefinitely?

That's one of the larger questions facing *The Universe*, as well.

I remember two years ago, walking through Costco with my wife, when we came upon a snazzy-looking DVD box set—*The Complete Universe*—which included seasons one through five. "Well," I said to her, "I guess that's that. No more *Universe* episodes..."

Four months later, I was standing in an industrial ice freezer, handing my jacket to a shivering astrophysicist as we filmed an episode for season six.

Months later, when the episode aired, I could almost still feel the chill. But now it was mixed with a tingle of excitement upon seeing everyone's hard work shared with the world. (That excitement and pride is a feeling I hope never goes away, no matter what I work on.)

New discoveries mean there will always be material for new episodes. Whether the economics of television production makes sense for those episodes to be produced is a question that is far above my pay grade.

I do know this: If the call comes, I'll answer. As a wise Jedi once put it, "Always in motion is the future," and if the future means another trip to the outer reaches of *The Universe*, well, that would be more than fine by me.

Notes

1. Within the confines of a standard NDA, of course...
2. Standard disclaimer: Here, I write to represent myself only; my thoughts and opinions are not necessarily those of A&E Television Networks LLC, HISTORY, Flight 33 Productions, or anyone else but me. Me! You hear me? Me!
3. Yes, I know they're now called just HISTORY. But that looks weird in a sentence.
4. "Alien Galaxies," "The Outer Planets," "Science Fiction, Science Fact," and most recently, "Deep Freeze." Catch 'em on iTunes!
5. Thanks, Mom, for years of bugging me about grammar.

6. The aqueducts appear to be made of material similar to a 6061 Aluminum/ Magnesium alloy; with 40% of the electrical conductivity of copper, it's suspected that the tubes may also be used for low-grade data transmission, although that has not yet been confirmed.
7. The famed "Summer Triangle"
8. LOST fans will recall it was via moon pool that one entered and exited the underwater "Looking Glass" station. Poor Charlie! But I digress...
9. Anyone who has visited, as I have, the "Secret Yosemite" or the "Incognito Yellowstone" knows our truly spectacular National Parks are closely guarded and protected, with tourists directed to nearby facsimiles which pale in comparison to the originals.
10. "Wine Spectator's" 2011 Wine of the Year, although they will bring you a bottle of runner-up Cabernet Sauvignon Napa Valley Kathryn Hall 2008 if you make a fuss about it.
11. The *Tajima* strain of Japanese Black, for those meatologists among you.
12. Also known as a Maitre d'Fromage. Really! He's the chap in charge of cheese, of course.
13. Thank you, Elton John and Tim Rice.
14. Yes, we do a lot of "Space - It'll Kill Ya!" What can I say? It gets ratings. And when we get ratings, we get to do more shows. More shows = more opportunities to cover real science. And this is a good thing.
15. Sadly, my pitch record is "0 for *The Universe*." Zippo the clown. A complete shutout. But maybe someday...
16. In my head, this moment is accompanied by an angelic version of the famed THX "Deep Note" Sound; your epiphany soundtrack may vary.
17. There's no CGI fakery involved in the steam breath Alex Filippenko generated during our interview inside the Union Ice facilitiy in Van Nuys, CA. Alex also had to borrow my jacket and gloves for the shoot so he could answer my questions without chattering teeth.
18. In the business, we don't call this "Pulling a Watson & Crick," but we should.
19. Practice, practice, practice!
20. Located in beautiful Kotzebue, Alaska. Although, with an average January temperature of 4° F, this probably would have been a pretty good location for "Deep Freeze." Next time...
21. Rocky Mountain National Park. I feel like we're able to use acronyms now, you and I. Am I wrong?
22. I actually thought this.
23. I know, I know; why stop now?
24. I don't want to call them out by name, but if it happened to you, you'd be pee'd, too. (Camera recording format joke!)
25. It should also be noted that when I started doing these shows, not all that long ago, an associate producer would send you mountains of books, purchased just for you, or checked out from a local library. This is no longer the case; almost all research is done online now. I miss the books.
26. It's a bad habit; I'm trying to get better/tighter on my first drafts.
27. Got to leave a little extra fat on the bones; it'll really help when the bosses start whacking out large chunks.

28. With several seasons of shows in the bank, you can also pull from an extensive library of already-generated animations. This helps you focus your budget resources on creating new animations that are way too specific to have been previously made.
29. This description of the post production process allows me to give an unreserved shout out to Flight 33's foxy post producer, and yes, I called her foxy, because she's my wife and the mother of my two beautiful children, and we're on a search for truth here, and she IS foxy...

Chapter 5

Narrative Alchemy: From Vision to Visual

Alexis R. Gambis*

Filmmaker, Scientist, Artistic Director of Imagine Science Films, Brooklyn, New York 11215
***E-mail: agambis@imaginesciencefilms.org**

Structured in the format of a scientific paper, I outline my philosophy on visualizing science from the perspective of a scientist-turned filmmaker and as the founder of *Imagine Science Films*, a non-profit organization committed to promoting a high-level dialogue between scientists and filmmakers. At the core of my philosophy, I propose that scientific truth does not need to be bent or embellished to make a screenplay exciting. Rather, scientific observation can be an effective springboard to cinematic imagination. By breaking stereotypes, humanizing scientists and mixing scientific material with personal drama, scientific concepts can be transformed into a narrative alchemy.

Introduction

As I sat in a 'Feature Film Screenwriting' class, my turn came to present a one-page synopsis of my feature, *My Geneticist* (Figure 1). The film was inspired by the story of Calvin Bridges, an American scientist known for his seminal research about fruit fly genetics. I spoke about the story of Calvin who worked with Dr. Thomas Morgan in the early 20th century at Columbia University. My presentation left a few students with puzzled looks. After class, a fellow student came up to me and asked: "Why do you only make films about science?" As a scientist turned filmmaker, science is encoded within my DNA, so it's only a logical next step of my personal evolution.

Figure 1. Still from the short film My Geneticist written and directed by Alexis Gambis (2012, 14 mins) and produced in the Film Graduate Program at New York University Tisch School of the Arts. Story: While at first Betsey feels closer to her father than she ever has, she ultimately discovers secrets about him and his laboratory that will transform their relationship forever.

So, what is a science film? A science film is a coming of age story about a scientist applying his findings in the lab to his personal challenges outside the lab. A science film is about basic research. A science film follows the creative deconstruction of nature and evolution. A science film is not genre-specific. A science film is an alchemy of real-life events and experiences. A science film is about understanding human existence. Above all, a science film explores the same fundamental questions about the world that are asked at the beginning of any scientific study.

Background: The Chemistry of Vision

Isaac Newton once wrote, "If I have seen further it is by standing on the shoulders of giants." In a similar sense, many science papers begin with a brief summary of the work of those who have come before, and the principles upon which the work is based. It is fitting that Newton speaks of sight, as it is vision that is one of the fundamental principles upon which filmmaking is based.

Vision is so common that we forget that it is the driving force that helps us see and interpret the objects that surround us. *Photons*, particles of light, emitted by the Sun shoot towards Earth at nearly 300,000 kilometers per second. They reach Earth in slightly less than eight-and-a-half minutes. While most of these particles will be absorbed by water, land, flora, and fauna, some reflect off these surfaces, and eventually make their way to our retinae.

The visual process is a beautiful example revealing how light can produce molecular changes. Although eyes come in many shapes and structures throughout the animal kingdom, all visual systems share basic mechanisms that respond to the information about the environment. From flies to humans, eyes are populated with light-sensitive cells called *photoreceptor cells*. These cells are used to capture photons and then transduce the photons into an encoded electrical signal to the brain. Receptors consist of two types: cones to provide color information and sharpness of images, and rods that provide vision when light levels are low. The tops of the rods and cones contain a region filled with membrane-bound discs, which contain the retinal molecules bound to a protein called *opsin*. The resulting complex is called rhodopsin. Because rhodopsin absorbs primarily light within the green-blue portion of the visible spectrum, it appears violet; for this, it is also known as *visual purple*.

Once light impinges upon the retina, rhodopsin molecules undergo chemical changes, notably isomerizations, leading to rearrangements and the formation of molecular complexes. Changes in geometry initiate a cascade of biochemical reactions that result in an electrical potential difference that builds up across the plasma membrane. This induces an electrical impulse that is passed onto the brain via nerve fibers. The brain determines which nerve fibers carried the electrical impulse activated by light at certain photoreceptors, and is then able to construct an image.

Entering into a movie theater, the cones in a movie viewer's retinae become more sensitive and the rods are typically activated later. Color perception remains limited, as cones need more light than rods to work properly in the obscurity and rods do not provide color information.

During the film, like LEDs on a billboard, the rods and cones of the eye blink rhythmically with the visual dance, creating their own coding from the moving images on screen. The coded information is then sent via the optic nerve to the brain where processing, decryption, and, ultimately, interpretation takes place.

As the viewer exits the movie theater, the world appears staggeringly bright at first sight. The viewer's rods, exposed to low light during the running of the film, have become *saturated*. They turn off in these bright conditions, and it often takes a few minutes for the cones to begin functioning nominally again, and for normal vision to return. For the retina, the movie experience does not end in the theater and continues on for a bit longer. For the brain, the movie experience can go on for hours, days, years or more, depending upon the emotional impact the movie made.

In the early 20th century, film was even used as a scientific visualization tool—to study nature and animal behavior because of its intrinsic empirical quality. Time-lapse films *The Birth of a Flower* and *The Acrobatic Fly* brought mesmerizing images and scenery to the public, capturing the poetry of flowers and insects. For a modern science film, the visualization of scientific research provides a way to synthesize, explore, structure, and communicate scientific information to others.

Method: The Science of Narrative

Storytelling is universal in our world and its universality suggests that it is deep-rooted in our genes. From birth, we rely on narrative order for survival and for the construction of our own identities. While there is similarity in our cultural and genetic makeup, the combination of environmental and genetic factors makes us unique.

In *The Man Who Mistook His Wife for A Hat*, neurologist Oliver Sacks writes:

> …each of us is a singular narrative, which is constructed, continually, unconsciously, by, through, and in us—through our perceptions, our feelings, our thoughts, our actions; and, not least, our discourse, our spoken narrations (*1*).

Narrative is not only the stamp of our existence but it is wired in our brains, which in turn feed on it for maintenance and stimulation. Simply stated, story enables the human mind to explain itself-to-itself and to other minds. Cognitive neuroscientist Michael Gazzaniga states that the "*the human mind is disposed to creating stories or narratives*" (*2*). He speaks of the importance of narrative in mental health:

> From early childhood, we tell ourselves stories about our actions and experiences. Accuracy is not the main objective—coherence is. If necessary, our minds will invent things that never happened, people who don't exist, simply to hold the narrative together.

Gazzaniga further suggests that "preserving narrative continuity" is necessary "to fill in the gaps of memory." A person struggling with Alzheimer's disease feels pain and confusion when recalling details from their personal narratives. A child exclaims, 'I am not alone' after identifying with a character on-screen. Stories link the factual to the emotional, the specific to the universal, the past to the present.

How does one accurately capture 'the personal narrative,' though? It is so complex that one is not even aware of its finer resolutions, and, evidently, it is a challenge to replicate. Both the filmmaker and the scientist courageously attempt to decode and 'visualize' narrative. To capture narrative one needs to create narrative. The scientist or filmmaker creates his/her own version of reality, one that is enriched with their own subjectivity. Nevertheless, the visual imagery is always an approximation of the 'real' narrative.

A true depiction of the 'personal narrative' remains impossible with the existing scientific visual. Perhaps, there will come a day when we can fully represent all the facets that comprise human perception—a hypothesis that is for now in the realm of science fiction. With even the highest resolution microscopes, we are still making observations and inferences from an exterior perspective. Even with advanced brain imaging, neuroscientists are still predicting realities and inferring meaning to brain function.

Cinéma vérité, a term coined by the French anthropologist Jean Rouch, is the empirical practice of using a camera to attempt to capture truth and subtleties in moving life. Since the birth of the camera, observational documentary has been used as a scientific model to understand the world and to find patterns in nature. With modern breakthroughs in digital filmmaking, the camera further takes on a voyeuristic appearance. Surveillance-like cameras have enabled us to study the subject without the latter being aware of the presence that he/she is being filmed. Macro photography has also enabled us to probe into the microscopic worlds, where the smallest of the microbes become the larger-than-life performers on the big screen.

MIT researcher Deb Roy wired an array of cameras around his household to study the acquisition of language in his newborn child (*3*). He refers to his scientific study as "a piece of what is by the far the largest collection of home video ever made." On the screen, he presents recorded material as a disc array, resembling petri dishes. Each camera contains data about the environment, its inhabitants and activities. In his TED talk entitled the "Birth of a Word," Roy takes you through the epic journey of a child coming into the world and striving to create order in his environment.

Observations: The Truth About Fiction

Observational material is the raw data of science and filmmaking. From here onward, scientists and filmmakers may not always share similar views on how to use the empirical data.

Scientists strive to be innovative thinkers yet always aim for realistic and objective interpretations based upon their collected data. Still, the scientific approach is inherently subjective and creative. Furthermore, human error is inevitable. Filmmakers do not face similar repercussions to truth bending. They can take observational footage and deliberately reshape it. They can lift real-life events and adapt them freely. They thrive from fictional constructs. Jon Amiel, who directed the movie *Creation*, about the evolutionary biologist Charles Darwin, shares:

> Plot is consequence. Plot is Newtonian physics. Plot is action and equal and opposite reaction. It's true in science, too, that sometimes the one thing that you make up will better illuminate the truth that you're getting at than a painful accretion of facts.

However, not every filmmaker can do this artfully. If done poorly, the viewer may be turned off by "fakeness" of the whole cinematic spectacle.

Scientific filmmaking may sometimes be caught in between the worlds of the imaginary and the reality. Take for example the collection of science films by filmmaker-scientist Jean Painlevé. In his *Science is Fiction* vignettes, he presents real footage of marine life overlaid with personal anecdotes sometimes told from the perspective of the creatures themselves. With Painlevé, surrealism and dreams get mixed in with 'serious science.' What may seem like confusion between the real and the unreal to some, may also be perceived by others as the purest form of science cinema. If the mixture of fiction and nonfiction is done with careful precision, the viewer latches on to real material to fill in the gaps with his or her imagination. On that point, Jon Amiel explains why, initially, he was uninterested in directing *Creation*, and what led to his change of heart:

> I started to find the voices of these people reaching out to me across 160 years in a most extraordinary way. As a parent, I started to connect to Darwin as a parent. As a husband, I really started to connect with Darwin in his marriage—the marriage of two people who loved each other dearly, but held profoundly different views on the most important topic for them in their lives. As far as she was concerned, his beliefs would prevent him from entering the Kingdom of Heaven with her, meaning they would be separated in eternity, yet they had this extraordinarily intricately-intertwined relationship. So I said, "OK, listen, these are my terms. One, I don't want to do a period movie—it just happens to be a story that's set 160 years ago. Two, I'm not interested in a chronological story—he was born here, then he went on the Beagle, and then twenty years later he wrote this, and 35 years later he died. Not interested. I am interested in doing a psychological portrait of a man at a specific time in his life. If we could do that, if we could use anecdote and dream and look at the way… at an extraordinary mind in an extraordinary state of crisis, at a specific juncture, and if we can do that in an emotive and associative way, I'd be interested to try that.

Amiel adds, "We're making movie for cinema, not a PBS TV show."

Filmmakers may opt to fully leap into fiction, attempting the reenactment of 'the observed.' However, constructing fictional narratives that appear seamlessly truthful is a difficult task. There are always several factors at play: sequencing ideas, using language coherently, shifting attention, and relating to other people. Experimental techniques, often involving post-manipulation of the image, may help change rhythms giving momentum to the story. However, narrative should always be held in place with a strong foundation.

Successful narratives often also withhold certain critical story points. Didactic films inclined to teach or lecture others often present information that feels stale, leaving little room for imagination or excitement of the mental processes. Viewers take more pleasure in the cinematic experience when engaging with the narrative created through images, uncovering clues and piecing them together; all the more captivating if they are misled or surprised. In the 1920s when movies by the Lumière brothers were first shown at soirées, salons and cafés in Paris, the audience had no idea what to expect. As trains grew in size on the screen, the audience would jump out of their seats in a mixture of fear and excitement, convinced that the train was coming right at them.

Nowadays, it takes more effort to excite the average cinephile, as he or she is typically blasé from all the fast-cutting visual effects seen on television and in theaters. Nevertheless, film originality may still be achievable by recycling old techniques, presenting never-before-seen perspectives with the advent of new digital technology (ex: microscopic filming) and by crafting unique visual metaphors.

Results: Science in Fiction—Applying the Scientific Method

Science in narrative filmmaking should be evaluated at all stages of development: script, production, post-production and exhibition. Not only does one need to ensure scientific accuracy during the script-writing stages, but also monitor its integration in acting, production design, and editing. The scientific paper with *Abstract, Materials and Methods, Results and Discussion* sections provides a good checklist for addressing all the facets of science portrayal in film.

The scientist should always feel human. Sometimes, this can be achieved by working with real scientists. This allows for mutually beneficial interactions between actors and non-actors and room for improvisation. While the emphasis should be on the scientist's everyday life, a breakthrough may occur. However, it should feel like the culmination of a laborious process rather than a 'quick fix' discovery *à la CSI Miami.*

It is best for the scientific theme to come second to the emotional arc or else the film may be perceived as having a scientific agenda with an imposing directorial hand. The science, however, provides support to the plot and strengthens each character arc. In many ways, the scientific process is akin to character growth, as both evolve through a series of happenstances and accidents, ultimately leading to some measure of understanding of the world in which we live.

Discussion: Science and Film at a Glance

Science and film have had tumultuous affairs over the last century with conflicts, compromises, and passions. In today's culture, the definition of a science film is stereotyped, often narrowed down to science fiction movies, cable channel nature shows, or news-like documentaries.

For as long as film has existed, scientific intrigue has also been expressed through fantasy. In films like Melies' *Trip to the Moon* and Fritz Lang's *Metropolis*, science becomes science fiction with futuristic, imaginary and speculative turns and twists. Einstein's Theory of Relativity or the birth of intelligent machines brought us to places where time travel is possible and human-designed robots invade the planet. The film industry has been continuously stimulated by science to generate fantastic worlds, and to explore fantastic concepts (Figure 2).

Jon Amiel shares:

> I think science doesn't need to be dramatized. Science is drama. Properly understood, and properly inhabited, there is intense drama inherent in almost every scientific idea. You just have to find it. There's the drama of what's going on, why there's tension between two magnetic poles, or the collision of two particles. We use the terminology of catalyst

and currents and charges and things endlessly in our talk about drama. Between massive events that are happening way beyond our Galaxy, or tiny events that are happening every time you switch on a light, there is an essential drama inherent in that if you but look for it, and if you but find a way to characterize it.

One could argue, that film with a strong visual and 'sexy' storytelling capacity has served as an educational tool making science more accessible to the public. It has instigated curiosity for the youngest and the oldest of us. I will be the first to admit that 80's science-fiction movies on TV brought the aspiring scientist out of me.

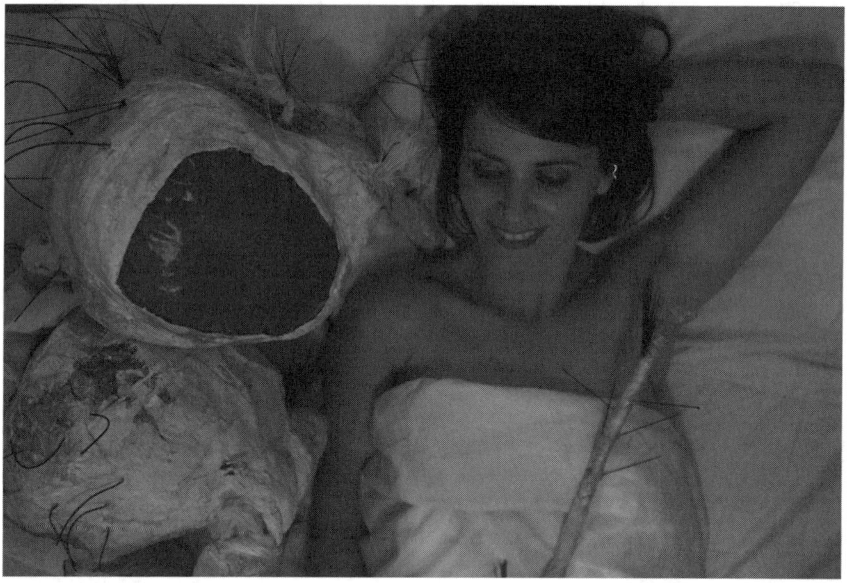

Figure 2. Still from Courtship written and directed by Alexis Gambis (2011, 12 mins) and produced in the Film Graduate Program at New York University Tisch School of the Arts. Story: Scientist-turned fly Lucien and call-girl Victoria speak about the genetics of sexual behavior in bed.

Film has also been helpful in raising the ethical boundaries and philosophical undertones of science. In *Gattaca*, we are placed in a disturbing and terrifying not-so-implausible future where humans are discriminated according to their genetic makeup and receive genetic enhancements at birth to favor their success. In *2001: A Space Odyssey*, the robot Hal 9000 turns against its own creator and takes control of the space ship. With these films, we are pushed to think about the bigger questions posed by scientific discoveries and let our imaginations go wild.

Unfortunately, there is a darker side to the relationship between science and film. Extrapolated science in films does not always ground itself in credibility. These films leave the realistic portrayal of science and scientists, entering the realms of the ludicrous and favoring stereotypes and caricatures. It seems that it is often the dark and powerful side of science, or science and technology gone awry, that appeals to filmmakers. Hence, with millions of people rushing to the big screen, film has the potential to promote misconceptions, damaging stereotypes and outright falsehoods about science and scientists. *Godsend* is a good example. It tells the story of an 8-year-old boy who is brought back to life by cloning. With a story not grounded in credible genetics, the film becomes manipulative and phony. What we are left with is a freakish film that takes an issue of topical interest from the headlines and grafts a wildly histrionic reaction to it, contributing to the ill-informed public that wants cloning, genetic engineering, and the quite beneficial issue of stem cell research banned on the premise that it is killing unborn souls or that it might produce races of three-legged mutants. The everyday scientist suddenly becomes a delirious, mad and 'unethical' scientist who mischievously wants to abolish the human race using pipette tips and stem cells.

The science film language seems to have fallen into two schools of thought. The first is the news-report model defined by a sense of rigor and protocol. The second is the science fiction model, which reacts to the so-called precision of science by revolting against it and inventing implausible far-fetched scenarios.

Rather than eliminating amazing science fiction and dense science documentaries, the hope is to widen the scope of science film by exploring new ways of incorporating science in film. It should be noted that the scientific process is in its essence 'inexact' where models and mechanisms are always faulty and riddled with exceptions. Also, the scientific protocol, similar to a recipe, is always personalized by a scientist and not as rigid as one may think: it contains shortcuts, tips from previous users and creative add-ons represented by an elaborate constellation of arrows and scribbles on the protocol sheet.

Finally, it is also important to recall that breakthroughs were often happenstances or fortunate accidents led by creative minds thinking outside of the box.

Future Work: Science New Wave

The last decade has witnessed a promising effort to change these preconceived ideas about scientific filmmaking and to demonstrate through practice that captivating science films can be made without resorting to clichés or bending scientific truth. And most importantly, as with any other topic, the science needs to be woven into a story.

An increasing number of film schools and foundations are working to influence the next generation of filmmakers to create more realistic science-based stories and to challenge existing stereotypes and biases about scientists through visual media. The non-profit institution *Alfred P. Sloan Foundation* has been at the forefront of these initiatives; it has provided awards, grants and high-profile platforms to increase public understanding of science to a wide non-specialist audience through different forms of communication, notably film. The Science and Entertainment Exchange, an initiative launched by the National Science Foundation, has fostered great dialogue between scientists and Hollywood filmmakers. Both have been involved in the production stages by matching directors with science advisers and also organizing 'science of the movie' discussions. There have also been open forums and Internet websites, notably the Museum of the Moving Image Science & Film site, which provide regular news about opportunities and events at the crossroads of science and film. It also encourages future scientists and filmmakers to become involved by hosting a database of science short films from student filmmakers.

Television and radio have also shown growth in science communication. AMC's hit show *Breaking Bad* demystifies chemistry by picking as its lead role, a renegade chemist, who is a professor and also a drug dealer. *New York Public Radio*'s popular radio show *Radio Lab* uses sound design as a way of connecting science with personal stories.

A fresh new wave of science communication has further been propelled by the increasing number of public events and festivals. The *World Science Festival* swings by New York in the summer every year, bringing fairs, exhibits, screenings, and discussions celebrating science and art. Contemporary art museums are presenting more science-related exhibitions. The Museum of Modern Art in 2010 and 2011, exhibited 'Design and The Elastic Mind' and 'Talk to Me' that explored the relationship between design, technology and science with narrative-inspired installations. The Secret Science Club invites, on a weekly basis, expert scientists to speak about their scientific work and personal anecdotes to an eclectic beer-drinking crowd at the *Bell House* bar in Brooklyn.

Conclusion

In some sense, every film can be considered a science film, because science is embedded in our world and lays the fundamental rules that determine how individuals exist and interact with that world. Scientific thought and observation is inherently narrative lending itself to the film medium. We have over the last decades drifted away from the essence of scientific filmmaking. Science films used to have a slower pace, and were filled with real-life examples with simple yet effective visual imagery. With the advent of digital filmmaking, CGI and the breakthrough in scientific research, we have presented science as a fast-paced delirious ride, oftentimes losing the essence of its organic nature at heart. Science communication should have no agenda. It should not persuade, embellish or scare but rather present the ingredients that will spark the imagination.

References

1. Sacks, O. *The Man Who Mistook His Wife for a Hat*; Summit Books, a division of Simon & Schuster: Orangeville, ON, Canada, 1995.
2. Gazzinaga, M. *Your Stortelling Brain*, January 15, 2012. www.bigthink.com.
3. Roy, D. *The Birth of a Word*, December 1, 2012. www.ted.com.

Chapter 6

The Science Advisor's Journey

Kevin R. Grazier*

Hollywood Science Advisor, Sylmar, California 91342
***E-mail: kg_eureka@yahoo.com**

The Hero's Journey is a narrative form that recurs in myth, drama, and storytelling of all types and from all places and ages. Described originally by writer Joseph Campbell, The Hero's Journey details the stages of a typical epic adventure of a character archetype, the Hero, who goes out into the world and achieves great feats—meeting friends, rivals, allies, and enemies along the way. Cast in the role of Hero, a scientist enters the Kingdom of Hollywood, naive in its workings and inhabitants, with the noble goal of improving the level of science-related discourse in TV and film. Along the way the new science advisor must overcome challenges and adversity, while friends and mentors in the form of directors, producers, writers, even other science advisors, serve to guide and instruct. If the science advisor survives the experience, he bestows upon humanity television and film productions with a greater fidelity to the workings of the natural world, while simultaneously being transformed into a new being: the boundary spanner.

Introduction

Writer and scholar Joseph Campbell observed in his comparative mythology book *The Hero with a Thousand Faces*, that myths and epic stories from wide-ranging eras, geographical regions, and cultures, have comparable dramatic structures and pass through a similar sequence of phases called the *Hero's Journey or monomyth*:

> A hero ventures forth from the world of common day into a region of supernatural wonder: fabulous forces are there encountered and a decisive victory is won: the hero comes back from this mysterious adventure with the power to bestow boons on his fellow man (*1*).

The story of Luke Skywalker in each of the first three *Star Wars* films parallels the Hero's Journey, with George Lucas even publicly acknowledging the influence of Campbell's work on his own. The character of Neo in each of *The Matrix* films follows a Hero's Journey, as does Prometheus in Greek mythology, even Jesus Christ in the New Testament.

As viewers, the Hero's journey should be our journey as well—the audience wants to share the journey, to go along for the ride, to relate to the Hero. A gifted storyteller is one who keeps the audience immersed in the story: who successfully spins a tale striking a careful balance between being fantastic enough to be engaging, yet plausible enough to seem like it *could* happen.

For as long as stories have been told, storytellers have used natural world settings, events, and themes to give their stories a sense of plausibility, of verisimilitude. One of the oldest known works of literature, *The Epic of Gilgamesh* explores themes of life, death, and immortality, in which the weather, a forest, mountains, wild animals, even a colossal flood, compose more than just the backdrop: they are practically characters in the story. Reversing cause and effect, many myths spawned from attempts to explain the aspects of nature—stories were cobbled, and Heroes created, to fit observed natural phenomena. Every previous and subsequent observation of that particular phenomenon provided a grounding for the story. Snowflakes at the dawn of every winter reflected Persephone's grief when she parted from Demeter; the first buds of spring hailed their annual reunion. The white ring around every Loon's neck represented Kelora's shell necklace. Each earthquake resulted when Kashima let down his guard, and the giant catfish Namazu was free to thrash wildly.

Over time, beloved myths and legends fell prey to cool rationality. Although the primary mode of human storytelling evolved from oral retelling to stone or clay tablets to the printed page to the screen—be that television, film, computer, or smart phone—a primary goal of the storyteller is still to keep the audience immersed in the story. Screenwriters today continue the ancient tradition of grounding their stories using our modern understanding of the natural world: a discipline that we now call science. Yet the same science that shed light on the chariot of Helios, can torpedo an episode of *The Last Resort* if depicted incorrectly.

In television and film science inaccuracies, gaffes, or inconsistencies can undermine a story—particularly if the story falls within the science-themed or science fiction genres. Writer/producer Andre Bormanis explains, "The problem is that when you see something absurd it pulls you out of the story. You are not in the world of the movie anymore. You're outside of it, commenting on it and being critical of it, because it's silly." All disciplines within the natural sciences can be difficult to understand, particularly for non-scientists, and screenwriters do their best to translate the worlds of science into something to which television and movie viewers can relate. Kath Lingenfelter, writer/producer for *House, M.D.*, adds, "It's not easy to write for characters who know more than you do, believe me." Even when all the technical jargon and concepts are correct in a screenplay, the writer may have the added degree of complexity of portraying the culture of scientists, a culture to which they may have had little or no exposure.

In the days when learned elders were both keepers and purveyors of knowledge, quizzical listeners dared not openly challenge the content of a story, and there was little recourse to "fact check" the storyteller anonymously. Until fairly recently, even a trip to the library might be considered overly burdensome just to check up on last night's *Man from Atlantis*. The Internet has changed all of that. Tom DeSanto, producer for both the *Transformers* and *X-Men* series of movies, says, "With the Internet, and the amount of research that we can do immediately… it used to be you had to go to the library and pull books. Now with this magic portal into the collective human consciousness, we can do a little more fact checking."

Not only are libraries of information available with a few mouse clicks, there can be a ripple effect. Even if a viewer fails to notice a problem in an episode of their favorite show, if an error was there, somebody somewhere certainly noticed and posted it a few minutes later. Rockne S. O'Bannon, creator of many popular science fiction television series over the past three decades (*Farscape*, *Alien Nation*, *seaQuest DSV*), explains, "I think we're living in a very different world. The bar has been set higher. With the advent of the Internet, if people are interested/passionate in a show, they might be inclined to go online, seek out others who are also fans of the show, at which point they have access to the world. Others who are interested in the same show, may, in fact, have the advantage of some scientific knowledge and, therefore, it starts to encroach." If word gets out that a movie or television series is "stupid," that can have an impact on that production's ratings/box office, longevity, and, ultimately, revenue. The very same fans that might be hypercritical of the scientific flaws of a movie can be equally as vocal supporters if they feel the storytellers have acknowledged their intelligence, so there is incentive to get the science right.

How do Hollywood creatives make strides to satisfy these technically critical fans? How do we keep them invested in our stories? Enter the Hollywood science advisor! Science advisors (hereafter SA) have been consulting on Hollywood productions for decades (David A. Kirby wrote an excellent book about this entitled, *Lab Coats in Hollywood* (*2*)): Stanley Kubrick, for example, consulted a cadre of scientists and engineers in the Los Angeles aerospace community for his 1968 film, *2001: A Space Odyssey*. Increasingly, Hollywood productions are relying upon the services of real world scientists to add verisimilitude to their stories, to ground their worlds of fiction in real world science.

Though the use of science advisors is increasing, it's far from ubiquitous, and the multi-faceted roles that a science advisor can fulfill is generally a mystery to viewers and showrunners alike—to people both in and out of the entertainment industry. What do science advisors really do? How do they work? How do their contributions fit in to the storytelling process? Are they, essentially, science content copyeditors? How do I get that job? In short, what is the Science Advisor's Journey?

Although scholars have argued that it's unreasonable to expect real world stories to follow monomythic structure—chiefly because of the paucity of people who might be considered Heroes—occasionally tales of real life do conform to the pattern with surprising fidelity, and it turns out that the career of a Hollywood science advisor does just that (without having to stretch analogies to the breaking point). Using the Hero's Journey as a framework, and my career experiences to supply real world examples, we will explore the types of tasks, roles, and challenges that face a Hollywood science advisor. Although improving the level of science discourse in Hollywood is hardly heroic in contrast with fighting fires, performing surgery, exploring space, saving Earth from a black hole swarm, battling Skitters and Mechs, or destroying the Death Star, even Joseph Campbell would agree: everyone is the Hero of his own story.

During the Journey, the Hero typically encounters others—characters who fit into well-defined roles or archetypes—whose intent is to either help or thwart the Hero in achieving the goal of the journey. Included are interview excerpts with writers, producers, even other science advisors—friends and allies that I've met along the way. They'll assist in telling this story of long odds, long hours, conflict, and accomplishment: one filled with Mentors, Allies, and Shadows.

The science advisor's journey begins with…

The Ordinary World

> Most stories ultimately take us to a special world, a world that is new and alien to its hero. If you're going to tell a story about a fish out of his customary element, you first have to create a contrast by showing him in his mundane, ordinary world (*3*).

It can start anywhere from East Lansing to West Lafayette—with partners watching TV at home, friends gathered in a dorm room, a couple in a movie theatre, or a lone person flopped onto a comfy chair after a long day's work. Anywhere there is a lover of TV or film—not necessarily rich, not necessarily athletic, not necessarily powerful or influential—special only because that person possesses a love of science: there is a Hollywood science advisor waiting to be born. Many, likely most, of those who have consulted for television series or blockbuster Hollywood movies have begun their journey under similar modest circumstances.

The Call to Adventure

> The hero is presented with a problem, challenge or adventure (*3*).

Then it happens: "Oh that is *so* lame! We've known for ages that people use more than ten percent of their brains! Why didn't they consult with a scientist? I could have set them straight! Why didn't they ask *me*?" It's that less-than-perfect science reference that pulls the science-literate viewer out of the story. Much of screenwriting is about creating "moments": character moments, visual moments, emotional moments, moments that make the viewer say, "Oh, wow!" Science misstatements, especially those that are easily-prevented, make a significant fraction of viewers, instead, say "OH, PUH-LEASE!" In that instance, the viewer is yanked out of the story, and is no longer immersed in the writer's creative vision. Instead, that viewer instantly becomes aware that they are sitting between four walls in a room in the 21st century, looking at a screen, arms folded, feeling cheated.

Most of the offended viewers are sated simply by complaining to themselves, their partner, friends, or co-workers. Some (too many!) vent on the Internet. For a very few, this moment is the inciting incident that Heralds the start of their Hollywood journey..

The Reluctant Hero

> Often at this point, the hero balks at the threshold of adventure. After all, he or she is facing the greatest of all fears—fear of the unknown (*3*).

There are many and varied ways in which Hollywood can be overwhelming, daunting, and overwhelmingly daunting. Think back to the thespian star, the outrageously talented singer, or that writing savant in your high school or college class—peers with more talent than should be bestowed upon any single human being. Now imagine how many high schools and colleges there are in the United States, and imagine how many of those talents come to Hollywood to become a star. Each year. Every year. Now imagine how many roles there are to fill. One of these numbers is much larger than the other.

If young talent truly knew the size and scope of the entertainment industry, fewer would every attempt this Journey: to "make it" in The Industry. In any science advisor's Journey, there are obstacles, real and perceived, that will provide stumbling blocks—perhaps even convincing the budding SA to give up the journey voluntarily. Many who start the Journey turn away soon after this early dose of reality.

Once a young talent lands an agent, the road has been paved to some degree—at least their foot is in the door. The young talent still needs one more thing to work in his or her favor: that element of luck, (a.k.a. random chance) known colloquially in the Industry as a "big break." Science advisors do not have agents, so for a long time there were even fewer avenues into Hollywood than for other talents. How does an enthusiastic scientist breach the towering, impenetrable walls of ~~Mordor~~ Hollywood? Dr. Malcolm MacIver, a robotics expert from Northwestern University and science advisor for the TV series *Caprica*, shares, "The most common question people ask me is, 'How did that opportunity come about?' People are really curious: how would you ever get lined up with people from this apparently different universe?"

In my case, though, I did catch a "Break". In the late 1990's when I was a graduate student at UCLA, Paramount Studios accepted unsolicited manuscripts—scripts written by writers who did not yet have agents—for their

Star Trek series. I collaborated with a friend from undergraduate days on a script for the series, *Star Trek: Voyager*. Paramount cavalierly admitted that they received 3000+ scripts a year, and only a handful of writers had a good outcome. Still, seven months to the day after we mailed in our script, I received a call from Executive Producer Jeri Taylor's assistant telling me that they loved our script, but couldn't use it because it went in a creative direction they were unwilling to explore (but did three seasons later). We were invited to Paramount to pitch story ideas to the writers on staff.

Although my co-author and I never officially sold a story, one of the staff writers I met there, Bryan Fuller, would later make the introduction that helped me land my first science advisor job on *Battlestar Galactica*. About a year later, after the series *Eureka* had been greenlighted, the *Eureka* writing staff shared the Rock Hudson Building at Universal Studios with *Battlestar Galactica*. At a "getting to know you" lunch, Writers' Assistants Kevin Fahey and Eric Wallace were comparing notes on how they deal with technical concepts and dialogue in their scripts. An hour later, at my desk at the Jet Propulsion Laboratory, I got the call from Eric: the executive producers of *Eureka* wanted a science advisor, and wanted to know if I would be interested in helping out.

In some instances, Hollywood takes the initiative. Scientists are sometimes pegged as consultants because either 1) they have made a name for themselves with their science and are already famous or, 2) they know somebody in the industry. When Rockne S. O'Bannon wanted to consult with an oceanographer for *seaQuest: DSV*, he sought out the most well-known oceanographer on the planet: "My most direct experience with hard science in developing a TV series was *seaQuest*, where I had the incredible advantage of being partnered with Steven Spielberg, and by way of Steven Spielberg I was able to make contact with Robert Ballard, the very famous oceanographer: the fellow who went down and found the *Titanic*, and who has found all sorts of other things since. He was just a fabulous kind of visionary undersea explorer. So that [series] really had the opportunity to be grounded in very real science."

Planetary scientist Dr. Josh Colwell was doing research on comets at the University of Colorado, while his brother K.C. was the first assistant director on a movie entitled *Deep Impact*. Although the production staff already had consultants for the film, they kept calling Colwell for assistance; "They already had NASA consultants on board. Their other consultants were in the area of manned spaceflight, and not on the physics of comets and impacts. In addition to having a Ph.D. in planetary sciences, I had recently been doing research on both the physical nature of comet nuclei as well as the effects of collisions between comets and moons. K.C. took my answers back to the staff and prefaced them with something along the lines of, 'My brother is an astrophysicist and he says…' Before long the producers decided that it would make sense to have me on board as a formal consultant."

Often scientists are reluctant to heed the call of Hollywood—particularly in the cases where the entertainment industry seeks their collaboration rather than vice versa—out of fear that an association with Hollywood could prove detrimental to their careers. The fear is that once they work in television or film, their research will be marginalized and considered less seriously—deemed somehow "tainted"—by their peers, and that, somehow, they have "sold out." There is the fear of professional envy and/or reprisals. For as beloved as Carl Sagan is today, as much as he did to popularize science, and as much of a rush as there was upon his death to be the "next Carl Sagan," Dr. Carl Sagan was not treated kindly by many of his colleagues and peers while he was alive (*4*). About Sagan's relationship with Hollywood, and in popularizing science, scientist and science fiction author Gregory Benford writes, "Many scientists don't think much of such endeavors… Unless the culture of research science realizes that it may be a major stumbling block to its own popularity, we'll remain part of the problem (*5*)." Sadly, these are very real concerns that a potential SA must weigh before getting involved in Hollywood.

The silver lining is that there is less of a reason for professional envy these days, because it is far easier for a scientist, who is interested in helping to raise the level of science discourse in TV and film, to make that entry into the Industry. The National Academy of Science staffs a Hollywood outpost known as The Science and Entertainment Exchange. The Exchange maintains a database of scientists, along with their areas of expertise, and match science consultants to production projects: they are the equivalent of a talent agency for scientists. There is increasingly less reason for prospective science advisors to be reluctant Heroes from trepidation due to the immensity of Hollywood, or out of fear of professional envy. It's a game more people can play now, and another way scientists can get involved with Hollywood.

All of these scenarios really happen: the Big Break scenario, the "It is not what you know, it's who you know" scenario, the "Recruit the big name scientist" scenario, and the Science and Entertainment Exchange scenario. DeSanto summarizes, "You ask a hundred different people in Hollywood how they made it, you'd get a hundred different answers. Each person has their own Journey."

The Meeting with the Mentor

> By this time many stories will have introduced a Merlin-like character who is the hero's mentor. The mentor gives advice and sometimes magical weapons. Sometimes the Wise Old Man/Woman is required to give the hero a swift kick in the pants to get the adventure going. The mentor can go so far with the hero. Eventually the hero must face the unknown by himself (*3*).

Mentors can appear at any time in a Hero's Journey, and there can be many mentors along the way. By virtue of being a scientist, a budding science advisor is already learn<u>ed</u>. To land a gig and get continued work, one should be dedicated to being a learn<u>er</u>. People who describe themselves as professional scientists likely have far more than enough science wherewithal to do the job. Although Hollywood prefers science experts with doctorates, having a Ph.D.-level understanding of astronavigation may get you that first job on *Lost in Space*, but it is that experience—along with a mastery of the basics in several different fields—that will keep the job offers coming.

While waiting to break in, even if you have landed your first gig, be your own mentor and take the time to learn as much about the Industry as possible. Learn the roles of the director, writer, producer, and executive producer, as well as how these roles differ between film and television. All of this is available on Wikipedia and IMDb. Seek out mentors. Many colleges and universities have television and film classes, and if you live in the LA area, both UCLA and USC have formidable television and film curricula, and even if you don't live in LA, many of the courses can be taken online. (Hint: many of these classes are taught by successful writers, producers, and directors who are looking for new blood to work with or for them.)

More germane to the advisory role, take a class in screenwriting. Know why a writer's concern is a writer's concern. Malcolm MacIver recommends, "My central word of advice would be to come to understand something of the art of storytelling. I think it's really in the best interests of science advisors to learn something of the craft, as well as the constraints, of the people that they are trying to advise." As a science advisor, it helps you empathize with the writers and recommend options if, at some point, you have written a story (even better a screenplay) that has a beginning, a middle, and an end.

Bradley Thompson, a writer/producer with experience on several popular sci-fi television series, added, "On *Falling Skies*, when we got our first set of notes from our science advisor, they were laid out in a way that A) we could understand them, and B) he understood *why* we were doing what we were doing—what the story demanded. He said, 'This is how you can make this science issue a conflict between these two people. He understood character. [As screenwriters] what we're doing is talking about characters and cool stuff happening, and he could milk what he knew to give us cool stuff happening without us having to yank it out of a bunch of facts and figures." Caltech theoretical physicist Sean Carroll, who has consulted on several movies, adds, "The word is slowly being spread that scientists can actually help the creative part of the process."

Implied in Carroll's words is that the relationship between storytellers and their advisors is a symbiotic one. Understand that since the majority of writers, producers, and directors, have very limited exposure to the scientists and their culture, the science advisor may, conversely, be a mentor in somebody else's Hollywood journey.

Crossing the First Threshold

> The hero fully enters the special world of the story for the first time. This is the moment at which the story takes off and the adventure gets going. The balloon goes up, the romance begins, the spaceship blasts off, the wagon train gets rolling. The hero is now committed to his/her journey and there's no turning back (*3*).

In October 2003 co-creator and executive producer Ronald D. Moore gave a presentation about his soon-to-be-released reimagined *Battlestar Galactica* in Burbank, California, at a science fiction convention called Galacticon. It was the first time that clips from his new pilot were shown publically. Though difficult to imagine today, Moore's presentation was met by an indignant hostile mob, angered by the perception of Moore's arrogance: "How dare he have the audacity to change the show we loved so?" (That hostile crowd was lampooned years later in an episode of *CSI* entitled "Space Oddity"—penned by former *Galactica* writers Bradley Thompson and David Weddle, even featuring Moore in a cameo appearance as a hostile audience member.)

I was in that Galacticon audience. My reaction to Moore's presentation was quite different than the prevailing one: "THAT. WAS. AWESOME!" In the intervening years I'd kept in contact with Bryan Fuller from when I'd pitched stories to *Star Trek: Voyager*. Since they were fellow *Trek* alums, I emailed Fuller and asked if he would find out if Moore planned to use a science advisor on *Galactica* and, if so, how did I go about being considered? That's only a half-truth. The email went more like, "PLEASE BRYAN! Please get me in to see Ron!"

My interview lasted about five minutes. I was ushered into Ron Moore's office, and after introductory pleasantries,

> RDM: Wait, you look familiar.
>
> KRG: At Galacticon I asked you the only two polite questions you got all night.

Before I left, Ron handed me the series bible and scripts for the first two *Battlestar Galactica* episodes, "33" and "Water." I walked out of Ron's office, out of the Rock Hudson Building, and into the Kingdom of Hollywood.

Tests, Allies, Enemies

> The hero is forced to make allies and enemies in the special world, and to pass certain tests and challenges that are part of his/her training (*3*).

Although the advisor is brought in as "the expert", it is essential for the SA to cultivate friendships and alliances. My experience has been more with television than with film, and in television the most important person on a series for a science advisor to have in his or her corner is the showrunner—one of the executive producers—a term with which many outside the Industry may be unfamiliar (the showrunner's responsibilities are spread among several people on a film). Kevin Murphy, showrunner for *Caprica*, *Defiance*, and other series, explains, "A showrunner is usually a writer, often the creator of the particular series, and is essentially the CEO of the corporation. You can be an 'Executive Producer,' and it can mean anything; 'showrunner' means only one thing: it means that the buck stops with the person who has that title." By simple virtue that a series is looking to bring aboard a science advisor, the showrunner is very likely—though not always, as we shall see—an *a priori* ally and, likely, mentor.

Once a science advisor lands the first gig, one of the most important issues to resolve early in the process is, "Why am I here?" because oftentimes why the science advisor is there, and why they think they are there, are not synonymous. In some more manipulative instances, the SA is brought aboard solely to provide the production with a measure of legitimacy—a situation which, while not representative of the average SA experience, does happen on occasion. Understanding the situation from the start is the best way to manage expectations, and even in the best instances, textbook-perfect science is seldom an aspiration. <u>The science advisor is rarely, if ever, brought aboard to ensure that the production has perfect science, but rather to assist in the timeless tradition of grounding the story in the natural world, to make it seem plausible, to minimize the, "Oh please!" moments</u>.

If there is a conflict between story and science, story wins every time. Kevin Murphy elaborates, "Personally, from my perspective, I don't care whether or not it's actually true, I care whether or not it seems true. Because it's <u>all</u> fiction. In terms of *Defiance*, this world doesn't actually exist. My job is to convince our audience that it can and could exist. I think what's important is verisimilitude. The science does not have to be accurate, but the world needs to be immersive, and you need to believe that the science is real. If someone says, 'OK, I don't believe that that's properly grounded, I don't feel that they did their research, I feel that the writing has been lazy…' people check out of the experience."

Rockne O'Bannon agrees, "It's really a matter of allowing the science to create a foundation of reality that lets the characters be real in the work that they do, and therefore you're then free to color in the first and last acts and characters and emotions around that very solid foundation." Michael Taylor, writer/producer on shows such as *Star Trek: Voyager*, *Battlestar Galactica*, and *Defiance*, adds, "My priority has never been that we're a science show, it was for dramatic entertainment. We're not making documentaries. So I will take the amount of science that I need to help inspire and create something that seems more original, something we haven't seen before, but I'll leave it at that. I won't hang up on the details."

While, perhaps, counter-intuitive, experienced science advisors like Sean Carroll and Andre Bormanis agree with the storytellers. Carroll explains, "Over and over we found that by thinking carefully about science, construed very, very broadly, we came up with interesting new scenarios and interesting new ways to make the movie work. I think that these movies are not science documentaries. Sticking as closely as possible to realistic science, as we understand it in the real world, is not either feasible or desirable." Bormanis, who served as the science consultant on several of the *Star Trek* series adds, "Yeah, basically, story wins. You're trying to tell a compelling dramatic story, but that doesn't mean that science has to lose. It just means that the science has to be… adjusted. If you have to adjust it to the point where it's no longer scientifically credible, then I would fight for doing something differently with the science. You know, 'If this is the story you want to tell, then instead of this kind of approach to the science, what if we try this kind of approach?'"

Far from "selling out" to Hollywood, experienced science advisors have learned to improve the level of science in TV and film while working within the constraints of the system. Those constraints can vary wildly from show to show, however. Writer/producer Kath Lingenfelter, who worked on both the science fiction series *Caprica* and the medical drama *House, M.D.*, says that she thinks the science accuracy bar is set differently depending upon the series: "People understand *House* as being set in the real world and *Caprica* as being set in a fantastical world. They want to believe House exists in their world and is available to them, so we need to have the science as close as possible. We tried with *House*, because it's set in the real world, everything had to be grounded and relatable and real to people because it's very serious. You're talking about the human body, which is something we all share and something going wrong with it is a nightmare we can all share."

Lingenfelter continues, "A really good, and I think germane, example is resuscitative medicine: where the majority of the public, if they know anything about resuscitative medicine, it's from what they've seen on TV and films. I mean our attention spans are so short now that it's easier for us to watch somebody do CPR on an episode of our favorite show than to actually read a manual, so it's always going to be that shorthand. If we're going to be arbiters of that shorthand, then it should be right."

For the SAs who were attracted to the job in order to prevent the very same type of, "Oh please!" moments that initiated their journey—perhaps even hoping to use the shows on which they consult as vehicles to teach the public a lesson or two about science—this is a big test: the moment when the SA must come to terms with the stark difference between his or her goals, and that of the showrunner's. Working with writers, producers, directors, and performers on a Hollywood production is an amazing, challenging, and often highly rewarding experience. For an SA, however, the moment of realization that perfect science never was, and never will be, the goal, is a character test that derails many: often leaving them feeling bitter or cynical.

Sean Carroll counsels, "The important thing to remember is that you're servicing somebody else's work. You're not doing it yourself, and the goal is to help the director or writer or whomever make the best thing that they can make. Their goals might be different than yours." Bear in mind that a TV series or film is a highly collaborative effort. No matter what ends up on screen, "teachable moments" are one of the rewards that await the end of your journey.

"It's complicated," describes the relationship between the SA and his or her production and, like any complicated relationship, a thick skin is a useful attribute. This is especially true when the first realization hits home that his or her involvement may not be universally-appreciated by the staff writers, potentially even the studio or network, and that enmity has nothing to do with the SA whatsoever. In addition to the writer of record, many voices speak through a screenplay. When a writer pitches a story, he gets notes—recommended story changes. When he submits an outline, he gets notes. When he submits a draft, each and every draft, he gets notes. He gets notes from other writers, he gets notes from the showrunner, he gets notes from the EPs, he gets notes from the director, he gets notes from the studio, he gets notes from the network. On occasion, he even gets a note from the talent who has to perform the script. As an advisor, you represent yet one more person who will be giving notes and adding to the writer's workload.

In general, however, writers appreciate and often solicit notes early in the story process—though even that is highly writer-dependent. To some, however, the SA symbolizes the corporeal manifestation of an ancient adversary: "It is the dread exposition," says Bormanis. "Exposition is the term for dialogue that is explaining to the audience what is happening at that stage of the story. Exposition, in the long history of drama, has been considered the thing you most want to minimize. 'The ideal script has zero exposition.' This is what every screenwriting professor, for time immemorial, will tell you."

Compounding the concern, the SA represents science exposition. Science fiction novelist Robert J. Sawyer elaborates, "There is this presumption that anything that is expository, is death on television. And yet *Sherlock Holmes*, in the current iteration, is nothing but Benedict Cumberbatch doing fascinating exposition. *House* is nothing but Hugh Laurie doing fascinating exposition... *CSI* is nothing but Marg Helgenberger looking really amazing, and delivering fascinating exposition. But there is something about the general public reaction to science, separate from all other parts of human endeavor, that engenders knee-jerk reactions in a large percentage of the population. As soon as we come to... physics... chemistry... biology... it's like, 'If you start to go down that road, you'll lose the audience.'"

"As soon as you stop to give a science lesson that seems a little overt people are going to run screaming from their television," agrees Lingenfelter. Sawyer, whose novel *FlashForward* was the basis for the ABC series, further elaborates, "The field that I'm known for, 'hard' science fiction—science fiction where the science is integral to the plot and is rigorous in its execution and extrapolation—is a non-starter on television. [On *FlashForward*] We were told every time we use a tech term, 25,000 people wouldn't come back after the commercial. Do that, you've got five commercial breaks in an hour, you've lost over 100,000 viewers."

A specific example, one that I have encountered on every show on which I've consulted, is that some writers will insert a made-up technical term in an early draft of their script, yet when the SA suggest a "plug and play" replacement—a real term that fulfills the same role—the writer response is frequently, and frantically, "We can't say that, there isn't time to explain it!" One writer confessed, "I know I have [been guilty of doing that]. What I think is that it's basically to get the audience to notice that this is real, and if you're taking that extra step to create a world that's as realistic as possible, you want the audience to get the full benefit of that." Kevin Murphy concurs, "It's human nature to get very excited when something is real. You want to share that with your audience: 'No. No no no, you don't get it, this is real.' Sometimes science can be so wildly 'out there' that if you weren't reassured that this is real, it would sound like [BS]."

Sean Carroll has noted similar: "I have gotten a knee jerk reaction that a certain idea can't be used because it would require too much explanation, so I try to say that it doesn't require any more explanation than any other thing. I mean, if there's a line that would do fine, just say it and move on."

This point is worth mentioning because each time this occurs it is in a similar manner, and happens with alarming regularity. It is frustrating to writer and SA alike, and makes it clear that the two worlds, Hollywood and science, are still in the early stages of learning to communicate. Succinctly, for the SA the interpersonal challenges trump the technical ones early and often, and the Science Advisor's Journey, time and again, detours through Babylon.

Approach to the Innermost Cave

> The hero comes at last to a dangerous place, often deep underground, where the object of the quest is hidden. In many myths the hero has to descend into hell to retrieve a loved one, or into a cave to fight a dragon and gain a treasure. Sometimes it's just the hero going into his/her own dream world to confront fears and overcome them (*3*).

There are duel-use words that represent both a physical space and the collection of people who inhabit that space: a naval Ward Room, Scotland Yard, Wall Street. The writers' room for a television series also belongs in this category. Every TV series has a room, usually a conference room, called the "Writers' Room." This is a room where the writers, producers, even directors, can gather to collaborate on their show. When writers have ideas for which they want feedback, they will say, "I want to pitch this to the room." They mean to the other staff writers, the writers' room (Figure 1).

Figure 1. The author working with writer David Weddle in the writers' room on the TNT series Falling Skies. (Photograph by Bradley Thompson).

The new SA's job begins in earnest when he or she finally gains access to the room, either metaphorically or physically. What should the new SA expect? Perhaps the most common question people ask of science advisors is, "So, like, what? They just send you the scripts and you tell them what they did wrong and fix it?" The question is phrased like that, or very nearly, with jaw-dropping regularity. Sean Carroll agrees, "You do get asked that and you have to explain, that is not how it works. The reason why a lot of people in Hollywood are reluctant to talk to us is because they worry that what we're doing is copyediting and that's not very interesting. We have a long way to go before we overcome that stereotype."

Although the job title may be "science advisor" the tasks can vary dramatically from production to production. The key is flexibility. Malcolm MacIver elaborates, "Every interview I've done, the question is, 'So, what does this mean, what do you actually do?' I tell them, well, it's everything from correcting some clear errors, to coaching on some concepts, to helping integrate more interesting and exciting science and tech, to— it's all of these things, right?"

Carroll also makes the point, "Sticking close to the spirit of science and how science works will actually help you make a more dramatic and compelling story." Bradley Thompson agrees, "Well, it's that a lot of us in writing are not science-oriented, 'Oh gosh, we got to take chemistry in high school? This is terrible!' But what you find out when you're talking to [a science advisor]… is that science is a lot more interesting, and has a lot more possibilities, than you can think of if you don't know these things. That, 'Oh! There is a tidal effect with big big spaceships if you bring them really close to planets they might be torn apart.' This is something that we would not have known had we not talked to a science guy. Now, all of a sudden, there are possibilities for story on that."

The creative process involved in crafting an episode of television passes through several phases, and the writers and producers may solicit the SA's input at any, or every, phase of the process. Kevin Murphy summarizes how he and his staff tackle science on the SyFy series *Defiance*: "It's very 'free form jazz,' the process. The way we go about doing it, with our science advisor, is we write the outline, we send the outline, we get comments back. We write the script, we do the best we can, we fake it the best we can, and we get comments back. If there's a specific question, we'll pose that question, because that may be something that ultimately informs how we write the story."

On various television series, showrunners, producers, and writers have solicited my input for all the phases of the creative process at some point. For the series on which I worked the longest, SyFy's *Eureka* (five seasons spread over seven years), I was encouraged to provide feedback for every phase of the creative process at some point. *Eureka* represented a superset of all other consulting experiences: everything I've ever done for any series, I did at least once for *Eureka*.

Before Story Development

It's not uncommon that a showrunner will solicit the SA's advice before a season's story development process begins in earnest. Jaime Paglia, *Eureka* co-creator and showrunner, explains, "We invited Kevin into the writers' room at the start of the season to give us a PowerPoint presentation about some of the cool things that were happening in science, new cutting edge research and development stuff, just to see if there was anything in there that could be interesting to use as the germ of an idea."

Andre Bormanis shared a similar experience on *Star Trek*: "[Part] of the job was to provide the writers with the information they needed to do stories that revolved around some interesting idea from science or some astronomical phenomenon that we had not seen before. So part of my job was to stay abreast of new discoveries in astronomy and physics, biology, and try to incorporate those ideas into storylines."

Early in the Writing Process

Somewhere between the point at which the writers "break" a story—where they determine all the twists and turns—and submit a more detailed outline, can be an important point for an SA to chime in. Paglia explains that on *Eureka*, "We would also let Kevin know what storyline we had settled on for any feedback about how to make it more scientifically accurate at the story breaking/outline phase." It is at this phase, and the next, where it is particularly useful if the SA understands the writers, their concerns, and something of the art of storytelling—in order to incorporate specific, rather than general, recommendations within story notes (preferably giving the writers a choice of alternatives at the same time). Bormanis concurs, "Always try come up with that 'something else' before you tell them that the think they're trying to do is not really going to work."

During Script Writing/Revision

Paglia continues, "Then we would always have Kevin read the final scripts for scientific accuracy, to look at the dialogue for places to refine or clarify." Shy of being a copyeditor, the SA is able to offer very specific recommendations at this point for scientific misstatements in the dialogue. Timliness is key at this stage, however. Submit notes when a deadline is looming and the writer is under the gun, expect that you will likely get a chilly reception and, except in rare circumstances, the ship has sailed on the implementation of your note.

On some series there are conventions for writers to ask for SA help within the body of an early draft script. Bormanis explains, "I was responsible for much of the so-called technobabble, technical dialogue, in our scripts. I would often get scripts that had a line of dialogue with the word '[TECH]' in brackets, and that was my cue to fill in the blank with an appropriate-sounding technical term. So I would look at the dialogue, I would look at the context of the scene, figure out what it was they were looking for, think about the terminology we'd used in the past." Likely because there were so many *Trek* alumni onboard, in particular showrunner, Ron Moore, this convention was carried forward to *Battlestar Galactica* as well.

While Filming

About once or twice a season I would receive a frantic call from the *Eureka* set in Vancouver of the nature, "Is it pronounced 'had-ron' or 'hay-dron'?" Or occasionally, "We need equations—lots of equations!" For the season one episode, "Before I Forget," Vancouver asked me to provide three full boards of equations. To this day, I'm still unsure how, when I finally saw those scenes on the air, the equations wound up in my handwriting.

I have received calls asking to help with props. There were two instances on *Eureka* in season four where a producer asked me to write pages for science textbooks. In both instances the production team realized that the script required one of our characters to crack open a textbook. Since TV series today shoot in high definition, it is possible that viewers could literally freeze the frame, and render the text in the books legible. That text, therefore, had to be topical, accurate, and not copyrighted. Call the science guy.

Post-Production

There were even instances on *Eureka*, after the scripts were finished and all the principal photography finished, when Jaime asked me to comment on the scientific accuracy of visual effects. This was a rare opportunity, one not afforded most SA's—many of whom have been shocked at the disparity between how they envisioned a scene, and what ultimately wound up on the screen. "Really think about what you read on the page, and what that's going to look like on the screen," advises Andre Bormanis.

Clearly, the SA job is usually more than simply, "They just send you the scripts, and you tell them what they did wrong." A science advisor can be useful at any stage of script development and, used effectively, represents not merely left-brained constraints, but a right-brained contribution to the creative process.

The Ordeal

> This is the moment at which the hero touches bottom. He/she faces the possibility of death, brought to the brink in a fight with a mythical beast. For us, the audience standing outside the cave waiting for the victor to emerge, it's a black moment (*3*).

As a science advisor, you represent constraints to the writers' freedom to tell their stories. You build walls around their play area. Some look at the walls as opportunities for a different style of play, others feel claustrophobic and push back. There may be any number of reasons that your input may not be used, and often it seems like the better your input, the less likely it is to be included. It could also be that when the show you worked on so hard finally airs, is when you first discover that very little of your science advisory input has been incorporated.

There may be the temptation to feel that all your work was for naught: "Nobody listens to me." A better interpretation, though, is that you are in the same boat with the writers, and this is something they must get used to as well. During the story breaking process, a lot of ideas are thrown out in the room, and most thrown away. Many notes are offered, a fraction accepted, and your notes are just like anybody else's. What appears on the screen can even change after it leaves the custody of the credited writer. It has been said that a script is not a rule, but rather a "recommendation." Even when a script is "finalized," the director may change things, the talent may not be able to say the words as written, or simply may not like them. It's just a fact of life in the world of television.

The cure for angst in this type of situation rests almost entirely with the SA. Experienced writers learn not to get married to their ideas. If it is a good idea, and not accepted when offered, keep it—it might be worth exploring in a later work. This is another concern that the SA can address proactively. Malcolm MacIver counsels, "As much as you can put yourself in the shoes of the story maker, you can fashion your science advice in a form that's a) much more useful and b) much more likely to be listened to."

Despite your best efforts, despite your sage advice, despite your charming personality, some may be viewed less as a science advisor, and more as a science adversary. You represent another set of notes, science exposition, and constraints. Though the use of science advisors in Hollywood is increasing, they are still far from ubiquitous. SA's haven't been needed as much until recently. O'Bannon explains, "In the past the vast majority of the audience probably didn't have much of a foundation in what a specific science was in any particular show they're watching. It just had to pass the smell test. If it seemed real, if there was a sense of verisimilitude to it, then that was satisfactory." One showrunner for whom I worked said, "If I don't understand it, they won't either." So even today there are writers and producers who do not see the usefulness of a science advisor whatsoever.

"Thar be dragons," on the path of your journey.

Reward

> Having survived death, beaten the dragon, slain the Minotaur, her hero now takes possession of the treasure he's come seeking. Sometimes it's a special weapon like a magic sword or it may be a token like the Grail or some elixir which can heal a wounded land (*3*).

In every variation of the Hero's Journey, the story is riddled with strife and conflict. Yet there are still alluring and tangible reasons why so many attempt to "make it" in Hollywood. Doesn't every child fantasize about working in Showbiz? The experience of contributing to a Hollywood movie or TV series is a reward in and of itself (Figure 2), and will never be anything but a dream for millions.

As a science advisor, your contribution may pale in comparison to that of the talent, the writers, director, producer, script supervisor, writer's assistant, composer, editor, even the key grip. From a time and effort standpoint, you probably rank below the craft services people as well. Still your involvement can have very real benefits for the production. By circumventing potential,

"Oh please!" reactions from the audience, your input contributes to better story enjoyment for many viewers and that can, in turn, manifest as a better box office performance for a feature film, or another season pickup for a popular television series. The mere fact that the production included a science advisor in the first place is often an indication that a production is committed to excellence from top to bottom.

To a scientist consulting in Hollywood, seeing your name scroll past in the end credits the first time may not be quite as thrilling as discovering something new in your research—something nobody has ever seen or known before—but on "Life's List of Big Thrills," it still ranks pretty high. For that matter, so does the second time. As does the third. It really just never gets old. What also never gets old is hearing your words come out of the mouths of your characters if you've had the opportunity to write dialogue. It can be endless fun interacting with appreciative fans.

Figure 2. The author on the Vancouver, BC set of Battlestar Galactica. (Photograph by Bradley Thompson).

Receiving a paycheck with a popular television series as the payor is quite gratifying as well.

Resurrection

> The hero emerges from the special world, transformed by his/her experience (*3*).

Just like the false ending, common in the horror genre, immediately after your show airs is when Shadows emerge and some of your thorniest challenges begin. Shadows are character archetypes common to the Hero's Journey that represent villains, sinister forces, even the demons within each of us. Now is their time to party.

It should be clear that there are many places in the story creation process where science can slip through the cracks. The general public does not fully understand your advisory role on your series or movie, and neither does anybody outside of your series or movie. Fans of your show will assume that you had copyeditor-like control of the science content, and <u>any</u> mistakes are yours alone.

Now think back to the inciting incident that began your Journey. Recall when you were forcibly yanked out of a favorite television show or movie by a science gaffe. Now that you are on the other side of the Hollywood equation, an entire clone army of <u>you</u> now has you in their crosshairs.

Rest assured that 100% of the time when you make a science goof, and a fair amount of the time when you do not, you will be taken to task. Anybody who has worked on a television series or movie has experienced both the joy and frustration of Internet feedback. Tom DeSanto shares, "The Internet can be a great place. It can be a awful place, as well, if someone just gets on there anonymously and starts bashing something for the sake of a therapy session. If people are constructive, and people are sincere in trying to make something better, that's great. But if people are name-calling or [if they] go the lower vibrations of human existence, it becomes toxic. It's the nature of human beings; you can go to the Light or you can go to the Dark Side. Unfortunately, a lot of people choose to go Sith as opposed to go Jedi."

Still, let's put the problem in perspective. If you're fortunate enough that you can count among your problems that people complain on the Internet about your work—that Cally and Chief should have popped when exposed to the vacuum of space (no they shouldn't have), or there's no good reason why Vipers should shoot lead instead of light (yes there is), or there's no real difference between natural water and artificial water (you've got a point there)—then that means you are working on a production that fans are watching, enjoying, dissecting, and discussing.

Don't complain. You've made it.

Welcome to Hollywood!

Return with the Elixer

> The hero comes back to the ordinary world, but the adventure would be meaningless unless he/she brought back the elixir, treasure, or some lesson from the special world. Sometimes it's just knowledge or

> experience, but unless he comes back with the elixir or some boon to mankind, he's doomed to repeat the adventure until he does. Sometimes the boon is treasure won on the quest, or love, or just the knowledge that the special world exists and can be survived. Sometimes it's just coming home with a good story to tell (*3*).

We live in a wounded land. Just when topics like climate change, genetically modified food, stem cell research, and our environment are not only fixtures in the news but also topics of ballot measures, we live amongst a population full of individuals that have a difficult time comprehending the details. Tricksters, on both sides of many of these issues, eagerly sow seeds of further confusion. How do we expect to solve the problems of today, when the general public is simply unable to understand them in sufficient detail? Transforming our population into a scientifically-literate one is a start.

K-12 teachers and professors teaching undergraduate science classes need not shoulder this load entirely. Hollywood productions, have been referred to as the "unofficial curriculum of society" (*6*), and the number of people worldwide who watch Hollywood-made science-themed productions is titanic. Science education can STEM from multiple sources, both complementing and supporting the efforts of formal education.

What can Hollywood do? Although the level of science in Hollywood productions is improving steadily, science fiction and science themed shows will never have perfect science, and will never be perfect learning tools. Yet as the level of science dialogue and accuracy in Hollywood improves, STEM (Science, Technology, Engineering, Mathematics) educators will have an increasingly formidable battery of positive science examples to supplement their regular curricula.

Science advisors can play a big role here. Anybody who has worked in the industry has interesting stories to tell, and by sheer virtue of working in Hollywood, attracts an audience. Speaking opportunities—conventions, professional conferences, educational workshops, even DVD extras—often follow (Figure 3). Advisors may even have opportunities to share their stories in print. Each instance represents a great opportunity to share real science, enlighten the audience about the intricacies of Hollywood, and explain the complexities of marrying the two.

Although blending perfect science with compelling and imaginative storytelling may be next to impossible, nobody tells an enthralling story like Hollywood. Perhaps Hollywood's primary educational virtue will always lie where it has for decades: motivating the scientists and engineers of tomorrow by creating the types of strange worlds and situations that can only be visited through science and imagination today. Tom DeSanto believes, "I think that is where fiction writers can plant the seeds of those dreams of tomorrow. I mean you look at *Star Trek* and the influence that it had on the real life scientists of NASA, and I think that's where the heart and science really walk hand in hand."

Figure 3. "Hollywood Science" panel at 2012 San Diego Comic-Con . Panelists (L to R) are Jane Espenson, the author, Ashley Miller, Jaime Paglia, Jon Spaihts, and Zack Stentz. (Photograph by Phil Plait).

To some purists, the motivational element alone is simply not enough, and any science inaccuracy in a production is a transgression against knowledge. I disagree. The war for science literacy will likely be won in small skirmishes, not large-scale decisive battles.

Besides, do we really want to live in a world without light sabers, Godzilla, and superheroes?

The Master of Two Worlds

> Upon returning to the ordinary world after dwelling in the special world, the Hero faced the challenge of protecting himself and others from the power he had acquired. It's all too easy to be thrown into one extreme or another, either to rise too far above those he wants to communicate with or to sink too low into the mundane concerns of material existence. The hero who manages to avoid these extremes, however, becomes master of the two worlds (*3*).

After working for a season of television, or on a few movies, the science advisor may find that he or she has acquired a new skill and now speaks the language of Hollywood. In *Lab Coats in Hollywood,* Kirby identifies a new species—a sort of *scientia consuasor superus*—which he names a "boundary spanner":

> The benefits of familiarity with entertainment culture have led to the development of a recent type of science consultant that I refer to as boundary spanners. Boundary spanners are individuals with some scientific training who also develop extensive experience within the entertainment world. The boundary spanners' methodology involves their own consultation with appropriate specialists from whom they obtain and synthesize scientific information that they translate into the language of cinema.... Boundary spanners provide advantages because they readily move between the social worlds of science and entertainment (*2*).

Metaphorically a boundary spanner is a genetically modified organism whose scientist DNA has been spliced with that of a screenwriter. Increasingly, if an SA hopes to be an adviser on more than rare occasion, being a scientist—even a great one—is simply not enough. Bradley Thompson elaborates, "Part of the job of the science advisor is as a really heavy cultural translator. If you're trying to translate between two separate and wildly diverse cultures, say Feudal Japan and 20th Century England… What you've got to do is understand both cultures so that you don't get each other killing each other. If you understand both cultures, you can put things in a language that both understand."

If you are a science advisor, whether or not the journey is over here depends upon you. The hero's journey is cyclic, and as the science advisor's journey ends, the boundary spanner's begins. Armed with a new array of skills, knowledge, and experiences, the boundary spanner is prepared to face the new and bigger challenges presented by the most ubiquitous of Hollywood denizens: the sequel.

References

1. Campbell, J. *The Hero with a Thousand Faces*; Princeton University Press: Princeton, NJ, 1973.
2. Kirby, D. *Lab Coats in Hollywood: Science, Scientists and Cinema*; The MIT Press: Cambridge MA, 2011.
3. Vogler, C. *The Writer's Journey: Mythic Structure for Writers*, 3rd ed.; Michael Wise Productions: Studio City, CA, 2007.
4. Davidson, K. *Carl Sagan: A Life*; John Wiley & Sons, Inc.: New York, 1999.
5. Benford, G. *A Tribute to Carl Sagan: Popular & Pilloried.* http://www.skeptic.com/reading_room/popular-and-pilloried/ (accessed August 30, 2012).
6. Mooney, C.; Kirshenbaum, S *Unscientific America: How Science Illiteracy Threatens Our Future*; Basic Books: Philadelphia, PA, 2009.

Chapter 7

Consonance and Dissonance: The Art and Science of Film Music

Bear McCreary, Meghen Miles, and Kevin R. Grazier*
***E-mail: kg_eureka@yahoo.com**

Why does a film need music? Music plays many roles in film, but it is possible to categorize all of them into two primary functions: creating consonance or dissonance to highlight the film's emotion or narrative. While early composers who attempted to set music to film to create moods operated on instinct and worked largely by trial and error, composers today have a much better understanding of how to tailor music to elicit desired emotions from the audience. Starting with fundamental wave properties, we examine how simple two-note combinations can sound either consonant or dissonant, and scaling that to chords and harmonies, we explore the basic science behind how a film composer creates an emotionally engaging experience for the audience.

Introduction

Cinema represents a fusion of all other art forms operating in concert—drawing together disciplines as diverse as theater, dance, visual arts, sound design, and music. The need for most of these crafts is immediately obvious: writers craft the story and dialogue, actors, working with the director, find the unique voice in order to personify each character, and editors then assemble the footage into a meaningful structure. Each of these artists contributes towards making the fictional story more "real" for the audience and, just as special effects designers can set the story amidst strange and wonderful backdrops visually, film composers create a complex emotional landscape through which the narrative wends its way.

The vast majority of films follow a structure like the following: we meet a character during a time of consonance, events happen that turn their world upside down, creating dissonance, and at the end, one way or another, we return to a state of consonance. In music, consonance and dissonance can be achieved in many ways. Individually predictable rhythmic patterns can be layered on top of one another, creating a dissonant new pattern. Musical styles can be transplanted from one genre to another, increasing the dissonance between the music and the audience's expectations based on the visuals. Those methods are effective, but we will specifically examine the most commonly used method of achieving musical consonance and dissonance—harmony.

Consonance or dissonance is found in the relationship between two or more notes, a relationship that musicians call harmony. Certain combinations of notes are called chords, and given names such as major or minor that quickly describe their properties. The notes do not need to be heard simultaneously. Two notes in succession also imply consonance or dissonance, and their relationship is called an interval.

A consonant chord or interval is one that is considered *stable*, or pleasing, when played together. A dissonant chord or interval is considered *unstable* or unpleasant by most people, and is associated with transition. Indeed, dissonance often creates within the listener the desire for music to drive towards consonance: a "happy ending" as it were.

"Tension and release," "set up and pay off," "consonance and dissonance," whatever the terminology, this pattern repeats throughout a film to generate increasing excitement. Music is a filmmaker's most powerful tool to highlight these moments, as well as the transitions between them. Often, casual viewers don't even notice that a film has music, unless it is poorly placed and pulls them out of the experience. On the other hand, a film that can survive without any music is rare, and statistically negligible against the thousands of films produced every year that use music as a narrative tool. While many viewers may not fully appreciate the richness that a score adds to a film, they certainly notice its absence.

History of Film Music

One of the world's first cinema audiences saw a film without a score in January of 1896. Auguste and Louis Lumière's now-famous 48-second film, *The Arrival of a Train at La Ciotat Station*, features a locomotive charging at the camera. Legend has it that the audience screamed in terror because they believed the train was literally hurtling toward them. Why shouldn't they? Their entire life experience informed them that if they witnessed a train coming at them, the train was about to hit them. Their minds made sense of the visual information the only way possible at the time.

Those few were among the last general audiences in history to be tricked so effortlessly. As audiences saw more films, they quickly adapted and the young medium had to evolve technologically in order for movies to remain interesting and unpredictable. One of the first tricks filmmakers tried was to add music to their films. The technology did not yet exist to synchronize recorded audio to picture, so instead organists and pianists were hired to perform live accompaniment for each film presentation. Often with little or no time to view the film in advance, these musicians frequently improvised or performed classical pieces. The music was limited in how effectively it could foreshadow or comment on the larger narrative.

Nevertheless, these early film composers were the first musicians to experiment with synching music to film. When two lovers swooned into each other's arms, a major chord seemed to feel better than a minor chord. When the dastardly villain twirled his mustache, a fully diminished seventh chord made the audience cringe just a little more than a regular minor chord. Trial and error was commonplace, and because few were writing down what they were doing, musical traditions developed slowly. It was literally the prehistoric age of film music.

In the late 1920's, cinema changed forever with the introduction of synchronized recorded sound, most famously beginning with Warner Bros.' 1927 feature The *Jazz Singer*. From that point, music written for film could be more complex. With this new technology, every film would have a singlular musical score written specifically for it, in collaboration with the film's director and producers. A composer could watch the film in advance, experiment with different harmonic and rhythmic ideas, record the score and it would be forever married to picture. Every audience that watched the film would hear the exact same music, solidifying those musical ideas for all time. For film music, this was the equivalent of the invention of the printing press, and the evolution of the art accelerated quickly.

In the early 1930's, there was a gold rush for top musical talent. Composers such as Erich Wolfgang Korngold, Max Steiner and Dmitri Tiomkin were hijacked from the European classical scene and brought to Hollywood to apply their craft. Distinctly American composers such as Aaron Copland, Alfred Newman and Bernard Herrmann were also highly influential in this early era. By the early 1950s, they and their contemporaries had firmly linked cinema with the lush, brassy sounds of the Western European classical tradition. A new genre of music had been born: "Hollywood Music."

In the 1950s, the uniformity of this musical style began to disintegrate. Composers such as Elmer Bernstein and Henry Mancini introduced jazz, while Louis and Bebe Barron's groundbreaking score for the science fiction classic *Forbidden Planet* offered mainstream audiences their first taste of electronic musical synthesis. In the 1960s, Quincy Jones and Lalo Schifrin brought new cultural influences, while Leonard Rosenman and Jerry Goldsmith pushed the cutting edge of avant-garde experimental music into the forefront. Throughout the 1970s and 1980s, film music drew increasingly from popular music, with composers such as Danny Elfman, Mark Mothersbaugh and Hans Zimmer transitioning from performing artists to leading film composers.

Today, film music draws equally from all disciplines of music. The European classical influence remains, but is consistently melded with elements of electronic, rock, jazz, world, and experimental genres. Today's film and television viewers experience a wider array of music than any audience in history.

Properties of Sound Waves

Despite vast stylistic differences, nearly all film scores share a common goal: to heighten the dramatic impact of their film. Today, film composers draw upon wildly disparate styles, instrumentation, harmonies, and rhythms; yet, at a fundamental level they create moments of consonance or dissonance the same way, using the same basic tools. All of these moments create expectations for viewers of the film, priming them for emotional responses.

What happens within the mind of an audience member when they hear music? A successful film score is a case study in consonance and dissonance, on both the micro and macro levels. On the micro level, in order to understand what makes a given harmony either consonant or dissonant, one must understand the acoustic properties of musical notes—in particular the overtone series. To understand that we must, in turn, start with a very brief overview of the properties of sound.

There are many different types of waves, yet despite their considerable differences, some principles are analogous among all the different types. In Panel 1 of Figure 1, we see a surface wave on water, and in Panel 2, an idealized wave. The point of maximum rise of a wave is called a *peak* or crest; and the point of the maximum depression, a *trough*. The maximum height, or maximum depression of a wave is called its *amplitude*. The distance from one peak to the next peak, or one trough to the next trough, is the *wavelength* of the wave, commonly denoted by the Greek letter λ. The number of waves that pass an observer—how frequently they pass—in a given amount of time is called the *frequency*. Higher frequency means higher pitch, and the unit of frequency is cycles per second or Hertz (abbreviated Hz).

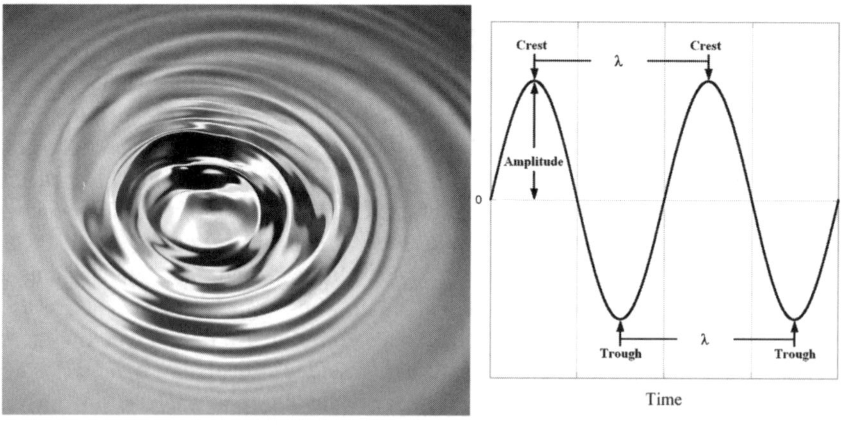

Figure 1. Surface waves in water, and the components of a wave. (Photo by Roger McLassus).

Sound, or acoustic, waves propagate very differently than surface waves on water. As a surface wave passes, the water level rises and falls—as a sound, or acoustic, wave passes, air pressure rises and falls. Figure 2 shows the relation between air pressure and the compressions (high pressure) and rarefactions (low pressure) present in the air molecules during the passage of a sound wave. What your ear senses is these tiny variations in air pressure, also known as sound. Both surface waves and acoustic waves are easily represented with a slinky spring or a telephone cord. Fix one end of the spring and oscillate the other; that wave is similar to a surface wave. Gather several coils of the spring together, release, and the "pulse" that travels the length of the spring (and back) is more similar to a sound wave.

Now to relate sound to music, we present a few simple experiments that anybody with access to an acoustic piano can perform. Go to a piano and play middle C (also called C4). That note is generated by a string vibrating back and forth at a specific frequency—a specific number of oscillations every second. The waveform for this tone (about 262 Hz) is represented in Panel 1 of Figure 3.

Move up an octave and play C5. This note is at twice the frequency, meaning that while on Panel 1, a lone wave has passed in one unit of time, for C5 two waves have passed. Play C6, an octave higher still (Figure 3, Panel 3), and the frequency is twice that of C5, and four times that of Middle-C. Two notes separated by an octave on a piano are separated in frequency by a factor of two. So, not only do C5 and C6 have higher frequencies than middle C, their frequencies are integer multiples of that middle C. When the ratio of the frequencies of two notes can be represented by small whole numbers, they are said to be in "resonance." With C5 and middle C, that ratio is two to one (written 2:1); with C6 and middle C, that ratio is 4:1.

Play middle C on the piano once again. Listen carefully and you'll realize that you're not hearing just one note. The strings in resonance with middle C will begin to vibrate as well, a phenomenon called *sympathetic vibration.* You're not simply hearing middle C, you're also hearing the next higher C, a G, yet another C, possibly even an E, and other notes ringing out quietly. The higher the frequency of the upper notes, the quieter they are, but they are definitely there. To amplify this effect, hold the sustain pedal (the one on the right) down with your foot. Now, loudly strike the C again. With the dampers off all the strings, the strings of the sympathetic frequencies will vibrate more readily. With middle C as the fundamental, or lowest, frequency, these upper sympathetic, or resonant, frequencies are called the overtone series (Figure 4), known to physicists as a harmonic series. Individual notes of the overtone series, higher in frequency than the fundamental frequency (in this case Middle C) and each a partial contributor to the series, are known as upper partials (also known as higher harmonics). An interesting property of human hearing is that if all the upper partials of the overtone series are played simultaneously, a listener has the impression that middle C is present when it is not (*1*).

The waveforms of Figures 1 and 2 are idealized in that they represent single frequencies, and have no higher harmonics. Such flawless audio signals are not common in our everyday lives, and certainly not in nature. If you try our little piano experiment you might think you can't hear these subtle overtones but, rest assured, you can. In fact, you've likely never heard a piano note without overtones in your entire life, and regardless of whether you can identify the overtones in a resonating piano string, you would certainly notice if they were suddenly gone. We all know the sound of a synthetically generated sound wave that contains no natural overtones: e.g., early videogame consoles, cheap consumer synthesizers, wristwatch beeps, sirens. We instantly recognize that wave forms generated by these devices come from an unnatural source because they are pure waves of a single frequency that have little or no overtone information. Every acoustic musical tone creates higher resonant overtones, whether originating in percussion, voice, wind, or string instruments.

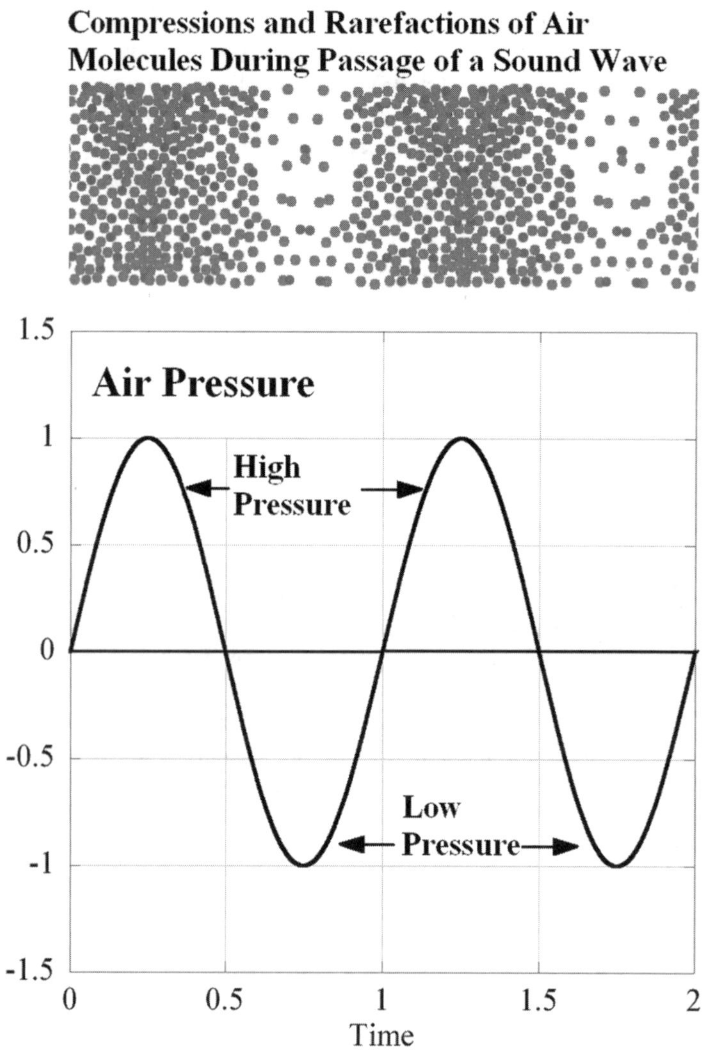

Figure 2. Compressions and rarefactions of a sound wave, and the relationship to air pressure. The wavelength and amplitude are normalized to 1.

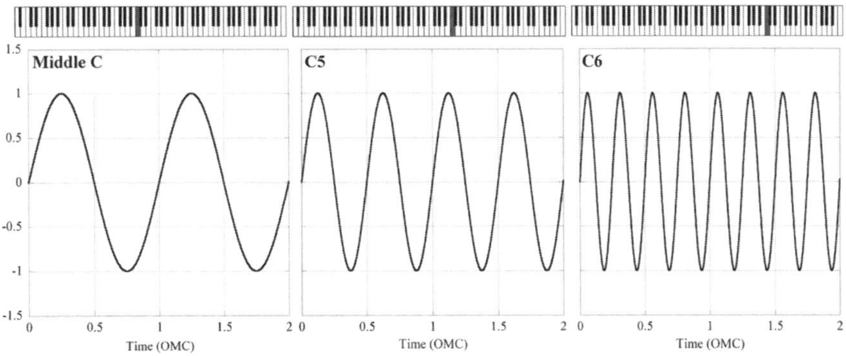

Figure 3. Middle C, C5, and C6 on a piano. Amplitude is normalized to 1, and the unit of time is in multiples of the period of one full oscillation of Middle-C (OMC), or about 3.8 milliseconds.

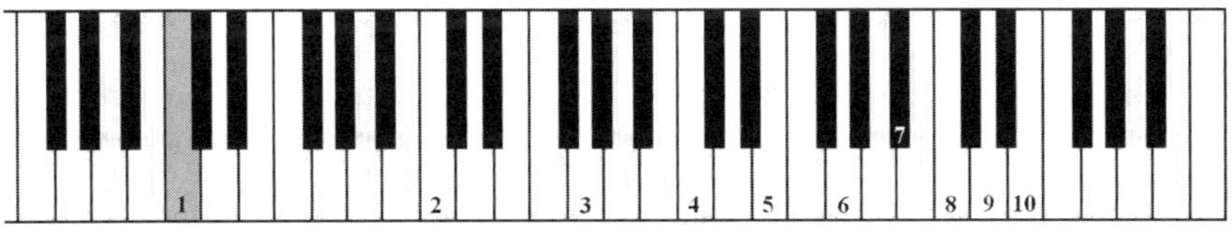

Figure 4. First ten members of Middle C's (highlighted) overtone series.

When a piano key is struck, many strings vibrate sympathetically, and your ear hears contributions from all of them. What occurs when multiple sounds, multiple notes, play simultaneously? Physically, when two waves pass over or through one another, they interfere with each other, meaning that if peaks from each wave overlap, the amplitude is the summation of the two amplitudes. When two troughs overlap, an extra deep trough is created. These are examples of constructive interference, but if a peak and a trough overlap, there is cancellation: a phenomenon known as destructive interference. When there are multiple wave sources (Figure 5, panel 1), the resultant amplitude at any given point is the summation, or superposition, of contributions of all the waves. Imagine a ship at sea. Choppy waves have a short wavelength (hence high frequency), yet small amplitude. They make the ship bounce around more rapidly, while swells are waves that have very long wavelengths with low frequencies and high amplitudes, and cause the ship to rise and fall much farther, but over much longer time spans (Figure 5, Panel 2).

Overtones contribute to the sound of a single piano key in much the same way that chop rides atop oceanic swells. Our minds reinterpret this upper-frequency information as tone, or what musicians call timbre (rhymes with "amber"). Timbre is like an acoustic fingerprint of an instrument, and a difference in timbre is why a clarinet playing middle C sounds different than an oboe playing the same note, why a viola and violin sound different playing their open D strings, and why an accordion sounds different than a choir.

Figure 5. Interfering surface wave, at small and large scales. Photos by Scott Robinson from Rockville, MD, USA and Kevin R. Grazier.

Figure 6 shows how the waveforms might appear for two hypothetical instruments. Even if waveform figures convey little information to some readers, the important point is that two instruments—with different timbres but both playing middle C—combine to form a sound that is still middle C, but with an altogether different tonal quality.

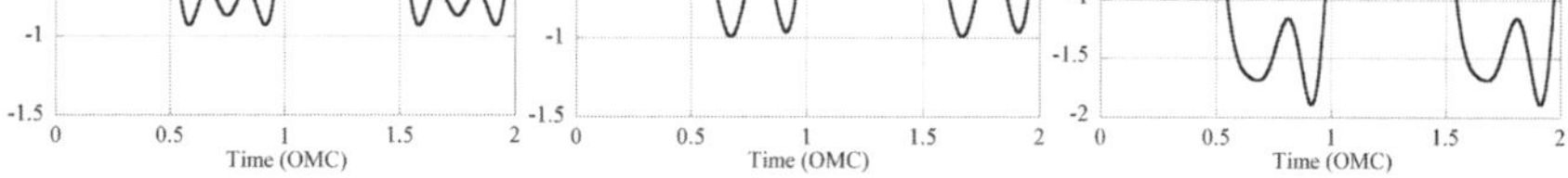

Figure 6. Examples of timbre.

Just like the overtones of a piano add to the fundamental frequency to give a piano its timbre, sounds or notes from multiple sources also interfere both constructively and destructively, with waves and swells from both sources combining to form a hybrid. Just as overtones modify the tone we hear from an individual piano string, two notes on an instrument, played simultaneously, similarly create a hybrid tone. When you hear music, the sound that your ear perceives at any instant is a complex result of contributions from every instrument, and oftentimes multiple types of contributions from individual instruments as well.

Consonant and Dissonant Intervals

In Western music, an octave has been divided into twelve notes, and musicians call the intervals between these notes *semi-tones*. The smallest interval, representing two adjacent notes, is the minor second, comprised of two notes that are one semi-tone apart. Larger than that, is the major second, comprised of two semi-tones, and so on. Physicists think of the relationship between two notes in terms of the ratios of their frequencies. The most common intervals and their ratio relationships can be seen in Table 1.

Table 1. Music Interval Names and Frequency Ratios. Ratios annotated with ∗ are for dissonant combinations, and the values approximate

Music Intervals		
Semitones	***Interval Name***	***Frequency Ratio***
0	Unison	1:1
1	Minor Second	16:15*
2	Major Second	9:8*
3	Minor Third	6:5*
4	Major Third	5:4
5	Perfect Fourth	4:3
7	Perfect Fifth	3:2
8	Minor Sixth	8:5
9	Major Sixth	5:3
10	Minor Seventh	9:5*
11	Major Seventh	15:8*
12	Octave	2:1

Certain intervals are more consonant than others, and can be defined by their simpler (smaller numbers) ratios in Table 1. These intervals are the unison (a note sounding with itself), the fourth (five semi-tones), the fifth (seven semi-tones) and the octave (twelve semi-tones).

These relationships are unique because when they are inverted they still create notes within the original fundamental's major scale. For example, assume a piece of music is in the key of C major and our fundamental note is, once again, middle C. To create a second above C, you would play a D, a note that is found in the C major scale. A second below C, however, would require a B flat, a note that is not in the C major scale. Try this experiment with one of the perfect intervals and observe a different result. From Middle C, a fourth above results in an F, and a fourth below results in a G, both of which are in the C major scale.

The unison, fourth, fifth and octave intervals are the only intervals for which this is true. Not coincidentally, these four intervals are the first in the overtone series and the simplest ratios of all intervals. To hear this in action, return to the piano, and play middle C and G5 (an octave and a fifth above) at the same time. The frequency of G5 is in 3:1 resonance with that of middle C: G5 is a member of the overtone series of middle C. The two notes played simultaneously result in a symmetric waveform, as we see in Figure 7, panel 1. This combination of notes, the resultant waveform, is said to be stable, since it repeats in a predictable way ad infinitium. The frequency ratio, however, does not have to be an integer multiple of the fundamental frequency (i.e., have the ratio, n:1, where n is an integer) for two notes to be in resonance. The ratio of the frequencies of G4 and middle C is in 3:2, and that is also considered resonant, consonant, hence pleasing. Western listeners find pleasing to the ear combinations of notes that result in waveforms that repeat over short time intervals. The most consonant combinations of two notes are those where the higher frequency is an integer multiple of the smaller. Also consonant are combinatinations where the frequency ratio can be represented by small integers.

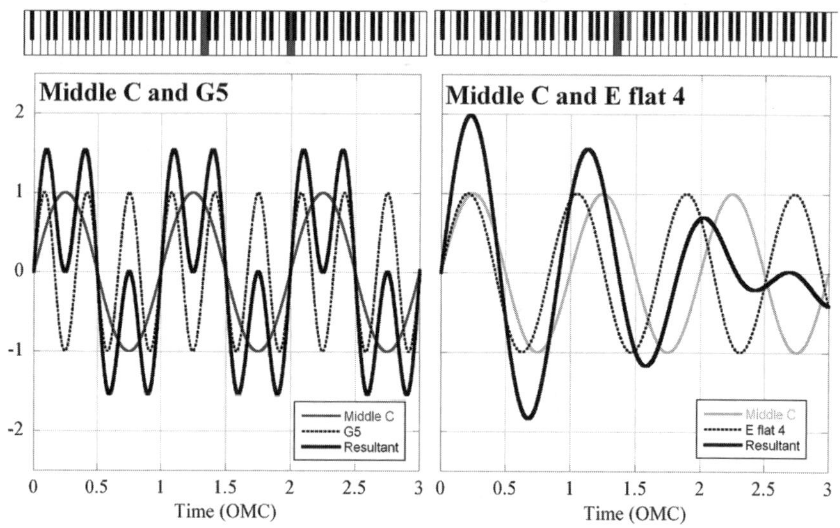

Figure 7. Two-note consonance and dissonance.

To hear musical dissonance, play that middle C again, but this time with the E flat just above it (a minor third). Although, technically, there is an E flat in middle C's overtones, it is very far up the series, meaning that it is very quiet and relatively difficult to hear. The combination of middle C and E flat (Panel 2 in Figure 7), results in an asymmetric, or unstable, tone—one that is ever-changing, never repeating. To the ear of Westerners this is dissonant: "An unstable tone combination is a dissonance; its tension demands an onward motion to a stable chord. Thus dissonant chords are 'active'; traditionally they have been considered harsh and have expressed pain, grief, and conflict (*2*)."

Two notes are all that is necessary to create tension, and there are intervals far more dissonant than a minor third. Return to your piano once more to play that middle C again. This time, add to it the first black key a semi-tone above it.

This clashing interval, the minor second, is among the most dissonant harmonies in Western music. Perhaps the most famous minor second in film music history is the shark theme composed for *Jaws* by John Williams, which begins with just two notes repeated in an accelerating pattern. Though the notes are heard in succession rather than simultaneously, the shark would simply not be as scary if his theme were oscillating a major third!

On the other hand, film music is rarely built from just two notes. Composers will stack notes into chords, where every note interferes constructively or destructively with every other note. In a good melody, the implied harmonic structure is clear—even though the notes play sequentially rather than simultaneously. The benefits of these acoustic properties for dramatic musical narrative are overwhelming. Composers select important events in a film and match harmonic structures to them, to maximize the appropriate amount of consonance or dissonance to any given scene. While any aspect of music can be consonant or dissonant, including instrumentation, style, harmony and rhythm, the manipulation of harmony is the most immediately effective way a composer can alter the mood of a scene.

Waveforms in the natural world condition us to recognize perfect intervals as consonant and less-perfect intervals as dissonant, foreign sounds. The importance of the perfect fifth has already been discussed, but Western music has subdivided the fifth even further. The frequency of Middle C is roughly 262 Hz, while the frequency of G4, a perfect fifth, is about 392 Hz. To put a note directly at the midpoint between the two notes in a perfect fifth is a note that isn't playable on a piano (just less than 327 Hz), because it does not fall in the overtone series. Instead, Western music has rounded this mid-point to the two nearest notes, resulting in the major and minor third. Going back to the example of C, the major third is a distance of four semi-tones to E natural. The minor third is only three, to E flat. By combining C, its perfect fifth G, and E major, you get a C major "triad" (a grouping of three notes). By shifting the E natural to E flat, it becomes a C minor triad.

Ironically, a minor triad includes a major interval and a major triad includes a minor interval, in both cases between the third and the fifth. So why do we hear a "major" sonority when we hear a C major chord, even though there is a minor third between the E flat and the G? The overtone series is the answer. We are conditioned to hear the lowest note as the fundamental and the upper notes as implied overtones. So, our minds instantly hear the C as the bottom tone, and the E and G as upper partials.

For proof that our brains are constantly looking for harmonic order to sounds, go back to the piano yet again. Play middle C with the E4 and G4 above it. That is a C major triad, clearly defined because the C is at the bottom and the other notes reenforce its overtones. However, those notes can be reordered so that the C is not on the bottom, and we still recognize it as a C major chord. To prove this, add the G3 below. Now, the G is the lowest pitch, at the bottom of the chord. Technically, a jazz pianist might be able to describe this sonority as a "G sus4 add6 without the fifth." That description, though, requires an unnecessary amount of mental gymnastics to rationalize this as a G chord. No, our minds will almost always recognize that this is still a C major triad, despite the fact the lowest note is a G, the fifth of the chord. Returning to the premise that "Resonance = Consonance," the ratio of the frequencies of C4 to G3 is 4:3; although this is not as pleasing to the ear as a higher G, these two are still resonant, and striking G3 also causes G4 and G5 to vibrate sympathetically.

Although the interactions of multiple sound waves—like those produced by musical instruments—can be quantified and categorized as either consonant and dissonant, the application of this principal to the art of film scoring is much more complex. The context in which composers and filmmakers introduce these sounds has a profound impact on how the audience perceives them.

Music and the Mind

The overtone series' emphasis on perfect intervals has dominated Western music for hundreds of years, which in turn, became the language of movie music. Yet, this interpretation of overtones is not the only way to organize notes, as can be heard in many examples of cultural music around the world. Listeners raised in the musical traditions of Nepal or India are culturally conditioned to appreciate harmonies and rhythms vastly different than Europeans'. "Hollywood Music" might sound very foreign to them—and create a completely unique set of emotional connections, based upon their cultural context. Most films, however, are targeted at a broad, global audience with the implicit understanding that they share a common familiarity with Western tonal music, so the overtone series is the foundation of the cinematic musical language.

The first five overtones in the series are the easiest to hear and are the most influential on music: the octave, perfect fifth, perfect fourth, major third and minor third. It is not a coincidence, then, that the soaring brass fanfares of *2001: A Space Odyssey*, *Star Wars*, *Star Trek*, or the original *Battlestar Galactica* are all structured around these intervals. We hear these intervals hidden within every note played by every instrument we hear. So, these melodies are building off an inherently familiar harmonic language. The audience hears that familiarity and their minds classify these themes as a consonant sound.

That kind of harmonic and melodic consonance implies strength and stability: the ultimate musical consonance. Audiences respond to these themes not only because they are orchestrated well and are based on appealing melodies, but because the use of intervals found in the overtone series reinforces overtones found in each of the individual notes. It creates a feedback loop of awesomeness and, used effectively, can leave the audience breathless—or at least with a serious case of the chills.

In fact the chills that listeners get from a particularly satisfying piece of music result when the brain releases a neurotransmitter called dopamine. "'Chills' or 'musical frisson' is a well established marker of peak emotional responses to music (*3*, *4*)." Building off roughly a decade of work on 'chills' induced by music, a recent Canadian study has shown that when subjects listened to music they found especially pleasing, their brains released dopamine. The release of dopamine is itself pleasurable, and it has long been known to be the brain's method of rewarding and encouraging survival behaviors like eating and sex. The study further found that even the anticipation of a pleasurable strain of music can elicit a dopamine response. Anticipation, or expectation, is bolstered in films via mechanisms of consonance and dissonance strategically employed by the film composer. Musical reward, as Dr. Robert Zatorre, a co-author on both studies, points out, is abstract, but nonetheless real: "These findings provide neurochemical evidence that intense emotional responses to music involve ancient reward circuitry in the brain," says Zatorre. "To our knowledge, this is the first demonstration that an abstract reward such as music can lead to dopamine release (*4–6*)."

The lead author of the studies, Valorie Salimpoor adds, "If music-induced emotional states can lead to dopamine release, as our findings indicate, it may begin to explain why musical experiences are so valued. These results further speak to why music can be effectively used in rituals, marketing, or film to manipulate hedonistic states. Our findings provide neurochemical evidence that intense emotional responses to music involve ancient reward circuitry (*5*)..." Music cognition researcher David Huron adds: "The findings suggest that, like sex and drugs, music may be mildly addictive (*7*)."

Let's return to our *Star Wars* example. The opening credits establish the film's well-known Main Theme, a brassy fanfare borrowing heavily from Korngold's score to *King's Row*. The theme spells out the overtone series as obviously as any film music since Kubrick's use of Richard Strauss' "Also Sprach Zarathustra" in *2001: A Space Odyssey*. The theme is heard once again in full during the film's closing credits. Aside from a handful of partial quotations, the theme is virtually absent from the entirety of *Star Wars*. The triumphant feeling generated by hearing that perfect fourth and perfect fifth stacked up in the brass is so consonant, it gives the audience such an intense satisfaction, that composer John Williams wisely chose to save it almost exclusively for the film's opening and ending, the parts of the film signifying narrative consonance. It remains the most memorable piece of music from the film, despite only being heard for a small fraction of the film's running time.

Consonance and dissonance apply not only to the narrative or the music, but also to the combined effect that occurs when a film is put to music. Psychologist Marilyn Boltz published a study in 2004 revealing that music that is congruent—consonant—to a film, enhances a person's memory of both the film and the music. Music that is discongruent—dissonant—to a film gets encoded into memory separately from the film; whichever component is focused on during viewing is the one that gets remembered (*8*). Furthermore, another study demonstrated that, while music alone can have an emotional effect, if it is paired with emotionally-discongruent film material—such as pairing a horror soundtrack with a benign pastoral scene—the emotional effect becomes neutralized (*9*). These additional findings support the idea that pairing a film with emotionally congruent music is what works: that the two work together to enhance perceptual context for audience members. The joint encoding of film and music (when the two are consonant) is what triggers mental imagery of scrolling, slanted words against stars in space when we hear that brassy *Star Wars* theme.

Consonance is bred from familiarity. The more familiar a sound is to our ears, the more satisfaction we derive from hearing it. For *Star Wars*, Williams wisely capitalized on the audience's built-in familiarity with classical music, and at the same time, did more to propagate mainstream appreciation of classical music than anyone since Leonard Bernstein.

Of course, there are many successful examples of composers deliberately deviating from familiar sonic ideas to effective results, including the 1968 *Planet of the Apes*, the 2004 *Battlestar Galactica*, the original *The Day The Earth Stood Still and Blade Runner*. These examples employed unusual instrumentation, exotic harmonies and obtuse rhythms to transport audiences to alien landscapes. Yet, a detailed look at any of those scores reveals that, more often than not, the harmonic language still boils down to harmonies built from the first five upper partials of the overtone series.

If musical consonance, can be defined as melodies or harmonies that reinforce the intervals heard commonly in the overtone series, the broadest definition of dissonance might be that it is anything that deviates from consonance. Musical dissonance has a powerful and immediate effect on the human mind, which tends to repel new sounds that can't be placed into a familiar, consonant, category. Dissonance in narrative is easy to spot: the protagonist enters into conflict with an antagonist, nature, external or internal forces. Almost without fail, these story arcs are scored with musical dissonance designed to excite the audience.

In 1913, the first audience for Stravinksy's *The Rite of Spring* was so shocked by the clashing harmonies and aggressive, angular rhythms that they nearly rioted in the streets of Paris. How awful that music must have sounded to them! As with all new sounds, though, people eventually got used to those clustered chords and jagged rhythms. A quarter century later, Walt Disney included *The Rite of Spring* in the animated anthology film *Fantasia*. Audience reaction was far more docile. Music that merited a violent reaction in 1913 barely raised eyebrows in 1940, but, it hardly takes decades for audiences to grow accustomed to a dissonant sound. It can happen over the course of a two-hour film!

The more the mind hears a sound, the less the it takes notice of that sound. We experience this in our everyday lives: the car alarm wailing outside, the buzz of a broken air conditioner, the blurring whir of heavy traffic. After a few moments, your brain realizes that the sound is not changing, nor is it important, and that noise gets filtered out of your concentration. We filter static sounds in order to be able to respond when something new is introduced to the situation. This phenomenon is a type of sensory adaptation we refer to as desensitization, or habituation. Without this ability, it would be virtually impossible to carry on a conversation with anyone in a restaurant, or to hear your cell phone ringing when walking down a busy street. While this type of adaption is temporary, very long-term habituation—of a few weeks or more—can lead to neuroplasticity, that is, anatomical changes in the brain. It is important to note that habituation is not the same as selective attention, though they can work together. Selectively attending to something involves choosing to focus on one stimulus, though many might be vying for your attention. Take a restaurant conversation for example. Habituation helps explain why the steady din of conversational noise from other tables disappears while you are holding a conversation at your own table. Selective attention helps explain how you are able to maintain focus on your conversation, despite new noises that might try to grab your attention, like a server rushing by, or a sudden outburst of song. Both principles are relevant to viewing films.

The principle of habituation applies in film scoring as well. The more music the audience hears, the more they recognize familiar material. In short time frames, for example, over the course of a film or even a long running television series, this principle can be used to a composer's advantage. Using themes to specify certain characters or recurring ideas is a marvelously effective way of allowing an audience to subconsciously connect events that are separated by time. This technique makes use of the co-encoding of film and music into memory that occurs when the two are in consonance. Hearing a theme again, the audience instinctively recalls the previous times they heard that theme and brings previous emotional weight to what they are currently experiencing.

Over long periods of time, however, say years or decades, audiences can become desensitized to music, especially when certain genres tend to use similar scoring techniques. The composer has to work harder to achieve the same emotional effects. The more movies a person sees in their lifetime, the more unusual sights and sounds they grow accustomed to. (If that first audience who watched *The Arrival of a Train at La Ciotat Station* were somehow transported to 2012 and watched *The Avengers*, they may all have had heart attacks and died of shock.)

Composers and filmmakers have entered a musical "arms race" to keep pace with audience expectations. Horror is the perfect genre to illustrate this principle. In the early and mid 1970s, genre classics such as *The Omen* or *The Exorcist* brilliantly used minimal music for maximum effect. Just the right dissonance was placed only where it was needed and audiences were terrified. Audiences quickly grew used to those sounds, so they had increasingly diminished effect. So, the use of musical dissonance increased, the orchestrations got louder, ensembles got bigger and filmmakers fell back on gratuitous stingers and jump cuts to terrify audiences. In just ten years, horror film scoring devolved into a mash of screaming orchestral sound effects, borrowing liberally from Ligeti and Penderecki, whose music was first used in film in Kubrick's *2001* to brilliant impact.

Anyone who's ever seen a slasher horror film knows what to expect when ominous low tones accompany a scantily-clad young heroine creeping cautiously down a dark hallway. We all brace ourselves for the inevitable sting as a monster leaps out of the shadows. Or sometimes just an alley cat. Either way, the music startles us out of our seats as much as the action onscreen. The effectiveness of this kind of cheap scare decreases with its use, both in a single film (short-term desensitization) and over the years (long-term desensitization). Audiences figure out that dissonant music means certain events are likely to happen. In the worst case scenario, this leads to the dreadful genre clichés that litter terrible movies. On the other hand, smart filmmakers recognize that audiences, consciously or not, will anticipate their musical cues and take advantage of this fact, using that anticipation to surprise them in unexpected ways.

A film score can only get so loud and frequent before it eventually saturates the audience in meaningless sound. In recent years, trends have begun to reverse. AMC's hit TV series *The Walking Dead* (Figure 8) is set in the well-worn "zombie apocalypse" genre. The score is surprisingly minimal, with long stretches of story passing before music enters. The dissonant chords are softer, more atmospheric, and work in tandem with the sound design to achieve the maximum emotional impact with the minimal amount of music. The series became a huge success and terrifies even its core audience of horror fanatics, many of whom have seen every movie in the genre. *The Walking Dead* represents a much-needed step towards creative disarmament within the horror genre.

Figure 8. The first author, composer for The Walking Dead. (Photo by Andrew Craig, courtesy of Bear McCreary).

Summary

There has been significant scientific effort over the years to categorize comprehensively the individual elements—pitch, tempo, rhythm, etc.—that make a piece of music sound like a particular emotion (*10*). These elements have been reliably categorized and replicated across studies, including some with cross-cultural participants. While there is much agreement that music written in the major mode, with a faster tempo and consonant harmony will communicate the emotion of happiness, for example, no film composer lives by the rule that minor chords are sad and major chords are happy. Communicating happy sounds just isn't the same as making someone feel happy. The former is the representation of emotion; the latter is its induction, and everything is modified by context. Human emotions are far too unpredictable and subjective to apply to general rules, at least when it comes to emotion induction. One can state with confidence, however, that a minor chord carries more inherent tension, more harmonic dissonance, than a major chord because it clashes subtly with each note's individual overtones. Film composers take advantage of this built-in tension to manipulate audiences.

Manipulation of consonance and dissonance in harmony is an essential component of film scoring. A conventional film score can be boiled down to its harmonic DNA and, time and time again, you will find more dissonant harmonies associated with scenes of increasing tension. On a subconscious level, audiences bring with them a built-in understanding that simple harmonies evoke positive emotions of familiarity and that dissonant harmonies create tension that wants to be resolved back to consonance. Theoretically, a composer could score an entire film using only two fingers and a piano: carefully placing dissonant and consonant harmonies against the right emotional moments in the film. Though, if the film became a hit, as you now know, this technique would likely work only once.

Of course, the art of film scoring is vastly more complicated and involves making many decisions beyond the application of harmony: the choice of instruments, ensemble size, rhythm, style and musical placement makes the art highly subjective. No two composers would score a film the same way, but all composers approach the task with the same set of basic tools. The most successful are the ones who guide the audience expertly along a narrative path and completely hide their tracks, so that the audience never realizes music has been manipulating their emotions throughout the entire film. The audience is simply lost in the experience of the darkened theater, enjoying every minute of the journey.

References

1. Schnupp, J.; Nelken, J. I.; King, A. *Auditory Neuroscience*; MIT Press: Cambridge, MA, 2011.
2. Kamien, R. *Music: An Appreciation*, 6th brief ed., 2008.
3. Salimpoor, V. N.; Benovoy, M.; Longo, G.; Cooperstock, J.; Zatorre, R. The rewarding aspects of music listening are related to degree of emotional arousal. *PLoS One* **2009**, *4* (10), e7487, DOI:10.1371/journal.pone.0007487.
4. Salimpoor, V. N.; Benovoy, M.; Larcher, K.; Dagher, A.; Zatorre, R. Anatomically distinct dopamine release during anticipation and experience of peak emotion to music. *Nat. Neurosci.* **2010**, DOI:10.1038/nn.2726.
5. Jha, A. Favourite Music Evokes Same Feelings as Good Food or Drugs. *The Guardian*, January 9, 2011, retrieved November 21, 2012.
6. McGill University. Musical Chills: Why They Give Us Thrills. *ScienceDaily*, January 12, 2011, retrieved November 21, 2012.
7. Sohn, E. Why Music Makes You Happy. *Discovery News*, January 10, 2011, retrieved December 10, 2012.
8. Boltz, M. G. The cognitive processing of film and musical soundtracks. *Mem. Cognit.* **2004**, *32* (7).

9. Blumstein, D. T.; Bryant, G. A.; Kaye, P. The sound of arousal in music is context-dependent. *Biol. Lett.* **2012**, *8*, DOI:10.1098/rsbl.2012.0374.
10. Juslin, P. N.; Laukka, P. Communication of emotions in vocal expression and music performance: Different channels, same code? *Psychol. Bull.* **2003**, *129* (5), DOI:10.1037/0033-2909.129.5.770.

Chapter 8

The Science & Entertainment Exchange: The National Academy of Sciences Goes to Hollywood

Ann G. Merchant*

Deputy Executive Director, Office of Communications, The National Academies, Washington, DC 20001
***E-mail: amerchan@nas.edu**

In response to numerous indicators, including findings from their own reports, that the scientific and technological building blocks critical to America's economic leadership are eroding at a time when many other nations are gathering strength, the National Academy of Sciences sought ways to inspire the next generation of scientists, engineers, and health professionals. In 2008, the Academy launched the Science & Entertainment Exchange, a program that seeks to leverage the power of mainstream entertainment media in Hollywood to create a synergy between accurate science and engaging storylines in both film and television programming. This "origins story" of The Exchange lays out the rationale for the program from the entertainment/education literature and describes the 1-800-I-Need-a-Scientist matchmaking service it provides to connect writers, directors, producers, and other members of the entertainment industry to science and technical consultants. It also enumerates a list of 13 key rules which have guided the growth of the program since its inception. As The Exchange looks to the future, it is now poised to ask some critical questions of its two most important audiences—the entertainment professionals it considers to be its clients and the many scientists, engineers, and health professionals who serve as consultants—the answers to which will help determine the impacts the program is having on the communities it serves.

As he told her that he loved her she gazed into his eyes, wondering, as she noted the infestation of eyelash mites, the tiny deodicids burrowing into his follicles to eat the greasy sebum therein, each female laying up to 25 eggs in a single follicle, causing inflammation, whether the eyes are truly the windows of the soul; and if so, his soul needed regrouting.

-- Cathy Bryant

Science and storytelling do not always mix. Uneasy companions, begrudging bedfellows, allies suspicious of one another when tossed together either by choice or happenstance. On the other hand, this oil-and-water relationship has always had the potential to be pizza and beer, peanut butter and chocolate, salt and caramel … essentially, two great tastes that taste great together, each benefiting from the flavor added by the other to be better than the one ingredient on its own.

Science as a discipline has bemoaned its inability to communicate effectively to lay audiences, to reach the "general public" (however you may choose to define that term) with the important messages it has sought to convey. Certainly, the National Academies has echoed remnants of this same sentiment in its landmark report *Rising Above the Gathering Storm*, which served as a call to action to the science and engineering communities that the time was now, the cause was urgent, and if we did not act, we were in danger of losing the "productivity of well-trained people and the steady stream of scientific and technical innovations they produce (*1*)." The Academies further warned that "without high-quality, knowledge-intensive jobs and the innovative enterprises that lead to discovery and new technology, our economy will suffer and our people will face a lower standard of living (*1*)."

We sent that message forth into the world to be repeated like a mantra by policy makers, decision makers, and influential leaders in the science and engineering communities around the country. Though the report was not released until 2007, we were well-aware of the efforts of the blue-ribbon panel convened to examine the issues—which included well-known names such as names Norm Augustine, Craig Barrett, Steve Chu, Shirley Jackson, Anita Jones, Josh Lederberg, and Chuck Vest—and were not ourselves untouched by the import that their conclusions were likely to have. Though the Academies does its best work advising others, it never hurts to take a page from your own playbook. We, too, wanted to do our bit to inspire the next generation of scientists and engineers who could see to it that our nation stayed strong and vibrant.

The Office of Communications at the National Academies is charged with communicating general messages of science and engineering as well as the work of the institution more specifically to a diverse set of audiences—in other words, people who *do not* read 592-page tomes like *Rising Above the Gathering Storm* (or even the Executive Summary thereto). We also oversee a host of programs that seek to communicate fairly broadly to the public. Our office is primarily engaged with informal science education and communication—what one of our reports calls the "overlooked or underestimated … potential for science learning in nonschool settings, where people actually spend the majority of their time (*2*)." Although relatively nimble at creating websites and producing collateral material designed to be reader-friendly and conversational, the declaration of urgency beginning to issue so clearly from both our own reports as well as from other influential sources called for something bolder, something surprising… something some might consider crazy-daring. Maybe… something like going to Hollywood. Or at least that is how we chose to interpret it.

At about the time that the National Academies Press was shipping copies of *Rising Above the Gathering Storm* around the country, Neil Gershenfeld, founder of MIT's FabLab at the Center for Bits and Atoms, described to us his experience providing expert input on the making of a Hollywood film. Neil enthusiastically conveyed to us how he gave feedback in the collaborative process and characterized the interaction as productive, both for him as a scientist seeking to ensure that strong messages about science were included in the film, and for the filmmakers, who hoped to enrich their narrative with interesting and unusual science-based details. Intrigued by the possibilities, but unsure of the true value in communicating via film or television, we turned to—where else?—the literature.

We know that television and film entertainment has the power to amuse and please its fans, or divert an audience's attention from reality. Mainstream narrative media can also convey messages and influence ideas. Research has shown that science, when incorporated into quality entertainment television and film portrayals, can change minds and overcome deeply entrenched opinions. But overcoming negative stereotypes about science and communicating its tremendous importance and appeal to a general audience is an enormous challenge. There is a bona fide communications strategy called entertainment education (EE), alternately referred to as edutainment, infotainment, or even enter-educate, that is used to affect behavioral and social change. EE is defined as the process of purposely designing and implementing a media message to both entertain and educate, in order to increase audience members' knowledge about an educational issue, create favorable attitudes, shift social norms, and change overt behavior (*3*). This very purposeful kind of entertainment has been used in highly effective ways, most notably in India and Mexico and in some forms of our own children's programming. In the 1980s Miguel Sabido, a vice president of Televisa, Mexico's national television network, used popular telenovellas to significantly shift behaviors on both literacy and birth control (*4*). And while Snopes.com says it just ain't so (or it just ain't proved, anyway) sources cite a fifth season episode of *Happy Days* in which the Fonz goes to the library to meet girls as the inspiration for a 500% increase demand in library card applications, which is exactly the kind of behavior we hoped to inspire (*5*).

Certainly we have evidence that entertainment media that portray scientists and their work in a positive light can have direct educational and socializing influences on audiences. Survey analyses, for example, show that some science-fiction portrayals appear to boost public perceptions of science. After controlling for education and other background variables, studies find that more frequent viewers of science-fiction television hold greater belief in the promise of science and are more supportive of controversial topics such as therapeutic cloning (*6*). In a study of the audience effects related to the 2004 blockbuster *The Day After Tomorrow*, viewers of the film, after controlling for education, gender, age, and political views, were significantly more concerned about global climate change, more likely to take action to reduce greenhouse gas emissions, and more trusting of government agencies such as NASA and NOAA (*7*).

One need only look at the commercial success of numerous television shows that incorporate STEM content into their storylines to see that Hollywood understands that science and engineering can actually be beneficial to the business of entertainment. Certainly medical dramas are a good example of this and shows like *ER* put very technical plots and dialog center stage (as opposed to the love lives of their characters) to draw audiences. Indeed, during the 1997–2000 television seasons, the Kaiser Family Foundation did some in-depth work to determine what viewers were learning from sitting down to watch one of their favorite shows. After surveying 3,500 regular viewers of the popular NBC hospital drama they learned interesting and revealing facts. It turns out that more than half the people that watched said they learned something important about their own health and that they were able to get enough information from the show to make personal decisions. One in five doctors even reported patients asking about diseases and treatments that they learned about from watching *ER* (*8*).

House, M.D. continued the tradition of putting the spotlight on medicine and practice rather than love and relationships while more traditional medical shows like *Grey's Anatomy* and the recently ended *Private Practice* did double-duty, using personal drama to humanize the medicine. Health storylines are also woven into the fabric of more traditional narratives in all sorts of shows, both comedies and dramas. We learned about Asperger's Syndrome from *Parenthood* (creator Jason Katims's own son has been diagnosed with Asperger's). We battled cancer with Lynette on *Desperate Housewives*, sitting with her during her chemo treatments and standing alongside her when she pulled off her wig to let her bald pate shine. The show *Army Wives* exposed a number of mental health issues onscreen and we shared Denise's concerns about Frank's mood changes after he returned home from Africa. Soap operas should also not be forgotten for their contribution to the public's understanding of HIV/AIDS. It was the long-running NBC soap *Another World* that showcased a soap character, Dawn Rollo, with AIDS.

The hit CBS procedural *CSI: Crime Scene Investigation* earned itself a place in the legal and science communications research literature with the now well-recognized reference to the eponymous "CSI effect." *Bones*, *NCIS*, *Body of Proof*, *Rizzoli & Isles*, and the *Law & Order* franchise all regularly feature forensic science as a key plot point while other shows like *The Good Wife*, *Criminal Minds*, *The Mentalist*, and others rely on forensics in a more limited but still important way. *Numb3rs* showed us that mathematics could be used to fight crime as well as any DNA test—and that the mathematicians were as sexy as the FBI agents who brandished Glocks rather than chalk. *Breaking Bad* demonstrated that chemistry (not just street chemistry, but honest to goodness book learnin' chemistry) can not only make first-class, grade A "glass" (crystal meth), it can also make great television.

These programs, and others like them, demonstrate that it is possible to pique public interest in science, scientists, and the scientific method when experienced through the lens of entertainment media. And that is just television! Feature films often rely on science, engineering, and technology to grab their audience's attention.

By some counts, as many as 8 of the 10 highest grossing films of all time are science-fiction flicks (*9*). And even at the low end, the count shows that half are sci-fi, so clearly the genre can get people into the theater. Science-based content can also be fodder for critical acclaim and artistic success. *A Beautiful Mind*, which won four Academy Awards, including the 2001 prize for Best Picture, focused on the life of John Nash, a Nobel Laureate in economics. Our pre-existing fascination with dinosaurs was brought to new heights when *Jurassic Park* hit the theaters. While we loved Drs. Grant and Sattlerand, those adorable kids, the Velociraptors clearly stole the show as the most memorable members of the cast, and there were a number of viewers out there who genuinely wondered whether someone might actually possess the technology to extract dino DNA from mosquitoes to make a real park! Under ordinary circumstances, the mating habits of the Emperor penguin might not be all that captivating, but thanks to a skilled Hollywood storyteller (and the fantastic voiceover from the talented actor Morgan Freeman), audiences of non-scientists can now explain the "journey like no other on the planet" of tuxedoed armies that trudge hundreds of miles across a cold, windy continent to complete their evolutionary mission.

Now that's not to say that things have always gone well. I'm not suggesting that entertainment media is somehow the holy grail of science communication that will solve all our problems as we sit back and let Hollywood do the heavy lifting required to change perceptions, influence opinions, or generate interest in, and public understanding of, science. Ask most scientists and you'll almost exclusively hear how badly Hollywood has done in depicting both the science and the scientist. You'll get the same from engineers of all stripes as well as most health professionals. Our oil-and-water problem may be here to stay for many in the STEM world, but the fact remains that that the public's interest in science often increases when exposed to science on television and on film (*10*). Indeed, films that feature STEM themes, particularly in the science-fiction genre, serve as an unintentional form of curriculum. Movies do not just entertain, they educate—whether we want them to or not. Researchers used the 2003 movie *The Core*, often trotted out by scientists as perhaps one of the worst examples of cinema-science gone wrong (and therefore one of the best examples of Hollywood's flagrant and egregious disregard for real-world science), to demonstrate just how much students relied on the viewing of the film to inform their study of geoscience. Students often referenced what they watched on-screen to inform their answers to classroom questions. Not only did students use examples and ideas that were expressed in the movie to explain their ideas, but they were more confident in their (sometimes wrong) answers because of the fact that they were able to reference a story that remained quite vivid for them (*11*). Another takeaway here is clear. The folks in Hollywood were sometimes better at communicating the science than the classroom educators, despite the fact that the latter science was far superior to the former.

So, what to do with this data? Faced with the finding that the guys in Tinseltown are doing a bang-up job of spreading the (sometimes supremely wrong) word on STEM topics so near and dear to our heart, we determined that it was better to join them rather than fight them. Accordingly, in 2006 Dr. Ralph Cicerone, president of the National Academy of Sciences, traveled to Los Angeles to meet with Jerry and Janet Zucker with whom he had a six-degrees-of-separation connection through a former student who worked for Zucker Productions.

Longtime Hollywood insiders, the Zuckers had relatively recently undergone their own very personal introduction to science. Some years earlier their daughter Katie had fallen gravely ill and was diagnosed with Type 1 diabetes. Their quest to find a medical solution to Katie's condition led them to a hospital room where she was administered her first dose of insulin. Her parents are fond of describing this emotional scene as the moment their daughter was returned to them feeling so much better. But they soon discovered that under the Bush by parental passion they followed the oft-trod path of smart, powerful people in Hollywood. They created a foundation and got political. Using a combination of connections, creative genius, and fight-for-right determination, they threw their hearts and souls into this mission, eschewing the business of movie making in favor of their very personal science-based cause. They did their homework (and drew on the access that administration of the early 2000s, long-term research into stem cell therapies—the most promising, long-term opportunities for people suffering from Katie's disease—were stymied by prohibitory federal laws. Incredulous, outraged, and fueled Hollywood clout can convey) to reach out to some very talented doctors and scientists doing incredible work who explained their research to them. Ultimately, the Zuckers and their foundation, CuresNow, were instrumental in the passing of Proposition 71, the law enacted in California to support stem cell research in the state.

Jerry and Janet, only casually interested in science before this journey, were hooked, both on the people they were meeting and the work those people were doing. They understood how much good science could do in the world and realized intuitively (without reading all the studies) the influence that the creative community could have on America's perceptions of science and scientists. And, yes, it did not hurt that all that science was just plain cool and smart and interesting to cool, smart, and interesting people like the Zuckers. So when Ralph Cicerone came for a visit, the timing could not have been better. Their fight to pass Proposition 71 was pretty much over but they were loathe to walk away from this new world they had discovered and hungered for ways to stay connected to science, and even more importantly, to continue to make a difference. They were primed and ready to join forces to create change.

The idea that scientists could act as technical consultants was not born the day that Jerry, Janet, and Ralph met in Los Angeles. This was not a unique notion and examples of other facilitated efforts already existed. Hollywood, Health & Society (HH&S), a USC program at the USC Annenberg Norman Lear Center, pre-dates The Exchange and is perhaps one of the most successful examples of this kind of effort, but HH&S focuses specifically on health storylines primarily for television where The Exchange's mission is more broadly stated to cover science, engineering, health, and medicine in film, television, and even video gaming. Other programs that had attempted to engineer a matchmaking effort similar to that of either HH&S's or our own had, for various reasons, never managed to gain much long-term traction. We thought long and hard about how the Academy's program might be different from those other efforts and created some goals and guidelines for ourselves. We believe that our model combined key ingredients and approaches to set us on the right path.

1. *Find a Hollywood partner with connections.* It never fails to amaze—and humble—me when my emails are answered, phone calls are taken, and invitations are readily accepted by some of the most important and recognizable names in science today. This has little to do with me. It is all about the institution for which I have very proudly worked for more than two decades. But while our Constitution Avenue address might contribute to our bona fides outside the science community, the National Academy of Sciences had insufficient name value or influence in Hollywood—a town that lives by name value and influence. We knew that it would be critical to have the help of an established Hollywood partner to get The Exchange off the ground. No one in Hollywood answers emails, takes phone calls, or accepts invitations on the strength of the NAS name and we knew it. The only way we had a shot at making this program work was to enlist a Hollywood partner. Janet and Jerry Zucker filled that role and worked tirelessly on our behalf for a full year to help create a strategic vision for how The Exchange would operate. They recruited a full roster of Hollywood board members that read like a who's who of important Hollywood names, which immediately conveyed legitimacy to the program. They helped plan—and fund—the 2008 symposium that launched the program. And they remain invested and involved in The Exchange to this day.
2. *Launch big (it's Hollywood).* As referenced above, we formally launched The Exchange with a full-day symposium in Los Angeles. We invited 300 people to come hear 18 scientists speaking on 6 different topics. Short TED-style plenary presentations were followed by individual "salon" sessions where participants were allowed to self-select into smaller discussion groups for a more intimate experience to learn more. The idea was to demonstrate the talent to which we could connect people and to give people a sense of the incredible stories that were naturally resident in science and engineering. We knew that we had put together a stellar lineup on our stage but we also knew that we were playing to the toughest crowd in the world. We had been advised by a number of knowledgeable people that we might not be able to hold onto the audience for a full day. Janet and Jerry believed in our plan—and gave us the strength to believe. They were right. Ultimately, the biggest complaint we received was that we offered the audience the chance to visit only two of the salons before concluding the day. The phones started ringing in the office the next day. More than 600 consults later, they have not stopped.
3. *Ensure that you have staff expertise in both science and entertainment.* Entertainment and science each have their own language and their own culture. We learned that early on when working with Zucker Productions to create the model for The Exchange and plan the launch symposium.

This experience made it clear that we needed to have representation from both communities in The Exchange's office. The program's LA staff has always consisted of a director and a coordinator and our rule has been that if the director's background and experience lies in entertainment then the coordinator's resume will reflect an education and job history in science—and vice versa. This rule has stood us in good stead and serves as a good internal check-and-balance system to keep us on the right track. We do not make "rookie mistakes" nearly as easily and can do things more quickly and smoothly based on the expertise that each person brings to the process.

4. *Locate your office in Los Angeles.* In some ways, it would have been easier to staff The Exchange out of our offices in Washington, DC. Certainly, I would spend a lot less time on airplanes. But I do not conduct the day-to-day activities of The Exchange, and consistent outreach efforts are critical for a program like this. Being available and accessible to the entertainment community are also important. Geography counts in other ways more subtle as well. Our Constitution Avenue address works just fine for the National Academy of Sciences but it would scream "outsider" to the entertainment community.
5. *Do not charge a fee for the service.* We do not charge a fee for services that are rendered. The connections we make, the events we host, even the scientists we fly in for consultations are paid for out of The Exchange's pocket. We want to keep the barriers low to using The Exchange and the benefits obvious from the experience. It is not always obvious to our clients, but The Exchange is supported by grants from foundations and other generous funders. We launched with internal funding to get the program off the ground and while we are not especially expensive as Academy programs go (about $460,000 per year in annual operating expenses), we are always in search of funding to keep the program going. We do not believe that the entertainment community will pay for the services we provide. In fact, it's not even clear that they could pay. Despite the reputation that Hollywood is awash in money, their budgets are often tight, and we've never forgotten that our most serious competitor is absolutely free—not always accurate or reliable, mind you, but free—and that's the Internet. Entertainers have free resources that they will happily access over a paid service. Although it's sometimes true that a long-term relationship with a science consultant might result in the studio paying that consultant, this is arranged directly with the scientist and The Exchange does not act as an agent and does not benefit financially in any way.
6. *Be quick and responsive.* Film and television is a fast-paced business (when it's not slow and ponderous). When a writer, director, or producer needs something, they generally need it pretty quickly. The Exchange is poised and ready to respond very fast to the requests we receive. For the most part, we work hard to fulfill a consultation request within 24-48 hours. This does not apply to our "think tank" meetings that involve flying in several scientists for a half- or full-day meeting to discuss a

variety of topics. These can obviously take longer to arrange. But for most of our consults where a few phone calls and emails will do the trick, we forge the connection quickly—lest Google start to look like the more attractive option.

7. *Do not discriminate between consults.* We take all comers, whether the query comes from a young writer or from an established director with an enviable IMDb filmography. We may match the young writer with a post-doc and the established writer with a more notable scientist—or not, depending on the requirements of the individual consult. The idea is to ensure that the query is addressed in the most efficient and effective way possible, leading to the best science content we can facilitate. From a very selfish point of view, today's unknown writer could very well become the next big star—with a hot science script.
8. *Complete several consults each week.* We strive to complete 3-5 new consults each week. The numbers count. We've clocked 600 consults to date and each time we bring a new client on board, it brings us closer to name recognition in the Hollywood community. The Zuckers made critical introductions in the early days of the program and continue to be supportive, but the real growth we've experienced has been achieved by the hard work involved in taking on a consistent stream of new consults, big and small.
9. *Hold regular events as part of an overall strategy.* Immediately following the launch symposium, we began planning our next event. Our goal has always been to have some kind of activity to which we can invite people every 6 weeks or so. These events vary significantly in size, complexity, and scope. We hold small salon events that are hosted by individuals, often in their homes. Janet and Jerry have been consistent salon hosts, welcoming 30-50 guests into their living room to hear from one or two scientists on topics ranging from string theory to 3-D fabrication. Intimate and personal, these events are so popular the last guests often do not leave until after midnight. We also do events in cafés and pubs, much like the model set by Café Scientifique, for larger (more boisterous) crowds. We host panel discussions and screenings for audiences of about 150. We sponsor exclusive lab tours for 6-12 people. We also do large symposia like our launch event. Regardless of the specifics, the idea is to remain a fixture on the calendar, to remain present in the minds of our current clients, to generate word of mouth as a way to secure new clients, and, of course, to use science to inspire new ideas for film and television. This has proved to be a highly effective way to achieve all these goals.
10. *Use science conferences as an opportunity to recruit new consultants.* As important as our LA events are in generating consults, outreach to the science community is just as important. Certainly the National Academy of Sciences and its sister organizations, the National Academy of Engineering, the National Research Council, and the Institute of Medicine have access to a vast number of members and committee members who could act as consultants. As indicated earlier, though, not

all experts are interested in acting as consultants and expertise is not the only qualification needed. Each time we present information about The Exchange at a conference or meeting of scientists or engineers, we receive expression of interest from individuals seeking to join the ranks of our consultants. These people may not be NAS members or have served on our committees, but they often make great consultants for The Exchange.

11. *Select consultants carefully.* We take great care to ensure that the scientist we select is the right person for the job. Many times, the questions that are asked are not so complicated or technical that it requires high-level expertise. We certainly have to ask someone with the right kind of expertise. If we are asked to provide input on aerospace engineering, we need to find an aerospace engineer. Oftentimes, however, that may be the easiest part of the selection process. The more complex role we play is in determining the right kind of personality for the job. Not all experts are created equally, at least not when it comes to communicating complex ideas in a simple, engaging, and creative way. We first look for people who actually want to do this. Not everyone does. Then we look for people who are good at this. Not everyone is. We interview all our consultants, and, before sending them on their first assignment, provide them with a set of instructions borne of our experience on what works and what does not. We either participate in the consult or check back with participants afterward to assess the interaction. I am pleased to report that we have made only one truly bad match, which is a track record we can be proud of in our 4-year history.
12. *Plausibility and authenticity are more important than accuracy.* I am fond of saying that we are not the accuracy police—rather, we are the plausibility patrol or the arbiters of authenticity, if you will. The moment we point fingers and proclaim Hollywood is WRONG is when they stop listening to us. We attribute much of our success to the first rule of improv: we do not say, "No, but…", we say "Yes, and…". Even when presented with very bad science, we advise our consultant to hold his/her tongue and refrain from the "No, but … that's wrong/that could never happen/that's not accurate" response. We counsel them to say instead "Yes, and … another idea/equally cool/you could even try." Essentially, make them a better offer. Because in our experience, the real science (the accurate science) is often better and more interesting and more fun. It's simply that when you preface it with a "No, but…" you turn people off.
13. *Be brave.* This was the first rule we set for ourselves and one by which we have lived ever since—though we have had to rely on it less as the program has become more accepted within the science community. Certainly there are those to whom The Exchange will never make sense. There are people who feel that getting in bed with Hollywood, those horrible lying liars, might not be safe. They believe that we should be the accuracy police and that our first order of business should be to get Hollywood to straighten up and fly right. Of course, that is in essence our goal. We just may not agree on the means by which we achieve the ends.

In order to achieve serenity in Hollywood, we at The Exchange have come to accept that story will trump science. Although film and television can and do educate, the writers, directors, and producers who ply their trade in Hollywood are not actually in the business of educating. They get paid to tell a good yarn. Their job description calls for making 'em laugh and cry, getting audiences to root for the good guys and boo the bad guys. The ultimate goal is to sell tickets, encourage downloads (legal, of course), and turn on televisions. The truth is that strict adherence to scientific accuracy can sometimes get in the way of these things; even weigh down the story and make it less entertaining.

If you can hook someone with just enough good science to entice a person to learn more from legitimate sources or captivate an audience with a fetching character who inspires imitation, thus swelling the enrollment in college science classes, then we can in good conscience sacrifice some accuracy for the greater good. Because even when the science is bad, there are opportunities to engage audiences in a conversation. Steve Jobs many times referenced the inspiration he drew from the *Star Trek* television episodes he watched as a young man. The touch-screen controls used with such dexterity by Lieutenants Sulu and Chekov, as well as by Commander Spock and Lieutenant Uhura, were all models for the iPhone screens. The PADDs (personal access data displays) used by numerous *Enterprise* personnel were the progenitor of the iPad. Were any of these technologies "real" or "accurate" when the show aired?

We feel very fortunate to have been welcomed in Hollywood. The fact that the service is free does not hurt but we believe it is also in large measure because we have been genuinely helpful. Though we worried that we might sit idle, waiting for the people to call, the fact is that since our launch in 2008, the phones have not stopped ringing. We did not have to take out ads in the trades or book billboards on Hollywood Boulevard. Turns out that science sells itself—largely because science can help the entertainment industry sell its story to its audience (*12*). We have called on skilled science communicators to work with our Hollywood clients and they have spread the word.

We have numerous anecdotes to which we refer when referencing the influence that The Exchange has had on science and engineering content in film and television. We have documented the use and growth of the program based on a variety of carefully recorded measures. For instance, we have recorded the number of consultations we have completed, the number of people who have attended events we sponsored, counted the number of science consultants entered into our database, and ticked off the number of customers we can claim as having used our services, paying especially close attention to repeat customers. But these output measures do not measure impact. We want to dig deeper to evaluate The Exchange by measuring the impact the program has had on its two core audiences—those working in the Hollywood entertainment industry and the scientists who consult for us.

With a generous grant from the Alfred P. Sloan Foundation, we now plan to evaluate the effectiveness of the network we have developed to understand both the influence that The Exchange is having directly on its client base as well as the onward influence its clients are having on the entertainment community in general. This evaluation will allow us to develop strategies for expanding, where appropriate and necessary, the reach of the program further into both the science and entertainment communities. If we are, for example, to scale up the program, we must know more about who is taking part and why—and perhaps more importantly, who is *not* taking part and why. We believe that there are some key questions that must be posed to our key stakeholders. Some of those questions may focus on the processes that The Exchange employs in the execution of its activities. Others may be more squarely directed to the perceived outcomes of the consultations and events sponsored by The Exchange in order to ascertain whether our efforts are well received and result in sufficient change to have been worthwhile. We are interested in both.

Some examples of the questions we want to consider include exploring whether entertainers have increased their interest in science as a result of their interactions with The Exchange. Have these entertainers changed their perception of what it means to be a scientist or to practice the discipline of science? Is The Exchange helping in the effort to alter mainstream stereotypes about those who work in science? Equally of interest is how the science consultants have been affected by their work with the entertainment community. Have their perceptions of scientific accuracy in narrative storytelling been altered by their interactions with Hollywood? Do they feel that their efforts have resulted in genuine impact? Has the experience of consulting on entertainment media been generally positive? Has the experience somehow fostered a greater appreciation for science communication among the science consultants? Has the experience influenced how they think about their research?

It remains to be seen what genuine impacts we have had to date or where this still-expanding program goes from here. But we feel gratified when we see that Jim Kakalios's (*13*) Science of *Watchmen* video on YouTube has received almost 1.8 million hits, knowing that The Exchange engineered the match between Jim and the filmmakers—and that he has parlayed this relationship into an opportunity to reach (and teach) more people about physics than he could ever manage by just working his day-job in the classroom. We cannot help but be pleased that Sean M. Carroll influenced the writers on *Thor* to recast Natalie Portman's character as an astrophysicist rather than the nurse that she was originally meant to be. Not, for the record, that we have anything against nurses but there just are not a lot of women astrophysicists and the world needs to *see* more of them if we *want* more of them. We are equally excited each time we hear that one of our science consultants is hanging out with one of our Exchange clients … just because they've become friends.

Acknowledgments

Support for The Science & Entertainment Exchange comes from the Alfred P. Sloan Foundation, the California Endowment, the Gordon and Betty Moore Foundation, Howard Hughes Medical Institute, Research Corporation for Science Advancement, and other generous funders. More information about The Exchange can be found online at www.scienceandentertainmentexchange.org.

References

1. National Research Council. *Rising above the Gathering Storm: Energizing and Employing America for a Brighter Economic Future*; The National Academies Press: Washington, DC, 2007; Executive Summary.
2. National Research Council. *Learning Science in Informal Environments: People, Places, and Pursuits*; The National Academies Press: Washington, DC, 2009; pp 1.
3. Singhal, A.; Cody, M.J., et al. *Entertainment-Education and Social Change: History, Research, and Practice*; Lawrence Erlbaum Associates, Publishers: Mahwah, NJ, 2004; pp 5
4. Sabido, M. Entertainment with Proven Social Benefit. http://www.miguelsabido.com/eeeng.html.
5. From the Desk of Norman Lear: The Fonz, Drunk Drivers, and Trash, November 13, 2003. http://www.ema-online.org/from-the-desk-of-norman-lear.
6. (a) Nisbet, M. C.; Goidel, R. K. Understanding citizen perceptions of science controversy: Bridging the ethnographic-survey research divide. *Public Understanding Sci.* **2007**, *16* (4), 421–440. (b) Nisbet, M. C., et al. Knowledge, reservations, or promise?: A media effects model for public perceptions of science and technology. *Commun. Res.* **2002**, *29* (5), 584–608.
7. Leiserowitz, A. before and after *the day after tomorrow*: a u.s. study of climate change risk perception. *Environment* **2004**, *46* (9), 22–37.
8. Entertainment Education and Health in the United States. *Kaiser Family Foundation: Issue Brief*; Spring 2004. http://kaiserfamilyfoundation.files.wordpress.com/2013/01/entertainment-education-and-health-in-the-united-states-issue-brief.pdf.
9. The Numbers – Movie Records. http://www.the-numbers.com/movies/records/#inflation_adjusted.
10. Barnett, M.; , H.Wagner; et al. The Impact of science fiction film on student understanding of science. *J. Sci. Educ. Technol.* **2006**, *15* (2), 179–191.
11. Barnett, M.; , H.Wagner; et al. The impact of science fiction film on student understanding of science. *J. Sci. Educ. Technol* **2006**, *15* (2), 179–191.
12. Kirby, D. *Lab Coats in Hollywood: Science, Scientists, and Cinema*; The MIT Press: Cambridge, MA, 2011.
13. Science of Watchmen. http://www.youtube.com/watch?v=zmj1rpzDRZ0.

Chapter 9

Making the Science of TV Meth Crystal Clear

Donna J. Nelson* and Melodie R. Lettkeman

Department of Chemistry, University of Oklahoma,
Norman, Oklahoma 73019-0390
***E-mail: djnelson@ou.edu**

The US requires a sufficient future scientific workforce in order to address many needs including national security, economic development, and technological strength. To achieve that, we must win those young minds effectively. Therefore, we should ensure that young people's exposure to science through mainstream and popular media is filled with positive and accurate science, lest negative experiences dissuade them from the field. I consider myself part of that mission in my work as a science adviser for AMC's *Breaking Bad.* While my fears of affiliation with illicit drug production mirrored those of my science colleagues, overcoming those fears was essential to bring accurate science to the small screen.

The Drug-Free Image We Seek To Preserve

Scientists are typically conservative and cautious. It is the nature of our work to be so. We possess knowledge of very scary and dangerous topics, from the production of illegal substances to the hydrogen bomb, so we are careful not to look as though we are part of any potential wrongdoing.

Trying to preserve that image is important to me, knowing how important it is to science. The way science and scientists are portrayed in film and television can have a serious impact on what course future scientists take. I have spent the last ten years collecting data that demonstrate women and minorities are underrepresented among science and engineering faculty at research universities.

While socio-economic status plays a significant role in this disparity, another reason this issue exists may be because they receive the wrong message about what constitutes being a scientist (*1*). The way these fictional stories and characters are portrayed—good or bad—has real world consequences on the science fields that we should be mindful to help make accurate.

When I first considered *Breaking Bad* creator Vince Gilligan's request for scientific advice on the show, this idea haunted me (*2*). In *Breaking Bad*, an organic chemistry teacher produces methamphetamine in order to secure his family's finances before his lung cancer kills him. It is the best methamphetamine in the southwest United States because he uses his chemistry talent to make it nearly 100 percent chemically pure. What kind of message will it send, I thought, to advise a show that presents science teachers as illicit drug producers? My colleagues raised the same concern when I broached the subject with them.

Not until I viewed the show did I realize that *Breaking Bad* presents the drug life as an awful existence. Because the protagonist's actions are presented as desperate, not glamorous, I resolved my dilemma and contacted Vince to volunteer. To benefit the scientific community and inform the general public, I wanted to help make this show scientifically accurate.

From the Mind of a Real-Life Chemistry Professor

While reading an issue of Chemical and Engineering News (C&E News), the membership magazine for the American Chemical Society (ACS), I came across an interview with Vince Gilligan. In that interview, Vince requested the advice of one who was part of a "chemically-inclined" audience, as the show's budget did not allow for a paid chemistry advisor. Later, he told me that he was researching much of the scientific information for the show on the Internet. He made it clear that he wanted the show to be as scientifically accurate as possible. Of course, this is also a value of my own; I cannot stand the thought of someone watching a show, being confused because what they've learned in the classroom does not match what is presented on television. Once the show began to win awards, it was clear that it was going to be a hit, whether it portrayed good science or bad science (*3*).

I threw my hat into the ring and called Rudy Baum, Editor-in-Chief of C&E News, to let him know I was interested. Vince emailed me shortly thereafter, asking if I could come to California to meet with some of the show's representatives. I happened to be traveling in the region with my son and we stopped by to visit them in Burbank.With *Breaking Bad*'s popularity on the rise, I figured Vince would be too busy to see me, so I would meet briefly with an assistant and be sent on my way. Such was not the case; I was whisked into a room with a storyboard. My son and I positioned ourselves with our backs to the board, thinking that Vince's group would not want us to see their secrets.

I expected to discuss the work for less than an hour, assuming I would answer a few science questions. Instead, Vince and the writing staff bombarded me with questions, first on what life as a scientist was like and how we become interested in the field. Previously, he and his group could not get many scientists to speak with them. It was clear that they wanted contact with a real scientist.

We talked for hours, and I was treated to a terrific lunch with the entire team (Figure 1). I learned they often checked for science accuracy by using the Internet and Wikipedia articles. They also received help from Drug Enforcement Administration (DEA) officials in order to reproduce realistic meth labs. Of course, the DEA could offer advice on the production of methamphetamine, but the writers wanted help with high school organic chemistry scenes and with some of the nuances of chemistry, knowledge I could provide.

Figure 1. Me (far left) enjoying lunch with the Breaking Bad team in Burbank.

Becoming AMC's Accomplice

I have received many requests from *Breaking Bad* producers and writers since that day in California. A particularly memorable one is a classroom scene of episode six; season two. Walter White was teaching a lesson on alkenes and their nomenclature, and the writers needed guidance on alkene terminology. I gave them a one-paragraph summary of some classes of alkene groups—monoalkenes, olefins, dialkenes, trialkenes, polyalkenes, etc. This proved helpful, and they asked for examples of material which Walt might write on the blackboard. I sent them several pages of drawings, and they reproduced one page of my structures on the blackboard (Figure 2). This would be the first time my work would appear on camera.

During this scene, which occurs about 11 minutes into the episode, we hear Walt listing off those classes of alkenes (monoalkenes, diolefins, polyalkenes . . .) and discussing how difficult alkene concepts can be. (This thought seems funny to me now, because thereafter, I was asked to help them simplify words and concepts, so actors could pronounce them easily.) In the alkene nomenclature scene, Walt then explains how carbon is "at the center of it all" and walks right past—wouldn't you know it—a diagram that exactly matches the drawing I sent to them as a result of their request!

mono-alkenes

$$H_2C{=}CH_2$$

$$H_2C{=}CH{-}CH_2{-}CH_3$$

$$H_3C{-}CH_2{-}CH{=}CH{-}CH_2{-}CH_3$$

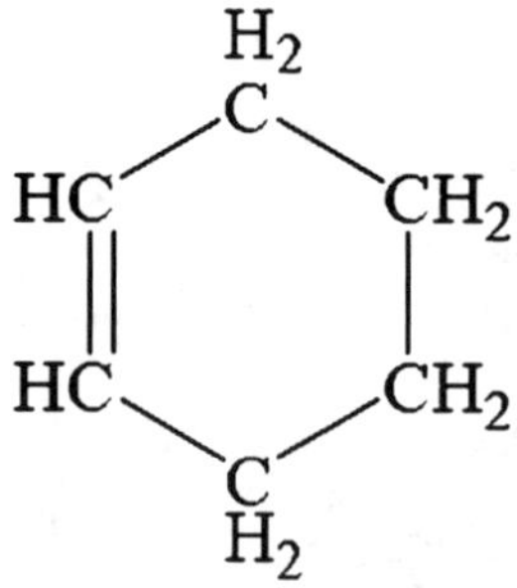

Figure 2. The structure I sent to the writers that made it into the show.

Working in television science is quite a different world from real-life science. Consulting for *Breaking Bad* gave me a chance to learn about a completely different community. We favor simpler words, but we would not use an unscientific word just to make the science easier to say. The entertainment industry terminology is different as well. In science, we operate on the metric system, so that the standard is the same across the world. But, the work for *Breaking Bad* is done in the United States, so we must translate to what is commonly used by the US audience: the US customary units (sometimes jokingly called Fred Flintstone Units, FFUs, by scientists). This disparity manifested itself in a somewhat odd way in episode seven of season one, when I helped the writers calculate how much meth Walt and his partner, Jesse, could make starting with a 30 gallon drum of methylamine.

The exact question was: Using the P2P method, how much meth could one synthesize from 30 gallons of methylamine—*in pounds*? Upon reading this, I laughed for a number of reasons. (A) First, in academia, it is almost unheard of to use such a large quantity of a chemical; scientists try to conserve as much as possible. In research, we run our reactions on the smallest possible scale in order to minimize the cost of reagents, equipment and disposed byproducts. In teaching, we minimize the size of reactions in order to maximize safety to students. (B) In academia, professors would never discuss making drugs, practical methods to do so, or calculate synthetic yields of the processes. (C) Finally, no one in the scientific community measures in pounds, and the calculations *Breaking Bad* requested had to be done using the metric system, then converted to FFUs.

Another laugh came later: In order to calculate the amount of product made, I needed to know the exact synthetic route the producers and writers wanted to use and to determine from the literature the percent yield of that synthetic route. The first step of this route is determined, but the second step (the reduction step) can be accomplished by using one of many possible reducing methods. Therefore, I emailed them the list of reducing reagents and asked which to use for the second step. They chose the aluminum-mercury method, because it is easy to say and would keep their actors happy. I found this hilarious; scientists never choose a reagent for a reaction based on its ease of pronunciation. I have selected reagents based on their ease of use, ability to create the most reproducible results, cost, safety, and highest yield, but never based on the ease of pronouncing the reagent name.

In a discussion with writers, my use of the word "precursor" to describe methylamine in the P2P synthesis also became a part of the show. When the writers heard the term, they immediately asked me what it meant, and then they incorporated it into the script. After Hank, a DEA agent tracking the new super-meth supplier, finds out that such a large amount of methylamine has been stolen, he says "Thirty gallons of precursor—that big a score, they're going to wind up stepping on some toes." This is just another example of the attention to detail which *Breaking Bad* writers take in order to have accurate science on the show.

When the Cameras Stop Rolling

Of course, work on an award-winning television show does not stop when the episode has aired. Since I began helping the *Breaking Bad* producers get the science right, I have been busy with additional related activities.

Recently, I fulfilled AMC's request for an article examining the validity of chemistry moments in the show's first five and a half seasons. One of my undergraduate researchers, Melodie Lettkeman, and I poured through the episodes to find six stand-out moments, including the use of thermite to break into a chemical warehouse and making explosives that look like meth crystals. We found a humorous discrepancy in the use of thermite. Walt and Jesse use Etch-a-Sketches in order to obtain the aluminum powder for the compound. In the scene, only about ten Etch-a-Sketches are visible. Our calculations revealed they would need nearly 200 of these toys in order to create as much thermite as was shown on camera in the plastic bag! The first part of the article is at the AMC website currently, and the second part should be posted to the AMC website in the summer of 2013 when the network begins airing the last half of the fifth, and final, season.

I also had the good fortune to visit the set in 2011, where I was invited to do a cameo as a nursing home attendant (Figure 3). Unfortunately, the scene was cut, but the process of being filmed enabled me to experience in a small way what actors do on the set.

Figure 3. Dressed as a nursing home attendant for my cameo on May 12, 2011.

Through visits to the set, I've developed relationships with the cast and crew (Figure 4). I learned how realistic the makeup and costumes are. I remember Aaron Paul, who plays Jesse, looking as if he was fresh out of a particularly awful fight. It was just makeup, but it was disturbingly realistic (Figure 5). The experience revealed that the crew on *Breaking Bad* cares as deeply about scientific details as all other aspects of the show.

Figure 4. Producer and Writer Moira Walley-Beckett, myself, and Executive Producer Vince Gilligan during my first set visit.

Figure 5. Aaron Paul with a few fake bruises and me during a set visit May 12, 2011.

Besides the cameo work, I have other wonderful memories of my time on the set (Figure 6). We had a laugh when I showed up with an Oklahoma State Bureau of investigation shirt during another visit in May 2012, when the security guards repeatedly asked me to which unit I was assigned. On another occasion, I gave Vince Gilligan a T-shirt from the University of Oklahoma chapter of the Student Affiliates of the American Chemical Society. On its back, it said "Chemistry. We do stuff in lab that would be a felony in your garage." He thought it fit his show so well he sent me a photo of himself wearing it.

The popularity of the show has given me the opportunity to share the work of science advisors with a broader audience. I've been interviewed by many newspapers, shows, and blogs. I've spoken on NPR's Science Friday and Smart Planet. I have been filmed for a number of television shows and appeared live on Good Day Sacramento. Locally, the Oklahoma Gazette and the Oklahoma Daily have written about my work on *Breaking Bad.* I've spoken to many professional, civic, and fraternal organizations as well.

A community concerned with factual science in popular media has been growing rapidly. The 2011 President of the ACS asked me to create a Hollywood Chemistry symposium for its spring meeting, which was held in Anaheim, CA. The thought was that this meeting was physically close to Hollywood, so that it might be relatively easy to convince producers, writers, and other science advisors to participate. We brought together these people, creating the first such bridge between the ACS and the entertainment industry. Science advisors from across the spectrum came to speak with students and other members of the society about the important service science advisors offer (Figure 7).

Figure 6. Bryan Cranston, Aaron Paul, and me during a set visit May 26, 2012.

Figure 7. Moira Walley-Beckett, me, and Kath Lingenfelter at the ACS National Conference in March 2011.

Some participants of the first symposium in Anaheim were the following: (A) Moira Walley-Beckett, who spoke about producing and writing for *Breaking Bad*; (B) Kath Lingenfelter, who discussed her personal scientific background and her writing for *House MD;* (C) Kevin Grazier, who discussed his extensive advising activities for *Zula Patrol, Eureka*, and *Battlestar Galactica*; (D) Jaime Paglia, who discussed his experience as co-creator and executive producer of *Eureka,* and (E) two speakers who delivered an important underlying message about science advisory work. Mark Griep from the University of Nebraska-Lincoln and Sidney Perkowitz from Emory University, who discussed the impact of Hollywood science on the scientific community. (F) I discussed how a televised portrayal of chemistry affects the next generation. This particular phenomenon has long been of interest to me because of work I have been doing since 2002 relating to trends in the involvement of women and minorities in science education ((*4*, *5*)).

The event was so successful that we were asked to put on a second one later that year titled, "Science on the Hollywood Screen" (Figure 8).This time, at the request of the ACS president, we worked hard to make sure we had a speaker from *CSI* in the lineup, as that was one of her favorite shows. We brought in writer and producer Corinne Marrinan, who opened with an important discussion on *CSI*'s effects on its viewers. She noted that while the show changed expectations on real *CSI* team's capabilities—sometimes in a negative way—it also inspired interest in the field of forensic science. She discussed the importance of writers and producers accurately explaining science while being entertaining enough to hold an audience as seasons and years go by. She noted the importance of advisors in the process and how, in CSI, they work with all departments very closely. The science in *CSI* must be accurate and explained without being condescending to the audience, she said.

Figure 8. The group that comprised the "Science on the Hollywood Screen" panel.

I presented an updated version of my *Breaking Bad* discussion at this conference. We also brought in the director of The Science and Entertainment Exchange at the National Academy of Sciences, Marty Perreault. Her discussion was humorously titled, "Damn It, Jim (Cameron) - I'm a screenwriter, not a chemist!" Next, writer and producer Jane Espenson taught us that great chemistry can make great television. Then, another *CSI* writer and producer, Aaron Thomas from the New York spinoff, *CSI New York*, brought a writer's perspective on including science in the script. He mentioned that the practice introduced a lot of anxiety. There is pressure to get terminology and concepts correct in television, especially from those in pertinent communities.

Breaking Bad's Positive Impact on Science Culture

I believe that popularizing science can have a positive impact on future US scientists if approached in the right way. First, science in television should be as factual as possible. This was the focus of Griep's discussion, "Using Hollywood movies to teach chemistry formally and informally: Chemistry in the movies." If a student learns a concept in class, he or she should be able to find that concept applied correctly in media.

Second, science should also be seen as fun. If viewers see science as a stuffy, lonely existence, they will feel less inclined to become scientists. The departure from this is profound in *Breaking Bad*; the viewer watches Walt do awful things with his chemistry, but he also does very unconventional and interesting things with it. He breaks into a highly secured chemical storage plant using material from Etch-a-Sketches. Though the scene is highly dramatized, the concept is sound.

Chemistry saves Walt and Jesse in a later episode from being stranded in the desert due to a dead RV battery. Using his fundamental knowledge of batteries, Walt is able to fashion a battery using items found around the vehicle.

Through these examples, it is clear science does not have to be boring. When students see science used to help lock up dangerous criminals or effect amazing accomplishments creatively, they can be inspired to pursue science in order to see if they can similarly push the scientific envelope.

But image is a critical player in the work science advisors do. This was the reason colleagues initially warned me to avoid consulting for *Breaking Bad.* It is the same reason I advocate for accurate science, which can only be accomplished with the guidance of professional scientists. When more science advisors step forward and reveal the good that comes from scientifically accurate media, then more scientists can feel comfortable advising producers, even if it means their name may be attached to illegal drug production!

Conclusion

Before I stepped into the *Breaking Bad* offices in Burbank, my mind was clouded with concern that people would associate me, and other scientists, with undesirable activities. The fear made others advise against stepping forward when Vince Gilligan called for help. This fear, though valid, should never detract from the importance of consulting for mass media. When I opened my mind to the potential benefits of science advising, I found fulfilling work that taught me as much as I taught others. I also found that television producers are working hard for accuracy, and that makes advising truly rewarding.

References

1. Nelson, D. J. *Nelson Diversity Surveys*. http://faculty-staff.ou.edu/N/Donna.J.Nelson-1/diversity/Faculty_Tables_FY07/FinalReport07.html (accessed September 27, 2012).
2. Kemsley, J. Novel TV show features chemist making crystal meth. *Chem. Eng. News* **2008**March3, 32–33.
3. AMC TV. *Breaking Bad Awards*. http://breakingbad.wikia.com/wiki/Season_1#Awards (accessed October 24, 2012).
4. Nelson, D. J. A National Analysis of Diversity in Science and Engineering Faculties at Research Universities, 2004. http://faculty-staff.ou.edu/N/Donna.J.Nelson-1/diversity/briefings/Diversity%20Report%20Final.pdf (accessed July 17, 2013).
5. Nelson, D. J.; Brammer, C. N. A National Analysis of Diversity in Science and Engineering Faculties at Research Universities, 2007. http://faculty-staff.ou.edu/N/Donna.J.Nelson-1/diversity/Faculty_Tables_FY07/07Report.pdf (accessed July 17, 2013).

Chapter 10

Getting the Science Right: Tales from the *Eureka* Writers' Room

J. Paglia*

Co-creator/Executive Producer of *Eureka*, Universal Cable Productions, Universal City, California 91608
***E-mail: jaimepaglia@gmail.com**

A humorous insider's look at the painstaking process of balancing science with fiction in the making of a sci fi television series. Specifically, the collaboration between science advisors and the Hollywood creative community. Note: No actual scientists were harmed during the making of this chapter...Well, maybe just one.

Let me begin with the fundamental question, "Why am I here?"

This is not intended as some existential query about my life's purpose. Nor is it a pontification on the human condition as I see it. I mean, why am I, a theatre arts and English literature major, co-editing and contributing to this esteemed science anthology?

Unlike my co-editors and the majority of our fine authors, I do *not* have letters like, M.D. or PhD. (or in some cases, both… clearly, pathological over-achievers) following my name. I did not spend years of my life mastering physics, or chemistry, or the physics of chemistry (is that a thing?). But I know quite a few people who did, including my father (which is the literary equivalent of saying, "I'm not a doctor, but I play one on TV."). Donald E. Paglia, M.D., is now professor emeritus of pathology and hematology after 50 years on the faculty at U.C.L.A. where he has pioneered research in iron metabolism, ATP deficiency, and lots of other sciency stuff. Maybe I didn't follow in his footsteps, but I spent many summer vacations hanging around his lab, playing with beakers, and generally driving him crazy ("Do not touch the centrifuge!"). Frankly, it galls him to no end that I have been welcomed into this inner-sanctum of scholarly writers, because instead of going to medical school and getting a *real* degree, I decided to skip all the fancy book learnin' and just make up stories about science instead.

In 2004, I co-created the Syfy (then Sci Fi) Channel original series, *Eureka*. Some have called it ground-breaking. Some have called it genius. Okay, yes, by "some," I mean my mom. Still, the show set records with non-relatives when it premiered, and found a devoted following of people who liked a little sci fi mixed with their character drama; a little peanut butter with their chocolate, as it were (Note: this metaphor will be important later). For the uninitiated, *Eureka* follows an every-man U.S. Marshal named Jack Carter who stumbles onto a remarkable town that doesn't appear on any maps. At least none that aren't classified "Top Secret" by the Department of Defense. In our mythology, it was founded by Albert Einstein and Harry Truman shortly after WWII, and designed to be a place where the most brilliant minds in science and technology could live and work with the best supports and quality of life. Einstein wanted to ensure we would always be on the leading edge of scientific discovery, and this town is now decades ahead of the rest of the world technologically. However, the scientific method being what it is, these remarkable geniuses frequently create chaos with that technology. This is where Carter comes in. He is the man with the average I.Q. who has a unique ability to see the forest for the trees. Where the brainy populace often becomes absorbed in the myriad potential causes of scientific anomalies, Carter's layperson perspective and investigative mind allow him to find logical solutions. As a result of helping save the day in the pilot episode, he is assigned to be Eureka's town sheriff. Sheriff Carter is our eyes into a remarkable world of eccentric geniuses, extraordinary technologies, and endless possibilities, all set against the backdrop of a quintessential small town in the Pacific Northwest.

As the "Jack Carter" of this anthology, I'm here to illuminate how we incorporated real science into *Eureka's* science fiction, how we navigated the precarious line between sci fi and magic, and how, over the course of seven years, our little science fiction dramedy found a devoted following of both sci fi fans, and fans of science.

The concept of *Eureka* was deceptively simple. Carter was a fish out of water lawman from the big city, now stuck in a small town of eccentric geniuses whose entire mission was to push the envelope of scientific discovery at the top-secret research facility, Global Dynamics. Each week, the brilliant townsfolk would create something potentially disastrous, and Carter would have to put the lid back on Pandora's Box. It was an Earth-bound, character-driven dramedy with sci fi elements. Every episode had a relatable human conflict that was grounded in our characters, and the sci fi problem of the week would be an extra catalyst for the drama. It had all of the human problems you'd find in any small town, but with the additional toy box of sci fi to amp up the stakes. Using our mathematical expertise, we devised an elaborately complex equation that would become the foundation of our show:

Small Town Issue + Big Science = Big Problem

(This concludes the math portion of our chapter.)

The initial challenge was finding the right balance between the serialized character storylines and the episodic science fiction elements in our show. In particular, how best to incorporate sci fi A-stories for Sheriff Carter and company to solve that would be conceptually interesting, emotionally dramatic, and visually compelling. Additionally, we wanted them to thematically complement the B-story character conflicts, not distract from them. We intended the science of the show to be a love letter to our favorite sci fi tropes while giving them a *Eureka* twist. The term "*Eureka*-fied" was born in the writers' room and quickly metastasized to the studio and network whenever we were pitching stories.

(This concludes the medical portion of our chapter.)

In order to be able to "*Eureka*-fy" a sci fi story, we knew we had to ground it in real science. After all, if you're creating a show about a town filled with scientific geniuses, you'd better get the science right. That's where our science advisor (and co-editor of this anthology), Dr. Kevin Grazier, came in. In addition to being a bona fide JPL rocket scientist, Kevin had been consulting on the critically acclaimed *Battlestar Galactica*, our sister show at Syfy. He later became a member of The Science & Entertainment Exchange, a group of science experts that was formed to unite scientists with Hollywood producers in an effort to improve the depiction of science in movies and television. Our show would become something of a poster-child for the organization and we utilized our advisor's expertise at many stages in the story-breaking process. Before each season, Kevin presented an overview of the latest research trends in fields ranging from astronomy to zoology. During the story-breaking process he gave feedback on the science of the arenas we were considering. After scripts were written, he vetted them to be sure our tech talk was accurate without being pedantic. Every episode was a balancing act of story and science. The more plausible the sci fi device, the easier it would be for an audience to suspend their disbelief and become invested in the character drama. We wanted to push the boundaries of scientific discovery, but always stay grounded in the theoretically possible. That would prove to be a moving target.

Science vs Magic: The Good, The Borderline, and "Seriously?"

The first thing we drafted was a document we humbly titled, *The Eureka Manifesto*, or series bible. In addition to detailed descriptions of the show's concept, look, tone, world, town history, mythology, and characters, it delineated the kinds of stories we would and wouldn't do. The primary rule was that we never cross the dangerous line between sci fi and magic. We wanted to respect our audience and their investment in the show. Most viewers are eager to go along for the ride with you and suspend their disbelief. But there is a tipping point; that moment in a story where you push the boundary of plausibility a smidge too far and the audience says, "Well, that couldn't happen." Then you've lost them. One moment, they are happily lost in the world of your characters, the next they're yanked out of it by a logic issue. They go from being invested to being insulted. Sci fi fans are incredibly passionate, intelligent, and loyal. But if you take them for granted, they will let you know (Twitter can be brutal). The tricky part is that the line between investment and insult is different for everyone. Sometimes you have a story that lends itself to hard science. Other times, it requires a bit more latitude. As we were finding the sweet spot for *Eureka*, that pendulum swung fairly wide. The challenge for any sci fi show is knowing when the story is strong enough that the audience will stay with you even if you're pushing the boundaries. This is further complicated by the very real challenges of writing to strict deadlines, budget constraints, and production limitations that affect your ability to execute an idea well. Following are three examples of episodes we did in Season One that represent the range of that plausibility variable, from the good, to the borderline, to, "Seriously?"

The Good

Early on in the process, we presented our network head, Mark Stern, with two-dozen "*Eureka*-fied" story arenas. Mark commented that one area of science he had no interest in exploring was human cloning. He felt the ethical issues had been done to death and it wasn't all that visually interesting.

We took this as a personal challenge.

How could we do a cloning episode differently? How could we find a unique take on the ethical questions it raised? Most importantly, how could we prove our network head wrong? (Note: Do not prove your network head wrong.)

Approaching the concept from a relatable human perspective, we imagined a scenario where the story would begin with the funeral of a couple who had died in the previous episode, Walter and Susan Perkins. Then at the end of the teaser, a woman who looked exactly like Susan would show up at the sheriff's office from out of town, decidedly *not* dead, and quite confused about receiving an invitation to her own funeral. Carter and company then discover that this is Walter's ex-wife. The *real* Susan Perkins. After their divorce, Walter moved to Eureka, and, missing her desperately, used his research to clone his wife, re-building her, cell by cell, from a sample of her DNA. The *new* Susan who recently died looked identical to the original, but on a cellular level, her body was only eight-years-old.

However, the real moral dilemma of our story wasn't about Walter cloning Susan. As our network head pointed out, that territory had been well-mined within the sci fi genre. (Note: Do not prove your network head right). Our focus was on Brian, the seven-year-old child Walter had with Susan's clone. As a result of his parents' untimely demise, young Brian was now orphaned. This is a boy that the original Susan has never met. Yet, biologically, he is her son. Does she have any moral obligations to care for him? Does she feel any connection to him when she sees their family resemblance? Will she decide to stay?

We wanted the science behind the cloning to be as accurate as possible so it wouldn't distract from the main point of the episode. Our advisor explained the theoretical process by which Walter could have used Susan's DNA to grow her into a full-grown adult. And we loved that this would make her biologically only a year older than her son. Creep-tastic. The science was simple but plausible, allowing us to tell an emotionally complicated story about the bond between a mother and son.

The Borderline

Sometimes, you have an idea for an episode that is just so much fun, it's worth pushing the line further into the "fiction" side of science fiction. Such was the case with an episode called, "Blink." The concept was grounded in a very relatable issue: work performance. We wanted to tell a story that dealt with the incredible pressures the townsfolk were under to produce results in their research projects at the Department of Defense-funded Global Dynamics. Two teams of scientists were pitted against each other for the same government grant and their deadline was looming. Stressed and exhausted, one of them decides they need a chemical advantage. With their genius I.Q.s, we figured the researchers at Global Dynamics could take performance-enhancing drugs to a whole new level (mind you, this was *years* before Lance Armstrong). What would a doping scandal look like in Eureka? We came up with "Blink," a neural enhancement drug taken through eye drops that would speed up the synaptic firing in people's brains, allowing them to think faster and more efficiently. Except one of the researchers "cooks" the compound into a super-concentrated version that speeds them up on a cellular level to the point where they can actually *move* super-fast, making them a threat to Carter and the town. This is where we began to push the boundaries of believability.

I called our science advisor and gave him the pitch for our teaser: A researcher is in his lab, working late on a project. He's surrounded by data, computer screens, books, stressed-out, clearly exhausted. Then he pulls out this tiny container of eye drops and places a drop in each eye. We push into his pupil as it dilates, deep into his brain where we see synapses begin rapidly firing. The scientist has renewed energy and begins processing information at an amazing rate, flying through books, typing with incredible speed. Then something goes wrong. Inside his brain, the synaptic firing turns into a storm of electrical impulses. Clearly, he has overdosed. A strange splashing sound begins to echo in the background as we slowly pull out of his eye and see he is now outside, running in slow-motion. His face is euphoric. And then we pull out further until we reveal that the splashing sound is actually his slow-motion footsteps because *he is running across the surface of a lake.* As we return to real time, he streaks across the surface of the water in a blur and slams into the side of an anchored boat. Carter would be called out to what appeared to be a boating accident, except he realizes that the deceased struck the boat, not the other way around.

Our science advisor's response was, "Cool!" Until I asked him, "How fast would the guy have to be moving to run across water without sinking?" Kevin asked me to give him ten minutes to do some rough calculations and he'd call me back.

True to his word, my phone rang exactly ten minutes later. Like many scientists I've known over the years, Kevin's brain tends to work much faster than his mouth can manage (though it makes a valiant effort). This requires one to pay very close attention to keep up, particularly when speaking over the phone. Kevin gave me his rapid-fire findings:

> Kevin: Okay, I've been doing some research. As you may know, the Basilisk Lizard, or "Jesus" Lizard, in South America can run across water, but relies on its webbed hind feet and tail to maintain buoyancy and balance. Clearly, humans lack both physical advantages. So, for a man of roughly one-hundred-eighty pounds to run on water for any kind of distance, he would need to be moving at an incredible rate of speed. Calculating his weight, surface area of size-eleven feet, and water density, that would be roughly 10 to the 18th power, or a quintrillion miles per hour, which would quickly peel off the subject's skin, flay the muscles from his bones, and cause him to disintegrate, probably after bursting into flames.
>
> Me: So… you're saying that scientifically it's… improbable.
>
> Kevin: Affirmative.
>
> Me: (beat, considering) Okay... I'm gonna do it anyway, and your name will be in the credits as our science advisor, so how fast should I say he's running in order for you to be able to sleep at night?
>
> Kevin: (heavy sigh, then) … Six-hundred-thirty-two miles per hour.
>
> Me: Done!

Okay, maybe I made up the quintrillion figure, but it was something with a whole lot of zeroes. In any case… now armed with a hard number, we considered the other physical tolls such super-speed would take on the human body. The amount of energy required to move so quickly would burn an enormous amount of calories. The friction would generate heat, causing severe dehydration. We reasoned that people who were using the drug would have to consume massive quantities of food and liquids to counter the effects. This later became part of Carter's investigation when he noticed the team of doping researchers bingeing in the cafeteria. These were small details that supported the concept of super-speed, but they proved significant in helping ground the story in science.

In the final version of the episode, we sadly had to lose the "running-on-water" scene for budgetary reasons. That conversation went something like this:

> Me: Hey, can we shoot a guy running on water and sink a pleasure boat?
>
> Studio Executive: You're adorable. I just want to put you in my pocket.
>
> Me: So, that's a yes then? (click) Hello…?

It was not a yes. We settled on a car accident instead. Admittedly, this was not as shocking as the opening we had originally conceived. But it still made for an exciting episode with great visuals and a unique investigation for Carter to solve. In the end, "Blink" successfully walked the line between science and magic, and became a fan favorite.

"Seriously?"

That brings us to what I consider to be the most egregious example of loose science from Season One: "Primal." A perfect storm of factors led to its creation. We were late in the season and running out of scripts. Ideally, you want five or six polished scripts in hand when you start filming a 13-episode season because it takes much longer to write them than it does to shoot them. It's a six-to-eight-week process to take an episode from concept to shooting draft, versus only seven days to complete principal photography. Once the production train leaves the station, it burns up nearly a script a week, until you find yourself laying track in front of it, doing whatever is necessary to keep the engine from careening off the rails. Under the best circumstances, we would have been racing to make it to the finish line. Then we had the bad luck of getting a first draft from a freelance writer that we had to abandon. The problem with the script was not just in the execution. It was flawed in concept. But we had run out of time and had to move forward with a page-one rewrite despite its significant shortcomings.

The theme of the story was about communication and emotional honesty in relationships. Two of our main characters were finalizing their divorce. Both wanted the other to reach out, but they suppressed their true feelings, which subconsciously emerged in destructive ways. Meanwhile, Sheriff Carter was trying to accept the end of his own marriage and the struggle to move on to a new relationship. It was a relatable and emotional armature to build on.

For the science A-story, we wanted to explore the future of nanotechnology, microscopic machines with remarkable applications. The idea was a researcher named Taggart had developed a "nano-bandage" that could synthetically replicate damaged tissue. You would simply apply the sand-like nanoids to the wound, and they were programmed to mimic the exact properties of skin, knitting together and healing the injury instantly. Meanwhile, another researcher named Fargo had developed a "mental mouse" that attached to the temple, allowing the user to navigate a computer interface using only their thoughts. In this case, the user was the man being divorced, Dr. Nathan Stark, the genius director of Global Dynamics and all-around egomaniac. (Note: This is where the chocolate/peanut butter mix I mentioned earlier becomes important.) The signal from Stark's mental mouse crossed frequencies with the programming in the tissue-replicating nanoids. Stark fell asleep with the mental mouse attached to his head, and the nanoids acted on the last mental directive they received from his subconscious dreaming. Once they realize what's happened, the content of Stark's dreams becomes critical in understanding what the nanoids intend to do. Stark reluctantly describes his dreams as wish fulfillment fantasies about his soon to be ex-wife, Allison… and some other stuff:

Nathan Stark: …The usual: power, success, control, over…everything.

Carter: (incredulous)You dream about world domination?

Stark: (shrugs) Not all the time.

This results in the nanoids forming dozens of angry replicas of Nathan Stark that are hell-bent on taking over the town. (Cue record screech sound-effect here.)

I know what you're thinking: *Huh?*

Even now, I see the promise of a compelling idea that was fraught with logic problems. Even if you buy into the programming error, why would the nanoids make multiple copies of Stark instead of just one? (Answer: Because multiple Starks are way cooler than just one.) Why would machines that were designed to replicate flesh also duplicate his Hugo Boss suit? (Answer: Because multiple *naked* Starks would be silly, not threatening. Though he is a very handsome man.) Why would Stark's subconscious regrets about divorcing Allison translate into a nebulous coordinated goal for his nanoid replicas to escape town and take over the world? (Answer: Because Carter had to have an escalating threat to defeat.)

The overriding answer was, *because we needed them to*, logic be damned. The show had to go on, and there was no time to start over with a brand new concept. So we wrote a new version of the script in record time, focusing on the emotional character stories of love and loss while grounding the science as much as possible. We knew we had broken our own cardinal rule. We crossed the line between sci fi and magic. I expected to be pilloried by our audience. Raked over the coals by reviewers. And I knew we deserved it.

Here is where I learned the most important truth about making a successful sci fi series: If the audience is invested in your characters, and you are true to them in your story-telling, they'll be forgiving. Just don't make a habit of it.

Was our science loose? Yes. Did we break our own rules? Without question. But did we let the sci fi plot device overwhelm the character story? No. If anything, because we knew we were treading on thin ice scientifically, we relied even more on the character dramas to keep viewers engaged. The science was admittedly far-fetched, but the emotional journeys were real. Was it our finest hour? Perhaps not. And I'm sure there were more than a few viewers who said those dreaded words, "Well, *that* couldn't happen." But they cared enough about the people in our little town of geniuses to keep watching. Undermining my entire argument about the importance of getting the science right, "Primal" has consistently been voted a *Eureka* fan favorite. I prefer to think of it as the exception, not the rule, though our science advisor disagrees with me. Kevin would argue that swarm intelligence is well-established in science fiction, most notably in Michael Crichton's, *Prey*. But *do not* get him started on the "artificial water" we created in "Shower the People."

The lessons from that first season of episodes shaped the entire seven-year run of the show. We always endeavored to tell relatable human dramas while keeping our science fiction grounded. Ultimately, it all comes down to telling a good story. But if you *can* get the science right, the story will be that much richer for the effort. And if you're really lucky, one day you'll be invited to speak as an expert on the subject.

Whatever you do, don't tell them that you're just making it up as you go along.

Science Reaches the Public

Chapter 11

Visions of the Past

Stephen Cass*
***E-mail: stephen.cass@gmail.com**

A chapter for a lay audience, discussing the relationship between the science and science fiction of time travel, with a focus on portrayals in movies and television shows. Various models of time-travel are discussed, including those consistent with the chronology protection conjecture and the Novikov self-consistency principle.

Time travel is a vexing thing to the curmudgeon. Unlike a science fiction staple such as, say, giant insects plaguing small town America—which can be ruled out on the basis of well-understood science—time travel is not so easily dismissed.

By time travel, we will be mostly be looking at the methods portrayed in movies and TV shows for sending either matter or information back into the past. While travelling into the future has its place, the ability to change the past has always had the most interesting dramatic (and scientific) possibilities.

We'll also mostly be confining ourselves to science fictional time travel. Of course, the idea of time travel predates science fiction by quite some margin, being a part of cultures around the world for many ages. In that context, time travel into the past was typically the province of information, with visions of the future being conveyed back to the mind of a prophet, as in the biblical Book of Revelation. Time travel into the future was more likely to be physical in nature, as in the Irish legend of Oisín, a man who visits Tír na nÓg (the land of the young) for three years; returning to Ireland he discovers, in best *Planet of the Apes* fashion, that three-hundred years have elapsed.

But, moving on from spiritual and mythological frameworks, lets turn to H.G. Wells and the moment in 1895 when the science and fiction of time travel started cross-pollinating, laying the ground work for how time travel is depicted and discussed to this day.

Physical time travel into the past had already been popularized in the 19th Century by tales like Mark Twain's 1889 *A Connecticut Yankee in King Arthur's*

Court. But the mechanism of this travel was either simply unexplained or ascribed to mystical or supernatural forces (*1*). It was H.G. Wells who established in the popular consciousness the notion that time travel could belong to the realm of science and technology, with his 1895 hit novel, *The Time Machine*. (Even the journal *Nature* warmed to the book, noting to its readership of the day that "the story is well worth the attention of the scientific reader, for the reason that it is based as far as possible on scientific data.")

Wells' influence can be gauged from the fact that to this day, "time machine" remains the term-of-art for science fiction writers and theoretical physicists alike to describe the mechanism that permits time travel, even when the mechanism in question is a geometric trajectory around an exotic cosmic object.

Yet, unlike much of the subsequent fiction on page and screen that has taken up the time travel gauntlet, Wells was not actually interested in exploring the implications of moving between the future and past. Rather than pondering what might happen if a time traveler attempted to change history, Wells' primary interest was in writing a social commentary. The bulk of *The Time Machine* revolves around taking the class divisions of 19th century English society to a grotesque end, with the evolution of humanity into the vacuous-but-beautiful Eloi and the industrious-but-brutish Morlocks in the year 802,701 A.D.

Indeed, when it came to the principles of his machine's operation, Wells simply cribbed them from an earlier 1888 short story of his called *The Chronic Argonauts*. Wells envisioned time as a fourth dimension akin to length, breadth, and height, and that, just as humanity had begun to travel in the third dimension through the mechanism of the hot air balloon, it was possible to move in the fourth dimension via another suitable device.

At the time, the concept of actual higher dimensions was a fresh current in mainstream thought, with their outlandish geometry popularized by works such as Charles Howard Hinton's 1880 essay *What is the Fourth Dimension?* Much as quantum theory is today, the science and mathematics of the fourth dimension were regularly co-opted by artists, writers, and the purveyors of the supernatural for their own purposes.

For Wells' purposes, identifying a fourth spatial dimension with time provided a nice veneer of plausibility. His idea may seem to anticipate Einstein's space-time continuum, but it must be said that the concept of four-dimensional time in *The Time Machine* is significantly different from what would be described in Relativity Theory a few years later.

In Einstein's space-time, the time dimension is not mathematically identical to the spatial dimensions, and space itself can be curved and warped. (For an example of what happened when the success of Relativity propelled this even more alien geometry into the popular consciousness, we must turn from H.G. Wells to H.P. Lovecraft, whose tortured tales of cosmic horror in the 1920s and 30s often featured "abnormal" topologies and "strange angles" that sorely tested the sanity of his protagonists.)

But although the idea of technologically-facilitated time travel owed its inspiration to genuine four-dimensional mathematics, it wasn't until 1949 that real science got in on the game. That year, Kurt Gödel published a paper that discussed the existence of *closed time-like curves* in Relativity's mathematical framework.

An object following such a curve could travel back in time. However, Gödel's particular solution to Einstein's equations required that the entire universe be rotating. As experimental evidence from astronomers indicated that the cosmos was doing no such thing, closed time-like curves were dismissed as physically irrelevant. Consequently, it wouldn't be until the 1970s that science would again take up the mantle of exploring time travel.

In the meantime, science fiction writers and television and movie producers were not so restrained as either Wells or the physics community. They delighted in exploring the various "What Ifs" of time travel regardless of the precise mechanism, merrily exploring scenarios that would later be formalized by scientists using terms like the *Novikov self-consistency principle*.

And, although time travel and its associated paradoxes were explored by literary writers in tales like Robert Heinlein's 1959 short story "—All You Zombies—" (in which a time traveler becomes his own mother *and* father), no warmer place for time travel could be found than in televised science fiction. Indeed, Rod Serling's prototype for the entire *Twilight Zone* series was a 1958 teleplay called "The Time Element," in which a man's journeys into the past eventually result in his erasure from the present.

Some of this warmth was, no doubt, due to the economic appeal of time travel shows. Studios could maximize their return on investments in period costumes, props, and sets otherwise sitting in warehouses, not to mention recycling already-paid-for footage from documentary or period productions. And with the penchant for novel-esque television serials like *Babylon 5* or *Lost* decades away, time travel made life easier for producers knocking out 20 to 30 self-contained episodes per season. What better way to return a show to the status quo each week than by whisking the cast away to another era (*2*)?

The most successful of these early time travel shows was launched in 1963, and is still running to this day: *Doctor Who* (*3*). Originally created as an edutainment show by the British Broadcasting Corporation to impart historical and scientific lessons, *Doctor Who*'s educational mission was soon sidelined as its dramatic popularity among children and adults alike soared. The show ultimately became well known internationally and the name of the titular Doctor's (*4*) iconic time machine—The Tardis—has entered the English language as a generic term for both a time-travelling device and anything that appears larger on the inside than the outside. The Tardis (the name is technobabble for Time and Relative Dimension in Space) has the external size and appearance of a 1950's British police box, but internally is a cavernous alien spacecraft.

For our purposes, the most important aspect of the show is its notion of *fixed points*. Essentially, while most of the time-space continuum is mutable (allowing the Doctor to vanquish malevolent forces), some events cannot be altered if the universe is to remain self-consistent. It's not clear if these events cannot be altered because it is physically impossible to do so, or if it's simply just a *really* bad idea to do so. The Doctor's only concerted attempt to alter a fixed point resulted in a character committing suicide in order to prevent the alteration.

We'll return to the idea of fixed points and self-consistency, but for now let's look at a second model of time travel, as exemplified by the 1966–1967 show *The Time Tunnel*. In *The Time Tunnel*, two scientists find themselves randomly transported through time and space, in each episode arriving just prior to some major event, such as the unfortunate mid-Atlantic juxtaposition of an iceberg and the RMS *Titanic*.

Attempts by the scientists to alter the course of these temporal milestones either fail outright, or actually ensure that events unfold as per the history books. In this model of time travel, it is possible to participate in the past, but not to change it.

These models anticipated the debate among physicists about the mutability of the past in the years that followed James Tipler's 1974 paper "Rotating Cylinders and the Possibility of Global Causality Violation." Tipler proposed that a spaceship in the vicinity of a dense, infinitely long, rotating cylinder could embark on a trajectory that would take the ship along a closed time-like curve without the need for Gödel's rotating universe. Although Tipler's theory would eventually be shot down, the theoretical cat was out of the bag. A cottage industry was born among scientists in minting paper time machines. Initially, most of these machines were based on Relativity Theory, but devices based on quantum mechanics have also been proposed.

For science fiction, the most influential of these theoretical machines has been the wormhole. Wormholes, under the rubric of *Einstein-Rosen bridges*, had long been thought about as creating a short cut between two points in space, but it was believed they were so fundamentally unstable that they would collapse as soon as a single particle entered one, rendering them useless. At least until 1988, when Kip Thorne, along with Michael Morris and Ulvi Yurtserver, published a paper in Physical Review Letters. Reading more than a little like science fiction itself, the paper not only postulated a method by which an advanced civilization could stabilize a wormhole, making it traversable, but also how such a wormhole could be used as a time machine, forming a bridge between two points in time.

Wormholes had an obvious charm for writers and producers seeking a breath of fresh air after decades of the descendants of Wells' original contraption of nickel, ivory, and rock crystal. A wormhole also has the advantage that its visual manifestation can easily be tuned to whatever budget is available—it can be depicted with an elaborate set and the full panoply of sound effects and computer generated graphics, or be little more than a glorified doorway (*5*).

True, visually similar portals and tunnels through time *had* appeared before Thorne's 1988 paper—examples include *Doctor Who*'s "Time Vortex" or the gateway in the 1967 *Star Trek* episode "The City on the Edge of Forever," but they weren't labeled as wormholes, and their nature was deliberately defined as vaguely as possible. In contrast, the theoretical wormhole revolution was a big green light for science fiction writers to get a lot more specific about the nature of their time machines. For example, over the course of the *Stargate* franchise's (*6*) run from 1994 to 2011, a relatively detailed set of consistent operating principles for the wormhole technology that formed the centerpiece of the franchise was developed, and frequently integrated into plot lines. While the wormholes were used almost exclusively for spatial journeys, several episodes played out Thorne's suggestion that they could be adapted for time travel.

Stargate's first foray into wormhole-induced temporal travel was a trip from 1999 to 1969, the same destination year that *Star Trek* chose for its first significant voyage into the past (*7*). That episode was ambiguous about whether or not characters could change history, but later episodes established that they could—"2010," a 2001 episode of *Stargate SG-1* set nine years into the then future, featured members of the 2010 SG-1 team sending a note warning their 2001 selves to abandon a scheduled visit to a particular planet. The 2001 SG-1 team dutifully deleted the coordinates of the planet from their wormhole control system, creating an apparent grandfather paradox. (The same premise was recycled for an episode of the short-lived *Stargate Universe* series in 2009.)

Indeed, as a rule of thumb, shows that regularly incorporate time travel allow the past to be changed, regardless of the possible creation of paradoxes. (If, however, a time travel plot is essentially a one-episode wonder on a show, an unchanging past model is more likely, as it guarantees no upsets to the established continuity.)

Of course, a vision of the past as mutable has more to do with dramatic purposes than any allegiance among writers and producers to a particular camp of theoretical science. After all, it's hard to sustain tension if the audience already knows the ultimate consequence of the cast's actions, which may explain in part why *The Time Tunnel* lasted for one season, while, at the time of writing this, *Doctor Who* is warming up for its 50th anniversary.

But although there's a general tendency to allow travelers the freedom to alter history, just how writers deal with the consequent potential for paradox varies wildly. This echoes the divergence of thought among scientists about how to handle the prospect of paradoxes—such as the grandfather paradox—that would arise from real time travel into the past. Broadly speaking, their thinking can be lumped into four camps:

1. Time travel is actually impossible *because* paradoxes would arise. Although solutions that permit time travel may exist within the mathematical framework of either quantum mechanics or Relativity, every such solution will have some problem that prevents its application in the real universe. In the early 90s, Stephen Hawking codified this line of thinking as the *chronology protection conjecture*.
2. Time travel *is* possible, *paradoxes* are not, and as any change would create a paradox, you can't change the past. It is therefore possible to travel to the past and interact with events, but it is impossible to alter history, i.e. time is fixed, and awareness of future events simply creates self-fulfilling prophecies.
3. Time travel is possible, and you *can* alter the past. However, the limitation is that you can't create paradoxes, such as doing anything would prevent the original trip through time, i.e. it's possible to go back in time and injure your grandfather prior to your parent's birth, but killing him will be physically impossible (unless someone happens to have frozen some of his sperm. Or clones him from his DNA. You get the idea).
4. Time travel is possible, and you can change whatever you like. It is possible to create apparent paradoxes because independent timelines can

interact; paradoxes only appear to be so from a point of view within a single timeline.

For obvious reasons, the first camp is rarely a foreground element in any science fiction, on-screen or off (a notable exception is a 1977 short story by Larry Niven, which borrows its title from Tipler's 1974 paper. In Niven's story, a galactic emperor finishes constructing an enormous cylinder in space, only to be prevented from using it as a time machine due to the immediate spontaneous nova of the nearest star (*8*)).

The second and third camps are based on the aforementioned Novikov self-consistency principle, a hypothesis put forward in the 1980s by Igor Dmitriyevich Novikov and later developed by Thorne and others. The third camp diverges from the second because it treats the period of time affected by time travel as something of a black box—it doesn't matter what happens inside the box, so long as the end result is the same. There may be many, many ways for the required end result—a traveler embarking on a trip into the past, or a specific signal being sent back through time—to be produced in addition to the pre-existing sequence of events (*9*).

Clearly, *The Time Tunnel* belongs to the second camp, as does the more recent *Babylon 5*. *Babylon 5*, which aired from 1994 to 1998, was an early example of the kind of highly serialized show that has become popular in recent years (*Lost*, *24*, *The Walking Dead*, et cetera). By being set in the 23rd Century, it avoided *The Time Tunnel*'s recurring problem of the audience knowing how historical events turned out.

Also, because *Babylon 5* was plotted in advance of production as a five-season arc by creator J. Michael Straczynski (who also wrote most of the episodes), the show was able to integrate an intricate time travel subplot perhaps more gracefully than any television show has before or since. Key elements of *Babylon 5*'s backstory, as well as several events that occur in the early seasons, eventually drive some of the characters to conduct an expedition into the past. In turn, the ultimate consequence of their actions in the past is revealed to be the very backstory and chain of events that led to their expedition in the first place, closing the loop.

TV shows whose productions are more open-ended than *Babylon 5* are necessarily constrained in how ambitious they can be in this regard. In particular, *Doctor Who*, which is technically in the third camp (assuming that its previously mentioned fixed points are truly fixed), gives itself a tremendous degree of latitude when it comes to temporal self-consistency. What is and isn't a fixed point is arbitrarily determined by the producers and writers as the plot demands; entire interstellar civilizations can be reshaped at the drop of a hat, but, for example, a small settlement of Ancient Rome must always be destroyed by the eruption of Mount Vesuvius in 79 A.D. In a classic example of turning lemons into lemonade, the show hangs a lampshade on the whole business, suggesting that the ability of the Doctor's species to instinctively know what is a fixed point and what is not is the very thing that makes them Time *Lords*, and why other time travelers are mere dilettantes. The only generally reliable inviolate observed by the show is the Doctor's personal time line.

However, in recent years, *Doctor Who* has tried to be more diligent about observing *Babylon 5*-style self-consistency, plotting out arcs on a season-by-season basis. For example, in the 2005 season following the show's lengthy off-air hiatus, the words "Bad Wolf" are a background motif somewhere in every episode—appearing as, say, graffiti, or the name of a television channel. At the conclusion of the season, a character is inspired by the phrase to interact with the Time Vortex and avert a crisis; after doing so, the character uses the Vortex to implant "Bad Wolf" throughout the history of the universe, saying "I am the Bad Wolf. I create myself. I take the words. I scatter them in time and space. A message to lead myself here."

When *Doctor Who* has attempted to reach beyond a single season it has generally fallen back on various *retcons*. "Retcon" is contraction of "retroactive continuity," and it means re-interpreting a previously accepted sequence of events as seen on-screen (*10*). For example, *Doctor Who* has had two major reunion shows, ("The Three Doctors" in 1973 and "The Five Doctors" in 1983) in which different regenerations of the Doctor, played by their respective actors, were brought together. As these adventures are not, of course, depicted in the original run of each regeneration, they are assumed to take place in between existing episodes, with care being taken during the reunions not to affect either the Doctor or his companions in a way that would be incompatible with established on-screen history.

Retconning can also be done by exploiting previously unexplained phenomena or character traits. For example, The Master, a criminally insane Time Lord, has been a recurring villain on *Doctor Who* since 1971. In a 2010 episode, "The End of Time, Part 2," the Master's insanity is shown to be due to a signal sent back through time to the Master's childhood by a tyrannical Time Lord seeking to escape his current sticky situation.

Of course, even with a show that has had decades to build up a rich mythology, such convenient hooks can be hard to come by. Realizing this, the creators of *Eureka* included a clever moment in their very first episode. *Eureka*, which aired between 2006 and 2012, features the sheriff of Eureka, a town that's an exaggerated version of wartime Los Alamos. The sheriff regularly protects the inhabitants from self-destruction due to various scientific misadventures (including several related to time travel). Arriving by car with his teenage daughter in tow in the pilot episode, the sheriff passes another car leaving Eureka, occupied by none other than himself and his daughter. The scene was never revisited until *Eureka* was unexpectedly cancelled. The producers were allowed to film an additional episode wrapping up outstanding plotlines—with the final scene showing the sheriff and his daughter passing a younger version of themselves on the way out of town.

Sometimes a production attempts to interact with its own past in a self-consistent way without any hooks at all. In 1989's *Back to the Future, Part II*, for example, Michael J. Fox's character must revisit the Enchantment Under The Sea dance that was the climax of the first movie. New scenes are interwoven with the old as he carefully avoids interacting with himself.

A particularly ingenious version of this approach was featured in a 1996 episode of *Star Trek: Deep Space Nine*. The lead characters found themselves cast back in time to the setting of one of the most popular episodes of the original series, "The Trouble With Tribbles." Skillful editing and digital effects seamlessly inserted the 1996 cast into, and between, scenes from the 1968 episode.

It's important to note that in these latter two examples (as indeed with most shows featuring time travel) temporal self-consistency is a *voluntary* arrangement, rather than the iron clad law of physics envisaged by Novikov, in which a time traveler could no more perform a non-self-consistent action than they could swim from Cape Canaveral to the Moon. Instead, the characters leave the past as untouched as possible because they fear alterations will create a new timeline, robbing them of a recognizable home to return to. This fear is due to the temporal butterfly effect, first described by Ray Bradbury in his 1952 short story "A Sound of Thunder," (made into a movie of the same name in 2005). In this story, hunters travel back into the past to shoot dinosaurs that are already marked for doom (*11*). Great pains are taken to prevent detectable alterations to past, but one of the hunters accidentally crushes a butterfly. The small death sends changes rippling through time, altering the present.

Similar to *Doctor Who*'s loose determination of what's a fixed point and what isn't, shows that worry about the temporal butterfly effect do still allow themselves considerable license in determining what's a detectable change and how much impact it has on the timeline, depending on the dramatic needs of individual episodes. For example, in one *Stargate SG-1* episode, a group of characters wind up several decades into their own future. Forewarned by their own memories and records, the denizens of that time drape sheets over everything in sight prior to the group's arrival, lest a single peek alter the course of history. Yet, on another *Stargate SG-1* episode, an extended and quite rambunctious trip into the past provokes only one detectable result: a previously empty pond acquires some fish.

The butterfly effect has become well known to television audiences, sufficiently so that it is now used as more of a background trope (like faster-than-light travel, or implanted computer interfaces) than the central conceit of plots. In fact, the butterfly effect is so well established in the minds of viewers that it is their default model for how a time traveler's actions might affect history. Shows that want to avoid the implications of the butterfly effect for their own storytelling purposes must find a way to let audiences know that something else is going on. For instance, in the pilot episode of the 2011 series *Terra Nova*, which featured a one-way gateway from the year 2149 to 85 million years in Earth's past, the creators found it necessary to include an entire scene's worth of exposition that involved characters standing around the (very ruggedly built) first probe that had been sent back through the gateway. Viewers are told that because no trace of the probe ever manifested in 2149, scientists concluded that the probe (and by extension everything that happens on the show) was actually in "a new time stream."

Thus *Star Trek*, *Stargate, Eureka*, and *Terra Nova* are all examples of the fourth camp of time travel, in which paradoxes are eliminated because timelines can exist independently of each other. Creating a new timeline does not alter the individual histories of time travelling objects, information, or people. On *Eureka*, time travel has permanently stranded characters in alternate timelines on several occasions, with the show following them by ruthlessly jumping wholesale over to a new continuity.

One show that has spent considerable time recently in exploring the consequences of independent timelines is *Fringe. Fringe* began airing in 2008 and, at the time of writing, is close to finishing its fifth and final season. The show features an FBI-sponsored team battling against a series of conspiracies of ever-increasing scope that employ beyond-the-bleeding-edge science and technology.

For the *Fringe* characters, the key event that must happen—if anything resembling their current timeline is to be preserved—is the dispatch of the constituent pieces of a fantastically powerful machine into the Earth's distant past. But within the boundaries of that constraint, they are perfectly willing to sacrifice entire time lines to meet their objectives.

What exactly happens to an existing time line after a new one is created is left vague, but internal evidence in *Fringe* suggests that different futures, each with their own histories, can co-exist (*12*) (although the point is moot on some occasions, due to planetary termination events occurring in some time lines).

How such co-existence might play out is explored by the 2007 South Korean movie *Project Makeover*. Although the time travel mechanism in the movie is essentially mystical in nature, it's still interesting for its unusual portrayal of what happens *after* the past is changed. In *Project Makeover*, the protagonist, an underachieving fashion worker, goes back in time to try to persuade her high school self from making a life-ruining error. In the past, she masquerades as a visiting cousin and gives her younger self her voice mail number. While the protagonist does end up changing this younger self's life for the better, when she returns to the present, nothing has changed, in contrast to movies like 1985's *Back to the Future*. As the character labors to improve her present day existence, she continues to receive voice mails from her younger self as that self experiences her changed timeline day by day. The past was indeed alterable, but, as the older protagonist's present is receding into the future at the rate of one day per day, the changes never catch up to her.

So far, we've been discussing the ways in which television shows and movies play out some of the real scientific questions about time travel. But there is one significant way in which the productions discussed so far disregard a limitation that even the most fervent scientific proponents of real time travel acknowledge. All proposed real time machines do not open a door to the entire history of the universe. Their reach can only extend as far as the machine's own existence in time: a traveler can not go back to a moment before the machine is ready for business.

One could argue in the cases of *Stargate* and *Star Trek,* as discussed above, that a broad interpretation of the time machines involved—an ancient wormhole network and the Sun respectively—means that the shows do respect this limitation. Their temporal horizon is just such a large epoch of time that the issue never arises. However, the wormhole mechanism requires that the wormhole be continuously open to maintain the temporal differential between each end, while in the *Stargate* franchise wormholes are created on demand, and can generally only stay open for 38 minutes. And true, *Star Trek* does rely on the Sun for its most iconic time travel adventures, providing a putative range of a little over 4.5 billion years. But the franchise regularly uses other time travel mechanisms as well, most recently in the 2009 movie reboot, where a newly created singularity creates a portal into the past.

But one movie that embraces the idea of a temporal horizon is *Primer*. This 2004 film was written and directed by Shane Carruth, who also co-starred in the movie. Shot on a shoestring budget, it is possibly the best technological time travel story since *The Time Machine* itself. Eschewing the need for the vastly advanced Relativistic technologies of shows like *Star Trek* and *Stargate* (or even Kip Thorne's 1988 paper, which postulates an "arbitrarily advanced" spacefaring civilization), in *Primer* two inventors accidentally create a small time machine while tinkering with some superconducting materials in their garage. The inventors soon scale their creation up to a human-sized box. A critical point about their machine is that if you want to travel a minute, an hour, or a day back in time with this device, you must start the machine running, and then come back a minute, hour, or day later. Then you climb into the box, and lie there for the required duration (*13*).

The protagonists of *Primer* are willing to put up with this because they have figured out a money making scheme. Each morning, they set the machine running for a few hours, and then clear out to a hotel where they spend their day noting stock prices, interacting as little as possible with the rest of the world. After the markets close, the inventors go back to the still-running machine and climb in, going back in time to just after their earlier selves depart for the hotel. They then proceed to rack up a profit trading on the stock market.

L.P. Hartley famously wrote in *The Go-Between* that "the past is a foreign country: they do things differently there." But the time machine in *Primer*, as well as the machines conceived by theoreticians, work precisely because things *aren't* done differently in the past. As Carruth told *The Village Voice* in an 2004 interview, the details of the *Primer* machine's symmetrical operation was directly inspired by the time-agnostic nature of quantum theory: "When you look at Feynman diagrams, there's really no difference between watching an interaction happen forward and backward in time."

Instead, the arrow of time in the universe appears to be established through entropy. As per the second law of thermodynamics, the macroscopic history of an isolated system is determined by the tendency of its microstates to be increasingly disordered, as disordered microstates vastly outnumber ordered microstates.

Primer echoes this concept. Despite the protagonists' attempts to maintain an ordered state of affairs, things spiral out of control as multiple loops through the machine begin to interact. While individual scenes are still coherent, the movie's overall narrative becomes disjointed as the inventors discover that they no longer fully understand what is happening and chaos threatens to overwhelm them.

Still, while *Primer* demonstrates a particularly close interplay between a scientific concept and the structure of a screenplay, for nearly 120 years the development of all time travel science fiction has owed a strong debt to our growing understanding of the nature of the Universe. Should future breakthroughs definitively prove Stephen Hawking and the chronology protection conjecture correct, time travel will join the canals of Mars or the swamps of Venus on the trash heap of scientific near misses.

In such a situation it's likely that some science fiction writers and producers will keep time travel around, and simply ignore the science, or posit some loophole (as is commonly done today with non-wormhole-based Faster Than Light travel). I suspect though that, over time, without a wellspring of real science to refresh the genre, tales of visiting the past would largely return to their mystical origins. With time travel's record of providing a rich vein of dramatic possibilities, out of which some of the most memorable on-screen scenes of science fiction have been mined, I can't help but hope that this is one occasion when a beautiful theory isn't killed by an ugly fact.

Notes

1. Still a popular dodge for writers not interested in inventing a technology, as in *Groundhog Day* (1993) or *Project Makeover* (언니가 간다, 2007), the later of which we'll return to when discussing some of the potential implications of time travel.
2. This also probably explains some of the attraction of that other great staple of TV science fiction—spaceship shows like *Star Trek* or *Blake's Seven*—in that they solve the same problem by visiting a different planet each week.
3. Albeit with a 16-year hiatus from 1989 to 2005, broken by one made-for-television movie in 1996.
4. Unusually for any science fiction show, the lead character of the Doctor is not human, but rather a humanoid alien known as a Time Lord. A convenient quirk of his species' biology, involving full-body regeneration, has permitted the part to be played by 11 actors to date.
5. Probably as a consequence of how scientists depict four-dimensional continuums in diagrams, the entrance to a wormhole is almost invariably portrayed on screen as flat portal. However, a real wormhole entrance would not be a two-dimensional circle, but a three dimensional sphere. The U.K. series *Primeval* and it's recent Canadian spin-off *Primeval: New World* deserves kudos for breaking with convention, depicting the "anomalies" that link different temporal epochs as multi-faceted spherical regions.
6. One theatrical movie, two made-for-television movies, one season of an animated television series, and a total of 17 seasons of three different live-action television shows. Even with such a sprawling franchise, *Stargate* maintained an admirable level of internal logical consistency throughout its run.
7. That episode was the 1966 episode "Tomorrow is Yesterday." An earlier 1966 episode, "The Naked Time," involved an experimental warp engine procedure hurling the *Enterprise* three days into the past, but it occurred in the closing moments of the episode and functioned purely as a *deus ex machina* to conclude the plot.
8. A similar idea of cosmic censorship was mooted during the construction of the Large Hadron Collider. Physicists Holger Bech Nielsen and Masao

Ninomiya postulated that the universe had such an antipathy toward the Higgs Boson that attempting to create one would result in ripples propagating backward in time, spawning events that would prevent the Higgs from being created. Thus, they predicted that the LHC would never go into operation. The theory was initially pooh-poohed, but resurfaced in the minds of many when the LHC suffered a major cooling malfunction during pre-activation testing. Then, during reactivation after that event, the LHC was shut down again after a bird dropped a piece of bread into just the wrong spot in a vital electrical substation. The LHC did, of course, finally escape the vagaries of fate and go on to discover the Higgs.

9. An analogy can be drawn here with statistical thermodynamics, where a given macrostate can be the consequence of any one of a very large number of microstates.
10. Of course, non-time travel shows also use retcons to explain away various production or casting issues, but viewers are somewhat less forgiving when this technique is used outside the time travel genre.
11. In a similar manner, in *Star Trek IV: The Voyage Home* (1986), Kirk and company set out to capture a pair of whales from the 20th century that they figure won't be missed.
12. For a detailed discussion of time travel in the *Fringe* universe, readers may be interested *in Fringe Science: Parallel Universes, White Tulips, and Mad Scientists*, (2011, Ben Bella) which contains a chapter on the subject by this author.
13. In an interview with the author for *IEEE Spectrum* in 2004, Carruth explained that the machine's operation was conceived specifically in reaction to earlier depictions of time travel "where you can arbitrarily jump around…however *Primer*'s machines worked, it was going to be something that you paid a price for."

Chapter 12

Constructing Crimes: How the CSI Effect Is Created

Stephen Cass,* Kevin R. Grazier, Bradley Thompson, and Corinne Marrinan
***E-mail: stephen.cass@gmail.com**

The so-called "CSI Effect," is actually a catch-all term for several different, and sometimes contradictory, perceived influences of the popular *CSI* television programs on juries, crime victims, and enrollment in undergraduate forensic study programs. We deconvolve the different aspects of the CSI Effect, and explore the degree to which each of these effects actually exist as well as their impact.

Thanks to all of the legal and medical dramas I've watched, I'm pretty sure I'm capable of winning court cases and saving lives.

-- Caprice Crane
--- Novelist/screenwriter/humorist

Perhaps no other genre of television exalts science as much as forensic procedural shows like *Crossing Jordan*, *Bones,* or, most notably *CSI: Crime Science Investigation* and its spinoffs *CSI: Miami* and *CSI: NY*. In almost every episode of these shows, a criminal is brought to justice through keen detective work and the application of a dazzling array of technologies. The immense popularity of these programs has given rise to the so-called *CSI effect*, actually a catch-all phrase for a number of different, and sometimes contradictory, changes in the expectations and behavior of juries, victims, and even criminals. The CSI Effect is also believed to be responsible for a surge of students taking forensics classes. In order to best discuss the *CSI* Effect, we should start with a firm understanding of how forensic procedural shows are created, and their relationship to real world forensics.

Forensic shows exploit the same principle that is at the heart of modern forensic science: Edmond Locard's *principle of exchange*. Locard, who worked in France between 1910 and 1966 (*1*), stated "every contact leaves a trace." In other words, it is impossible for a criminal *not* to leave some evidence of their presence at a crime scene, and conversely, the criminal will always carry away with them some material from the scene.

Of course, finding this trace evidence and using it to link a specific person to a crime can be tricky, if not impossible. For real investigators, this is a challenge they spend their careers trying to minimize. For a storyteller, it is a wellspring of drama. The more difficult the forensic challenge, the more heroic the sleuth, and the more suspense and tension that can be added to a narrative.

The first person to realize the dramatic potential of forensics was Arthur Conan Doyle, creator of the consulting detective, Sherlock Holmes. Sherlock Holmes was the hero of a slew of novels and short stories that appeared between 1887 and 1927, and his creation was inspired by an early pioneer in forensic pathology, the Scottish surgeon Joseph Bell. Such is the appeal of Sherlock Holmes that he made the leap to the screen within a few scant years of the invention of motion pictures (his first appearance was in 1900), and he's never stopped appearing since, with hundreds of television and film portrayals—indeed, as of this writing, there are two television incarnations of Sherlock Holmes in production on either side of the Atlantic (The BBC's *Sherlock*, which, while set in present day London, draws much of its inspiration directly from Doyle's stories, and CBS's *Elementary*, a more free-form adaptation set in New York City.) There's also a big screen franchise starring Robert Downey Jr. and Jude Law heading towards a third installment.

Although Sherlock Holmes' skills are explicitly described as being founded on patient scientific experimentation (for example, Holmes publishes a monograph on the distinguishing features of 140 different kinds of tobacco ash) science gets little time, either on the page or on the screen, in any Holmes' story. Instead, the narrative focus is squarely on Holmes' fearsome powers of direct observation and mental deduction, with the occasional bit of undercover work or gunplay thrown in for good measure.

Not so with the more recently minted forensic television dramas, in which science is so prominent that is essentially an additional character alongside the other members of the cast. The full force of physics, chemistry, and biology is brought to bear on the crime scene, revealing and analyzing trace evidence quite beyond the limits of Holmes' magnifying glass.

Of course, such shows are not documentaries. Producers and writers strive to portray science truthfully, but the highest priority is to tell a story that engages viewers and fits within the budgetary constraints of a television production. Dull shows don't get renewed, nor do those which fail to make a profit. The goal is to get the science as correct as possible without sacrificing story, or breaking the bank. This goal has some immediate and universal consequences for the portrayal of science and technologies on *CSI* and similar shows.

The first is that time is compressed; completely squashed. "We fully admit that we cheat the time factor," says Richard Catalani, co-executive producer of *CSI*, who also acts as the series' informal senior technical advisor. "We show what we want to show to tell our story, and we avoid things we don't want, like the amount of time it actually takes to look down a microscope and find semen on a slide.... We shorten some tests and we don't show all the steps of some tests, because some of them are boring."

Another fact of real life forensics that's also ignored in scripts is the issue of police budgets. On *CSI*, the investigators' laboratory is graced with a virtually unlimited collection of expensive equipment, and expenses are incurred that no actual police department would ever okay. For example, it's routine on screen to get a report of who has been calling a suspect's or victim's cell phone. In fact, the phone company charges for those queries, so it's only done when the police gauge that the case is significant enough to warrant the expense, and this parsimonious approach is true for many forensic tests.

The resulting situation in the United States is described in *Bodies We've Buried: Inside the National Forensic Academy, World's Top CSI Training School* by Jarret Hallcox and Amy Welch (2006, Berkeley Publishing Group): "Most police departments do not have their own [forensic] lab and must send their evidence to the state crime lab to be processed... These labs usually put a limit on the amount of evidence a CSI can submit for each case... most of the time, CSI's are allowed to submit only five to ten pieces of evidence." Now imagine a episode of *CSI* where the investigators box up a single suspected murder weapon, a couple of finger prints, and two shoes, and then wait days, weeks (or even much, much longer) for the results.

The last big universal liberty that's taken with the science on screen is the division of labor. Real crime investigation is often done by large teams. The person who conducts the autopsy, the person that does the blood work from the autopsy, the person who is investigating the grout from the tile on the floor where the body was found, the person who is doing mass spectrometry, the person doing DNA analysis, not to mention the people questioning suspects: all different, highly specialized people. On a television show, however, you can't have a recurring cast of hundreds, or even dozens.

One reason for that is because as mentioned above, it's not just police departments who worry about budgets, but network and studio executives as well. Actors are expensive, so casts are kept as small as possible, which means combining multiple roles in one person.

Other reasons for a smaller cast are grounded in the needs of storytelling. Shows want audiences to connect with characters, and with fewer characters, each individual cast member has more screen time to make that connection. Shows also want the cast to be heroes, so on *CSI* the characters do a lot of things that cops do (other forensic procedural shows try to get around this by integrating a cop into the team, as on *Bones*, where an FBI agent is teamed with a forensic anthropologist.) It's simply more dramatic to have the guy who figured out the identity of the killer to come charging out with a gun, and click the handcuffs closed, than for that guy to just hand a report off to someone else. *CSI* is about crime scene investigators. The stars have to do something active, and be heroic. If a police officer does the heroic stuff, that's not *CSI*.

However, it should be noted that there are many smaller police departments in reality that do not have a dedicated forensic investigation unit. In these departments, the crime scene investigator is indeed going to be the person who is also interviewing suspects and trying to make an arrest. The aforementioned National Forensics Academy was founded in large part to give these multitaskers a firm foundation in standardized crime scene investigation techniques, and allow them to create small labs of their own in order to process more evidence than can be handed off to a state lab.

So, forensic shows take liberties with the realities of time, people, and money on a general basis. On a more fine-grained level, writers often try to arrange matters so that things just happen to work out without breaking the boundaries of science and technology. For example, *CSI*'s investigators have access to perfect databases: a common scenario is that the investigators are hot on the heels of a suspect, and they put a DNA sample or a partial fingerprint in for matching against a database. Whoever is doing the matching will typically go: "Oh, we've got a hit!" and one suspect will have been identified. In the real world, things are much more complicated, with either a fair number of possible matches that have to be winnowed down by hand, or no matches at all because the suspect just isn't in the system.

The writers finesse this problem by doing things like giving the suspect a convenient history, say by having him have worked in a casino 10 years back. Employees of Las Vegas casinos are required to be fingerprinted for a work card, and those fingerprints *are* available in a database. So the investigators can get their match without the writers having to pretend that the fingerprints of everyone who lives in the United States are digitally recorded somewhere.

Another element of the nitty-gritty of real world forensic investigation that's largely sidestepped on *CSI* is the issue of negative results and the presence of false positives. Watching characters methodically plod through a series of dead ends doesn't work well within the temporal or dramatic boundaries of a crime show like *CSI*. Instead, writers find a way to give failures a storytelling purpose. For example, luminol is a chemical used by forensic investigators to perform a presumptive test for the presence of blood. Luminol can also react to things like detergents. So *CSI* might have a character who's initially a suspect in a murder because a luminol test on their clothing has a positive indication. A confirmatory test shows there is no actual blood present, and the suspect is released, but then the initial failure is recontextualized—with the investigators realizing the false-positive luminol test is evidence that the suspect washed his clothes—and it turns out to be a detergent that puts the suspect at the crime scene. An example of this occurred in the episode "The Accused Is Entitled" where a luminol reaction was a false positive because a golf bag containing a body had been washed with a detergent, then bleached. What gave the killer away was the the general streaky glow of the bag—as it would be for a garment washed in a detergent. Rather than having drops, drips, or smears of biological transfer, there would be an overall glow for the entire bag.

In part, *CSI* remains grounded in reality because the writers constantly look to actual cases and technologies for inspiration. They sift through crime literature and other forensic case studies, looking for a hook on which stories can be hung

For example, one *CSI* episode came out of a real case where a man had used a natural gas explosion to destroy a house and kill a woman. The murderer had used a timer and tried to make the explosion look like an accident. Investigators realized it was a case of arson-homicide when they discovered that a particular little piece of metal had melted. Under normal circumstances the surface of this piece of metal wouldn't have been exposed to high enough temperatures to melt, even in an explosion. So that led the investigators to look for tool marks, and those marks showed the gas line had been tampered with, exposing the piece of tell-tale material in the process. So when the writers find this kind of neat piece of actual forensics, the next step is "What are you going to do to make a story of that? Why did this person blow that person up?" Then the writers must work through how the fictional investigation will proceed, with questions like "Who can the investigators suspect?" Sometimes in the process of developing a script, the original forensic example that inspired a story can be overtaken or replaced, but it's a critical element in the process of origination.

The writers also mix and match ideas from different sources. In the episode that was inspired by the gas explosion described above, for example, the timer used to detonate the explosion was something found in a 1969 booby trap handbook for the U.S. Military, a "bean bomb." A bunch of beans are placed in a jar, and water is poured in. The cap has two electrical contacts that form a trigger; it takes about three hours for the beans to swell sufficiently to push the two contacts together, setting off the bomb. In developing this particular story, as with most others, the writers had a lot of input from Catalani and also *CSI*'s de facto research department, which is John Wellner and David Berman.

Wellner and Berman aren't forensic or police experts themselves, but they know a lot of people who are. Catalani explains how they are used in the scripting process: "They have people who know everything from timing of traffic signals in urban areas, to bouncing information off of satellites, to brain surgery—anything you can imagine, they have a personal relationship with some person that can answer those sorts of questions… we can go through the process of breaking a story [deciding the outline of the plot] and decide we want to get from *here* to *there* [in the story] using trace evidence that comes from, say, the process of making bricks." As no-one on the staff is a expert in brick manufacturing, a space is left in the script for the relevant details. Catalani then crafts a query that's likely to elicit the information that meets the needs of the script rather than a glop of brick-related trivia, and sends it to Wellner and Berman to pass on to one of their experts. "The next day, we have an answer. It's the just the best thing ever."

To investigate these carefully constructed crime scenes and suspects, *CSI*'s cast use a wide range of forensic technologies and methods. As noted above, the laboratory on *CSI* contains equipment well beyond the budget of any real local police department, but again the writers do look to the real world to try to stay grounded in reality. The rule of thumb is that if something has been done once in the world, or is reasonably within the bounds of possibility, then there is license to use it.

This means that technologies that are still essentially works-in-progress, such as facial recognition (which has had mixed results in field trials in recent years) can be featured on *CSI*. However, the show does make a point of retiring techniques if they've been rendered obsolete by the progress of real forensics. (Occasionally the writers have deliberately turned this notion on its head and revived an old technique for dramatic purposes; the investigators are denied the use of the latest equipment by various circumstances, forcing to them improvise.)

Of course, *CSI* is not intended to provide an education in how to build nefarious devices such as bombs, or how to conceal trace evidence from detection. Consequently, the writers and producers, having found out what *will* work in reality, sometimes find themselves in the situation of deliberately obfuscating their research. In these cases, plausibly similar, but deliberately inaccurate, criminal and forensic techniques are depicted instead.

This all means that the picture of science presented by *CSI* and other forensic shows has a complicated relationship to the real-world state of affairs. Some critics have ascribed this variance between fiction and reality to either laziness or the moribund intellects of the show's producers and writers; we hope the window we've opened onto the creative process above convinces you that that is an overly harsh and flawed assessment. Still, the fact remains that *CSI* is not reality, yet creates an authentic-feeling world that is avidly entered by millions of viewers around the world.

This situation has led to claims that the show has fostered exaggerated and misleading beliefs about forensics in the public consciousness, and this is perhaps the most notorious meaning associated with the *CSI* effect. Defense attorneys contend that the *CSI* Effect means that juries believe that forensic evidence is infallible (an incorrect belief, as the 2012 scandal involving an incompetent chemist in Massachusetts' state drug lab has recently reminded us.) On the other hand, prosecutors contend that it is now harder to gain convictions without presenting *some* kind of forensic evidence, even though the prosecutors contend that other (and cheaper) types of evidence, such as testimony from eye witnesses or confessions should be sufficient in many cases.

However, the evidence for these manifestations of the *CSI* Effect remains largely anecdotal, with studies showing little overall impact on acquittal or conviction rates. "I think that there is a *CSI* Effect," says Catalani, "I think people in the legal community use it when they lose a case as an excuse!"

Outside the courtroom, however, there is no doubt that shows like *CSI* have raised awareness of forensics among both victims and criminals. For example, in 2010, an 18-year-old woman was kidnapped and sexually assaulted by a serial rapist in the United Kingdom. When being transported in the rapist's car the victim—a fan of forensic television shows—deliberately pulled out strands of her own hair and spat on the seat to generate trace evidence of her presence there. This evidence was later discovered and used to convict the assailant. Conversely, there are reports of criminals attempting to clean trace evidence from crime scenes or victims. However, as—to put it bluntly—most criminals are not terribly smart, these efforts often fail.

The increased awareness of forensics has also lead to another positive outcome: the increased professionalization and profile of crime scene investigators. Not so long ago, forensic departments were often neglected within law enforcement agencies, with widely varying levels of training and competence, and little standardization in techniques, a situation that prompted the creation of the National Forensic Academy under the auspices of the U.S. Department of

Justice and the University of Tennessee (home of the notorious Body Farm) in 2000.

In *Bodies We've Buried,* Hallcox and Welch report that in the early days of the Academy they had great difficulty recruiting students (who all come from law enforcement agencies): "Before the TV show *CSI*, the vast majority of [police] departments called their investigators crime scene technicians, and funding crime scene schools was not a priority…We had to practically beg departments to take a chance on us and send an officer through the program. In fact, we could barely give away a seat in the class." Today, competition for places is high, and there is a multi-year waiting list.

Other forensic courses have sprung up to meet the demand from students inspired by *CSI* to pursue careers in crime scene investigation. Perhaps some are disappointed to discover how the day-to-day reality of investigating a crime scene differs from the heightened drama of a television show, but we suspect that most students take it in the same spirit that an aerospace engineer does when moving from the inspiration of watching a science-fiction show to the practical business of building a 21st century satellite or airplane.

For now, the viewing public's appetite for forensics shows appears to be still strong, so we anticipate many more seasons of *CSI,* and its fundamentally truthful portrayal of science as a valued servant of justice.

References

1. Maze, M.; Stagnara, D.; Fischer, L. P. *Dr. Edmond Locard (1877−1966), the Sherlock Holmes of Lyons* **2007**, *41* (3), 269–278, http://www.ncbi.nlm.nih.gov/pubmed/18348491 (accessed December 20, 2012).

Chapter 13

Cosmic Catastrophes in Movies

Joshua Colwell*

**Department of Physics, University of Central Florida,
Orlando, Florida 32816
*E-mail: jec@ucf.edu**

The cinematic appeal of a cosmic catastrophe is clear. First, it opens up a treasure trove of exciting and exotic visual effects opportunities. The threats are generally unfamiliar, and therefore potentially more scary, than mere terrestrial hazards such as earthquakes, volcanoes, and storms. The converse, of course, is that making the familiar scary (think about birds in *The Birds* or the beach in *Jaws*) can also make for particularly potent drama. But the unknown gives the filmmaker a particular liberty to invent and to change the rules of the game mid-stream by introducing new aspects of the unfamiliar threat. Hollywood versions of threats from space have taken plausible scenarios and exaggerated them to various degrees. To the extent that the stories excite the imagination, their scientific accuracy, or lack thereof, is secondary to their ability to make us think about our place in a hazardous universe.

Exterior Shot: Space. The blackness is deep and complete, interrupted only by the hard, pinpoint lights of the stars. The camera pans across the starfield to reveal the harsh and craggy profile of an asteroid. Cut to a peaceful scene on Earth where people go about their daily business, unaware of the disaster hurtling toward them at tens of kilometers per second. Or, instead of an asteroid, it's a comet, venting gases like a celestial steam locomotive, out of control and bent on destruction. Or, instead of a natural hazard, it's an invading armada of ships, piloted by bug-like aliens. Or robots. Maybe instead they are spores that will alter our DNA and turn us *into* bug-like aliens.

Impacts!

The most common version of the "cosmic catastrophe" is the threat of an impact of some celestial object (asteroid or comet) which may bring about civilization-ending destruction. The most prominent examples of this form in movies are *Armageddon* (asteroid) and *Deep Impact* (comet), both released in 1998. The timing of two major science fiction films with the same central premise (both movies featured a team of daring astronauts sent on a heroic mission to destroy the impactor before it arrives at Earth) being released within a couple of months of each other is less coincidental than it might first appear. In 1994 the comet Shoemaker-Levy 9 smashed into Jupiter leaving markings in the giant planet's atmosphere nearly the size of the Earth. The event left a mark on the headlines as well as on Jupiter, and movie producers took note. Gene and Carolyn Shoemaker, co-discoverers of Shoemaker-Levy 9 (with David Levy) were eventually brought on board the production of *Deep Impact* (in addition to me, and Chris Luchini from NASA's Jet Propulsion Laboratory) as "Comet Advisors."

Although the differences in style between the two movies is striking, there are certain similarities in how the real threat of a comet or asteroid impact was adapted to meet the dramatic designs of the filmmakers. The most obvious of these is the use of astronauts to plant a bomb inside the threatening impactor. In reality, of course, it's far easier and cheaper to send unmanned spacecraft. In any event, if you're going to blow up an asteroid the size of Texas (*Armageddon*) it isn't going to matter whether you bury your nukes 900 feet below the surface or not. In neither case will it make any difference

It's worth taking a moment to discuss the difference between asteroids and comets. Both classes of objects are remnants from the formation of the Solar System, debris that did not get incorporated into a planet and has instead languished in a few stable pockets of the Solar System for the last four-and-a-half-billion-years. The primary difference between the two is that asteroids are rocky and metallic objects which formed relatively close to the Sun and comets have significant amounts of ices due to their formation in the cooler climes of the outer Solar System. Most asteroids now are between the orbits of Mars and Jupiter, but there are significant populations of small asteroids whose orbits are close to that of Earth, making them the most likely celestial impactors. The long term reservoir of comets is beyond the orbit of Neptune, more than thirty times further from the Sun than the Earth. The more common smaller comets are not detectable at those great distances. We see them only after their orbits have been perturbed so that they make their way close to the Sun where eventually the warmer temperatures cause the ices to evaporate producing the characteristic wispy appearance. Thus, while they are fewer in number in our neck of the woods, it is more difficult to make very early predictions of a comet impact if it is a comet making its first visit to the inner solar system. Ironically, *Deep Impact*, the comet movie, gave the longer (and realistic) time frame from discovery to impact (a few years), while *Armageddon*, the asteroid movie, gave an unnecessarily (and ludicrously) short time frame of just a few weeks from the discovery of the asteroid until D-Day.

There are far more small comets and asteroids than large ones, but an impactor only a few km across will wreak global havoc. The Earth's relatively large mass is ultimately responsible for much of the damage because it is the Earth's gravitational pull that assures that any impactor will be traveling at least 11 km/s when it strikes the surface. In the case of the most likely impactor, the near-Earth asteroids (or NEAs), the impact velocity is not much greater than that speed due to the similarity of the NEAs' orbit to that of the Earth. A comet will most likely approach the Earth with a significantly larger relative velocity due to its more distant point of origin, resulting in a much larger impact speed and thus a much larger destructive potential for objects of the same mass. The kinetic energy of a 1 km radius asteroid with a modest density of only 1.5 g/cm^3 striking the Earth at 11 km/s is $3.8x10^{20}$ joules, roughly equal to the world's total annual energy consumption and more than 1000 times the energy of the most energetic nuclear bomb ever tested. Increase the size of the impactor by a factor of 10 and the energy increases by another factor of 1000. That is the energy of the Chicxulub impactor that led to mass extinctions and the end of the age of dinosaurs 65 million years ago. Like most things in *Armageddon*, their 900 mile asteroid was certainly overkill.

Thanks to the heroics of the astronauts in *Armageddon* that large asteroid does not end up hitting Earth. The most prominent recent movie version of a large-scale impact is the comet fragment in *Deep Impact* which has a diameter of about 1 km. On approach the comet is seen to approach the Earth at a very low angle to the surface of the Earth. A more vertically oriented trajectory is more likely; the glancing approach shown in the movie is so unlikely as to be inplausible. The effects of the impact itself are realistically depicted, as far as these things go: the impact event is depicted as an explosive event in the ocean followed by an atmospheric shock wave and a supersonic tsunami. The amplitude of the tsunami should decay with distance r from the impact site as $1/r$ as long as the ocean depth is constant, followed by a slowing of the wave and a commensurate increase in amplitude (preserving energy in the wave) as it approaches shore.

The most glaring physics error in the impact scene is the meteor shower into the atmosphere following the astronauts' destruction of the larger, 10-km, comet fragment prior to impact. This is shown as a harmless meteor shower. If all the pieces of the original object impact the Earth, they still deliver the same amount of energy to the Earth as if they arrived in one large chunk. The energy would be spread over a larger area, but at this scale that would still lead to global firestorms. Like with most movies, however, the timetable is not precise or clear: if the astronauts destroy the comet far enough in advance for the fragments to disperse over an area that is significantly larger than the cross-section of the Earth, then only a fraction of the impactor energy ends up being deposited in the Earth's atmosphere. Catastrophic fragmentation of a comet or asteroid would lead to fragment speeds of ~100 m/s. To disperse that material over ~10 Earth diameters then requires destruction on the order of a few days before impact in order to significantly reduce the destructive consequences. (*Seeking a Friend for the End of the World* begins with the failure of an expedition to destroy an asteroid 21 days before impact.) The best hope we have of avoiding a collision is to provide a relatively modest deflection to the path of the impactor as far in advance of impact as possible. This sort of timetable does not lend itself well to dramatic storytelling.

The immediate consequences of an impact of the scale of the Chicxulub impactor is a global firestorm and a prolonged cold and dark spell resulting from a global ash cloud blocking most of the sunlight. Another side effect would be a loss of atmospheric ozone due to reactions with atomic Nitrogen. The energy of the impact would dissociate enough N_2 in the atmosphere for N to react with O_3 producing O_2 and NO. Of course by this point, the movie has long been over. Regardless of the details of the threat from impact, the audience knows they will be catastrophic and the drama lies in the steps taken to avert disaster.

Aliens!

While impacts such as the one depicted in *Deep Impact* are a very real threat (one the scale of *Armageddon* will not happen; those only occurred during the epoch of planet formation 4.6-4.5 billion years ago), the other common cosmic threat according to the movies is one that, for the moment, exists only in our imaginations: alien invasion. While there are the occasional benevolent aliens (*E.T. The Extraterrestrial*, *Starman*), sinister aliens bent on taking over our planet (*Independence Day*, *Signs*, *The War of the Worlds*, *Falling Skies*) or, worse, our very bodies (*Invasion of the Body Snatchers*, *Alien*) are more common in the movies. There have also been invasions that are more nuanced, such as the shipwrecked aliens of *Alien Nation* and *District 9*. While no aliens have yet appeared on our cosmic shore, it is not an unreasonable premise to explore cinematically. Nor is it purely a speculative venture.

NASA takes the potential threat of alien microbes quite seriously. The Planetary Protection Office is charged with assuring that risks of contamination from any alien microbes that might reside in a sample returned from Mars or some other potentially hospitable abode are safely contained. The office is also responsible for assuring that terrestrial microbes do not contaminate any habitable environments beyond the Earth. This concern leads to the deliberate suicide plunges of spacecraft such as the Galileo mission at Jupiter and the Cassini mission at Saturn to ensure that they do not ever crash onto one of the moons of those planets that may harbor life and costly sterilization procedures for Mars landers. Europa at Jupiter, and Enceladus and Titan at Saturn are moons that have at least some of the necessary ingredients to be considered habitable (subsurface liquid water and, in the case of Titan, a hydrocarbon weather cycle). The *Andromeda Strain* features a deadly extraterrestrial microbe which, unlike the aliens in most alien invasion movies, acts without intent. It is merely a deadly microbe with unusual characteristics due to its alien origin.

Aside from the obvious dramatic potential (the bad guy may be very very bad with inhuman powers, or maybe the bad guy isn't a bad guy after all, but just a scary looking bug from outer space), the question of why there haven't been visits from extraterrestrials remains something of a puzzle, nicknamed Fermi's Paradox.

Enrico Fermi, one of the fathers of the nuclear age, reportedly posed the simple question to his colleagues, "if extraterrestrials exist, where are they?" It is not an idle question, and the answer is less obvious than one might initially assume. If a civilization reaches the point of being able to achieve interstellar travel, then even if it carries out its voyages between the stars at the modest speed of, say, 0.001c (c is the speed of light) that civilization should propagate across the Milky Way galaxy in only 100 million years. Each new colony established by the original technologically advanced civilization will, by definition, have the same technological tools available to continue the spread of the civilization to the next habitable planetary system. Given the age of the galaxy ($12\text{-}13\times10^9$ years) and the age of our own planet (4.5×10^9 years), Fermi's Paradox suggests that if such a civilization existed, it should by now have colonized the entire galaxy. The possible resolutions are a bit unsatisfying: (1) we are the first (or only) technologically advanced civilization in the galaxy, or (2) all such civilizations engage in something like Star Trek's "Prime Directive" of avoiding interference with other civilizations.

The first alternative suggests that life does not naturally evolve much past our current state. The number of potential habitats in the galaxy is staggering. In just the last decade the number of known extra-solar planets has soared to nearly 1,000, and the Kepler mission (*1*) has added thousands of planet candidates to that list and continues to find more. Given that our current searches are limited to a small fraction of the Milky Way and to discovering planets that have peculiar properties that make it easy for us to see them (*2*), it seems clear that the total number of planets in the galaxy may rival the number of stars ($\sim4\times10^{11}$).

The number of intelligent civilizations in the galaxy with whom we might, at least in principle, communicate, N_I, is estimated by the Drake equation, which neatly separates one wildly speculative number (N_I) into several somewhat less speculative numbers:

$$N_I = N_* f_P n_P f_L f_I L / T$$

where $N*$ is the number of stars, f_p is the fraction of stars with planets, n_p is the average number of planets in a planetary system, f_L is the fraction of those planets on which life evolves, f_I is the fraction of planets with life that evolve to an intelligent civilization, and L/T is the ratio of the age of the civilization to the age of the galaxy. These parameters are less and less certain as one goes to the right in the equation. When Frank Drake first developed the equation in 1961 only the first term was informed by any data whatsoever. We can now make a reasonable estimate that the product of the first three numbers is on the order of 1 to 100 billion. The next term (f_L) is in principle discoverable within the next decade or two through detection of chemical inequilibrium in the atmospheres of the now abundant population of known extrasolar planets. The product of the remaining terms, the fraction of those planets that currently harbor an intelligent civilization, is largely speculative, but could be as small as 1 in 100 million and still require that the second solution to Fermi's paradox be invoked.

The cinematic portrayals of aliens in movies generally involve a long list of repetitive fundamental errors in physics. While the Fermi paradox points out that interstellar travel is possible due to the long expanses of time available, this would require "generation ships" in which the travel time exceeds the lifetime of any one individual on board. Movies usually adopt a "warp drive" (*Star Trek* in all its various incarnations) or "hyperspace" (*Star Wars, Babylon 5*) shortcut to this problem so that characters can get from point A to point B in the galaxy and still participate in their own story. A variant of warp drive that accomplishes the same goal with slightly different technical justification is folded space and wormholes (*Dune*, *Battlestar Galactica*, *Star Trek: Deep Space Nine*). The most common exception still preserves the individual characters through a form of suspended animation or "hypersleep" (*2001: A Space Odyssey*, the *Alien* movies, *Avatar*).

The rise of computer generated images (CGI) has made it possible for filmmakers to develop less humanoid aliens (*District 9*), though they still frequently resemble a human with some extra appendages or protuberances (especially in *Star Trek*). One could argue that the diversity of life on Earth which shares a common biological heritage is greater than the diversity of extraterrestrial aliens seen in movies. Movie aliens also tend to arrive in spaceships that have an unspecified energy source and occasionally spend a lot of time hovering over cities with no apparent means of support against gravity (*Independence Day*, *District 9*, and (tongue-in-cheek) *The Hitchhiker's Guide to the Galaxy*, the television series *V*). These liberties, taken presumably for the dramatic and imposing effect of a technology far superior to our own, fall under the category of violation of the law of conservation of momentum. This is where we also find the "inertial dampers" of *Star Trek* and the ability of spaceships in most movies to acclerate to relativistic speeds with no negative side effects on the passengers.

Perhaps the most implausible aspect of the depiction of aliens in movies, however, lies not with the all-too-common violations of conservation of energy and momentum, but rather with the intentions of the aliens. They come to Earth (at what would be a tremendous investment of natural resources) to take our planet's resources when there are much more abundant and easily obtained natural resources for a spacefaring civilization in the asteroid belt, the Moon, and Mars. In *Signs*, for example, the aliens are undone because water is toxic to them. If there is one place in the Solar System you don't want to go if you have a problem with water, it's Earth. Most alien invaders would have a much easier time if they looted the abundant resources everywhere else in the Solar System where they would not only not have to deal with pesky humans, but also would be able to avoid the Earth's rather deep gravity well.

Even in *Star Trek* where there was a Prime Directive, akin to the second resolution of the Fermi paradox, James T. Kirk couldn't restrain himself from getting entangled in alien affairs. This puts us back at the first solution to the paradox, namely that our current capabilities not withstanding, either the value of f_l is precisely zero or L is a terrifyingly small number. (For us, one might argue it is currently at 100 years and counting, so that $L/T \sim 10^{-10}$.) With the notable exception of *Star Trek*, there has not been much cinematic science fiction dealing with a galactic culture of civilizations. Instead we are usually the target of a single invading species, whether it be as spores that invade our bodies or with preposterous ships that float threateningly over our cities. Many instances of space-faring civilizations instead place the action at a great distance from Earth either in time, space, or both (such as *Star Wars*' "long time ago in a galaxy far, far away"). The sobering implications of Fermi's paradox in the context of the

Drake equation suggest an uplifting subtext even to movies with horrifying aliens (such as *Aliens*): even if there are monsters out there, at least we are not alone, and perhaps we will make it in the long run to become space-faring ourselves.

The Sun

There are a number of potential cosmic catastrophes that have not been explored (or exploited) by Hollywood. (For a more detailed treatment of these than allowed by the space here, see (*3*) and (*4*).) The Sun, a mere 500 light-seconds away, can be pretty scary. In addition to its welcome radiant energy, it is continuously emitting high-energy protons and electrons that are deflected by the Earth's magnetic field. The aurorae near the Earth's magnetic poles are one manifestation of these charged particles as they funnel along magnetic field lines and impact the atoms and molecules of the upper atmosphere, causing them to emit photons in characteristic red and green wavelengths. However, the geologic record from rocks near the mid-Atlantic ridge shows that the Earth's magnetic field changes polarity on timescales of 100,000 to 1 million years. How abruptly this change occurs, and how much the field might weaken when it does occur, is not well understood, but the shifts are almost certainly too slow to be able to be captured in a typical cinematic timescale. The Sun occasionally belches a particularly dense and energetic stream of charged particles into space. These Coronal Mass Ejections (CMEs) pose a hazard to our electrical infrastructure even with the protection of the magnetic field. Combine a CME with a weak or absent magnetic field, and there could be a civilization-altering impact on the power grid. The TV series *Revolution* plays with the idea of a permanent loss to the power grid (which would not be the result of a CME), but there has not been a major production exploring the chaos of the initial disruption and its immediate aftermath. The movie *Knowing* features a deadly solar flare (and aliens), but does not deal with it other than as an end-of-world event.

The largest solar flare observed, known as the Carrington event after the astronomer who saw it on September 1, 1859, produced aurorae visible in the tropics and bright enough to wake up miners in Colorado. Telegraph service was disrupted, but there was no global (or even local) electrical power grid. A comparable CME today could destroy power transformers. Because these devices cannot be quickly replaced, there is a risk of loss of electrical power and all the devastating collateral damage that implies for medical services, climate control, and for food production, transportation and refrigeration.

The television series *Space: 1999* had the novel premise of a band of the ejection of the Moon from the Solar System, complete with a band of stranded scientists. While the Moon could not be ejected from Earth orbit by a nuclear detonation, as in the show, it is possible (though vanishingly unlikely) for it to be stripped from the Earth through a gravitational interaction with a sufficiently massive object passing through the Earth-Moon system on just the right trajectory. The series did not explore the effects of the loss of the Moon on those left behind on Earth. Some theories of the origin of life on Earth suggest that the lunar tides, much stronger in the past, may have played a critical role in producing shallow pools of water rich with nutrients and the chemical building blocks of life.

Movies and the Public

There are several aspects of cosmic catastrophes that can be accurately exploited for good dramatic effect. The threat can be huge (destruction of the planet or civilization) and is generally exotic. Frequently there is a date-certain for doom. Comets on a collision course have a well-known arrival time. They come with a built-in countdown clock to Doomsday. Think about how many movies feature a bomb ticking down to destruction. That inevitability of destruction with a certain timescale adds dramatic tension. The form of destruction may have great visuals. Both aliens and asteroids seem to arrive with big explosions. *Armageddon* had a barrage of precursor impacts to spice things up. While it is possible to do something to save the world, the steps required are both heroic and unfamiliar. Snipping the right wire on a ticking bomb with one second to spare is now a cliché (but that doesn't stop it from being used over and over again), but assembling a last-minute space mission to destroy an asteroid (*Armageddon*) or building caves to preserve some remnant of civilization (*Deep Impact*) are unconventional and grand visual spectacles, and thus heighten dramatic interest.

Nevertheless there are several obstacles to making a good movie about death from the sky. There is no tangible bad guy. The villain doesn't have a face (unless it's an alien one). The real solutions to cosmic threats are generally complicated, involve many steps, and take a long time. Take for example the case of the threatening comet or asteroid. The determination that the impactor is actually on a collision course would take an extended period of observations, with each subsequent observation shrinking the uncertainty about whether the object will actually strike the Earth. Not surprisingly, Hollywood typically shrinks this process down to a minute or two of a single scientist's calculations if it is even depicted at all.

The steps to be taken to avoid such an impact do involve a space mission, but there is no reason to send humans. We have become quite expert at designing and operating robotic spacecraft. In addition to operating a small squadron of rovers on the surface of Mars, NASA has sent spacecraft to every planet; sent a probe through the tail of a comet, captured some of its particles, and returned them to Earth; successfully impacted a probe onto a comet nucleus (the mission "Deep Impact," aptly named for its role reversal with the movie); worked jointly with the European Space Agency to land a probe on Saturn's moon Titan; and the Japanese Space Agency has captured particles off the surface of an asteroid and returned those to Earth. So, what would be done with the discovery of a celestial body on a collision course? Assuming the discovery is made with sufficient advance warning, an unmanned space mission would be launched to affect the orbit of the impactor to deflect it off its collision course. The simplest approach is to increase the reflectivity of the asteroid so that solar radiation pressure would alter its orbit. From the point of view of saving the planet, this technique requires action years in advance of the impact. From the point of view of Hollywood, this is rather dull and anticlimactic. Nuclear bombs can be put to use for deflection as well with better cinematic visuals.

The key to all these measures, like cures to cancer, is early detection. There are a number of telescopic surveys now in place dedicated to discovering near-Earth objects (NEOs). Because the objects are nearby, they are relatively bright, and therefore very large telescopes are not necessary to detect them. For example, the Lincoln Near Earth Asteroid Research (LINEAR) survey using meter-scale telescopes has discovered, as of late 2011, more than 200,000 objects including more than 2,400 NEOs. The Catalina Sky Survey (CSS) has been discovering hundreds of NEOs each year since 2005, also with modest telescopes. The challenge is the large amount of images and the data analysis, and then following up on discovered objects to determine accurate orbits.

While the Shoemaker-Levy 9 impact on Jupiter arguably spurred the production of two movies about cosmic impactors, it also helped raise the profile of this real threat in the political arena. While NASA is doing relatively little in the search for extra-terrestrial intelligence (SETI), surveys such as LINEAR and CSS were born in the aftermath of Shoemaker-Levy 9. This raises the question of whether and how the cinematic portrayal of these and other potential calamaties influences public opinion and even public policy. Do portrayals of events such as those in *Deep Impact* and *Armageddon* raise awareness of a real threat, or cast them as mere Hollywood fantasies in the public consciousness? Would a major motion picture depicting the aftermath of a global blackout due to a Coronal Mass Ejection spur governments to put safeguards in place, or at least create a contingency plan? Would it make a difference if the movie were more or less scientifically accurate?

My own anecdotal experience with students is that they make the connection between science fiction movies and the real world. While I have never been asked if *Spiderman* could happen, I have been asked about the verisimilitude of events in movies such as *Armageddon*, *Red Planet*, *The Core*, and various natural disaster movies. There has been an increasing amount of research on both the attitudes of the public toward science as influenced by movies and television (e.g. (*5*, *6*)) as well as interest in taking advantage of the cinematic material to teach science in the classroom (*7*). The "CSI effect" (e.g. (*8*)) is based on the idea that the attitudes of jurors toward forensic data is affected by the portrayal of the collection, analysis and interpretation of this data in television programs such as *CSI: Crime Scene Investigation* (see chapter by Cass, Grazier, and Thompson, this volume). There is also broad public interest in the possibilities of extraterrestrial life. Cinematic portrayals of aliens send the message that the galaxy is crowded with aliens and that, presumably, it is only a matter of time before some show up on our door. With the rapid pace of scientific discovery on many fronts, not just astronomical, the exposure of many people to new ideas comes primarily through popular media such as television and movies. As a scientist I wish those movies were more grounded in reality, which as we've seen is full of dramatic potential. But perhaps the most important thing is that the movies are making them think enough to ask the questions.

References

1. Borucki, W. J.; et al. *Science* **2010**, *327*, 977.
2. Fressin, F.; Torres, G.; Charbonneau, D.; Bryson, S. T.; Christiansen, J.; Dressing, C. D.; Jenkins, J. M.; Wakowicz, L. M.; Batalha, N. M. *Astrophys. J.* **2013**, in press.
3. Plait, P. *Death from the Skies*; Penguin Books: New York, 2008.
4. Tyson, N. D. *Death by Black Hole and Other Cosmic Quandaries*; Norton: New York; 2007.
5. Kirby, D. A. *Soc. Studies Sci.* **2003**, *33*, 231–268.
6. Kirby, D. A. *Lab Coats in Hollywood: Science, Scientists, and Cinema.* MIT Press: Cambridge, MA, 2011.
7. Efthimiou, C.; Llewellyn, R. *Physics Educ.* **2007**, *42*, 253–261.
8. Cole, S. A.; Dioso-Villa, R. *New England Law Rev.* **2007**, *41*, 435–470.

Chapter 14

Solving for *x* When *x* Is the Audience: A Case Study within Fan Culture

Carrie J. Cole[*]

School of Theatre, Film & Television, University of Arizona, Tucson, Arizona 85721-0003
[*]E-mail: ccole@email.arizona.edu

The essay examines the Fan, who occupies a space between the production of media and the scientists that inform it. The notion of a known audience is shown to be impossible due to the unstable definition of what constitutes a fan of a given franchise. While Hollywood may extrapolate a conceptual or ideal "fan" based on statistical analyses, they do so at their own peril. By using an intersecting framework of performance and fan studies as applied to ethnographic fieldwork, variables of audience engagement are brought to light.

I am not from the Hollywood part of this equation.

Nor am I from the scientific part of this equation.

But I'm the x for which they both attempt to solve.

More accurately, I am one x of many, for in the great calculation at the foundation of all Hollywood science fiction, x is a constant variable. The science advisor and the Hollywood writer together craft stories that maintain the integrity of both science and story while formulating equations to solve for x. Entertainment industry executives crunch numbers to determine all the variables of x that create a sound formula for projected revenue numbers. Everyone is looking for quantifiable proof that one particular theoretical set combining science and story will result in an exponent of x greater than hypothesized.

Yet x is a far more unstable variable than anyone cares to acknowledge. Ultimately, whatever equation gets devised, x is not just a designation for revenue streams or critical accolades. x is not a solid mass. x represents a multi-faceted, constantly evolving factor that often befuddles scientists and Hollywood insiders alike.

x = the Fans.

Exponents of *x*

The Fan can take on a monolithic, almost mythological characterization. Hollywood knows the Fan is out there, and a wealth of people try to analyze the Fan, to capitalize on its possibilities, only to find that each strategy for examination reveals different data sets—and leads to more hypotheses demanding more results. Philip Napoli, in *Audience Economics: Media Institutions and The Audience Marketplace*, summarizes this concept succinctly: "Human attention resists the type of exact verification and quantification that typify the transactions that take place in most other industries (*1*)." The Fan remains elusive. It's a Sisyphean task, in the end, to try to codify the myriad types of fans using a single equation based on a uniform conceptualization of the Fan.

When taken individually, a fan can confound the basic theoretical equation which teams of Hollywood writers and science advisors strive to solve: determining the best balance between the science and the story. This may be why it is easier to think of fans either in terms of the economic investment they symbolize as a market group on the one hand or as stereotypes that have been perpetuated since the early days of science fiction pulp magazines.

For decades, fans, particularly fans of science fiction have been fighting a stereotype that has been out of date for nearly as long. This is not to say that the pimple-faced young men in David Hartwell's seminal essay "The Golden Age of Science Fiction is Twelve" are not still a thriving and invaluable aspect of the fan demographic. However, Since Hartwell's essay was first published in *Age of Wonder* in 1984, the twelve-year-old young men he acknowledges as central to science fiction fandom have grown, and have been joined by a great diversity of gender, race, and ethnicity as technological advances and information networks have made both science and science fiction more accessible. As Hartwell recognized even, the fan of science fiction "may be your attorney, your dentist, your children's schoolteacher…happily living in two worlds at once (*2*)."

The geeky little twelve-year-old boy, whether he be the child of the 1920s eagerly anticipating Hugo Gernsback's next edition of *Amazing Stories* or the 21st century child counting the days until *Defiance* is released simultaneously as a series on SyFy and an MMO (massively multiplayer online game), still makes a compelling argument for x. However, our 21st century techno-savvy tween has been joined in the fan community by another x determinant: equally savvy tween girls. Each x defined thus far is likely to be either overseen or joined by x = dad, x = mom, x = aunts/uncles/older siblings…the variants are numerous. The fan community has become less like *Logan's Run* in which people get…purged…from the community when the sum of x's age equals 30. Today's fan community is a multi-generational, non-gender specific network of communities, and for those creating narratives to appeal across this network, an increasingly problematic mutable variable.

Some entrepreneurs in the Hollywood industry do capitalize on the variability of the Fan, and focus on specific demographics within fan communities. As Ashley Eckstein told *Forbes Magazine* when she launched the company Her Universe specifically to create merchandise for female scifi fans, "When I asked around I was told—mostly by men—that girls won't buy scifi products or merchandise made for them. I was told to just be happy with the men's size small. I did some research and found that 45% of New York and San Diego Comic Con attendees are female. And 50% of Syfy Channel's audience is female. And when you add in the fact that 85% of all consumer purchases are made by women, it made sense to launch Her Universe (*3*)." Eckstein limits the variable in her equation in order to capitalize on a specific outcome.

But while this limitation offers a nice, compact sample set for study, researchers extrapolate formulae based on this sample at their own peril, and "the result is a cultural system run by magic numbers, numbers that shape content, creation, and availability. The question then becomes: Who counts in the ratings (*4*)?" Each fan offers demographic variables, complicated by the diffusion of fan communities as digital media and multiple platforms widen access to entertainment, and their own eagerness and ability to enter into the creative conversation through fan fiction, fan art, and fan engagement at conventions.

When x = Me

I am a fan. In fact, my actions as a fan are what led to my inclusion in this anthology. It seems fitting then, to offer myself as an example of how the variables and heterogeneity of fan cultures have led to new, more inclusive theories of fans not as passive receivers of entertainment commodities but as a participatory culture which dynamically interacts with the focus of their fandom.

As I am part of the fan demographic Ashley Eckstein has targeted with Her Universe, my fandom includes my gender as a variable. However, I have most certainly passed the "golden age" of Hartwell's science fiction fans. I'm also an academic, which colors how I experience and examine my own fandom and my place in a community of science fiction fans. My area of scholarship creates another variable: as a theatre scholar and practitioner, my academic world often seems far afield from any of the sciences, and occasionally still eschews popular entertainment as a denigration of its aesthetic. So how on Earth did I end up here, intermingling with chemists, physicists, composers, and screenwriters?

Growing up, I was part of a somewhat binary education system: I could choose humanities or sciences, art or chemistry, languages or biology. I showed an early aptitude toward the humanities and from that point was never particularly encouraged to pursue science. What I remember of high school biology after the span of time is minuscule but vivid: my teacher's helmet of gelled-to-perfection hair (not science), getting detention for being tardy (not science), and the stench of grasshoppers that were not properly preserved for dissection (ah! Science!..ish). I know there was a fetal pig in there somewhere, but it does not resonate in my memories with equal footing.

Of chemistry, I remember both less and more. It was the first time I was cognizant of enjoying science. Admittedly, it had nothing to do with the experiments (Bunsen burners freaked me out) and everything to do with balancing chemical equations. I was not particularly a math person, but there was something both soothing and productive to balancing equations for me—it was like finishing a story left incomplete.

Ah. There it is: the story. For me, it's about the story. As a theatre historian, it's about the historical dramatic narrative. As a theatre practitioner, it's about translating the story from page to performance. As a dramaturg, it's about being an intermediary between the writer and the audience.

A dramaturg?

Yes, "dramaturg" sounds like some sort of nefariously brilliant science fictional entity. In fact, one of science fiction author John Brunner's novels from the 1970s, *The Dramaturges of Yan* (*5*), features a space-traveling dramaturg who refashions the histories of colonized planets into theatrical space spectacles—all in an effort to craft the best story, yet to the peril of the alien race and human colonists who, until the interference of said dramaturg, peacefully coexisted.

Machiavellian aspects aside, Brunner has a decent, albeit negative, understanding of what a dramaturg can do. As a dramaturg and as a fan, I'm interested in narrative and the world it creates. Build me a world, stay true to the natural, cultural, and social laws you've created for this world, and I'll likely accept whatever story you tell within those given circumstances. Defining dramaturgy in terms of structural, cognitive, spatial, and intuitive world-building makes the field more applicable beyond the realm of my home discipline and is probably the simplest definition I can offer of "dramaturgical fandom" that crosses the media of live performance, television, film, and video games. What seems problematic is that just as dramaturgs define their work in their field in specific ways, fans define their fannishness individually as well. There is no unified definition of either.

Variables of *x*

Yet it is the duality of my roles as a dramaturg and Fan that dynamically mark my "*x*" variable. I've admitted to a significant lack of science and mathematical knowledge, but it seems to me that $x = \infty$. Fans can be *anyone*. What good does that do scientists, writers, and scholars working toward the common goal of generating and cultivating a creative relationship with audience that is inclusive of science and entertainment?

In *Digital Fandom: New Media Audiences*, Paul Booth offers a definition of the Fan that narrows the parameters of *x*: "in traditional parlance, a fan is a person who invests time and energy into thinking about, or interacting with, a media text: in other words, one who is enraptured by a particular extant media object (*6*)." However, I quibble with his phrasing "extant media object." Arguably, the investment, the energy, and the rapture of the Fan are not aspects of a passive engagement with the *object* of their attentions. Rather, it is an active and interactive relationship that develops between a fan, a piece of media as *subject*, and the many tangential objects that orbit that subject.

Ultimately, *x* is so difficult to solve because the fan parameters are not determinately fixed. A more inclusive way to look at fans would be to acknowledge that:

- Fans are intertextual—and text goes beyond the written text and the performed text
- Fans are intermedial—they access their entertainment across media and across platforms
- Fans are experiential—they look for ways to experience the object/subject of their fandom through various forms of fan production

Defining Fans as intertextual, intermedial, and experiential acknowledges the complexities of fan culture Darshana Jayemanne explores in "Microstatecraft: Belonging and Difference in Imagined Communities":

> Fan culture, where a sense of community is generated around the reception and remediation of cultural texts, has developed its own extremely complex systems of belonging. Fandom is variegated not only along the obvious lines of which texts are appreciated and appropriated by a particular group, but also by the medium in which the text is expressed, the specificities of translation, the location of the fans, the engagement with or collection of peripheral merchandise and the particular historical narratives and self-imaginings of the group in question, amongst many other factors (*7*).

When *x* Meets Why

While science was not emphasized in my younger years, science fiction certainly held a prominent position on my family's bookshelves. As a voracious reader with an older brother and father who read science fiction with an equal fervor, I grew up reading science fiction. They also introduced me to scifi tv: with them, I watched early reruns of *Star Trek*, and the first iteration of *Land of the Lost* with its cheesy sphere-eyed Sleestaks and the Pylons housing multi-colored crystals that never quite helped the Marshall family out of that alternate, prehistoric universe. And, of course, I watched *Battlestar Galactica*.

My recollections of the first iteration of *Galactica* are fuzzy—in part because growing up in rural Maine, we only received a clear picture from the local ABC affiliate in the best of weather conditions. My memories scatter now like those of my biology class, but instead of my teacher's hair, it was the glistening coifs of Dirk Benedict and Richard Hatch and the booming resonance of Lorne Greene's voice that captivated me as I sprawled in front of the tv, closer than we were normally allowed, completely engrossed.

So when I first saw the promotional teasers for the reimagined *Battlestar Galactica* in 2003…I ignored it. I had to ignore it. I knew, standing in the center of my father's living room watching that promo, I would be equally engrossed, and my relationship with the show would be both torrid and enduring.

At the time, I was in the throes of writing about fan performance and production of an entirely different kind, focusing on nineteenth century historical feminist performance things. I had no room for scifi. More accurately, I placed an embargo on my entertainment while I worked on this study. I had to earn my tv breaks. I couldn't allow myself to develop a relationship with a tv show. I ignored the Cylons' call.

So I waited. I played coy. I held myself at bay until the opportune moment. Eventually, with my dissertation done, defended, and relegated to a backup drive, I surrendered. The 2007-2008 Writer's Guild of America strike, which held Hollywood in limbo for almost four months, seemed like the perfect time to catch up on the first three seasons of *Galactica* in one fell swoop. I became passionately enamored with the show. Succumbing to the series, I let episode after episode ravish me. I was addicted to it. I had made that indiscernible and inevitable transition from casual viewer to devoted fan from the moment I heard that ominous susurration of the dradis.

This was not the *Battlestar Galactica* I remembered—and that was part of its appeal. The re-imagined series did not look nostalgically to the past that I associated with the original series. This was a new look forward into a great dark void of terrifying possibility. The stories that captivated me in this new version of *Galactica* were no longer about the well-coiffed heroes (although the new aesthetics never hurt). Instead, I was transfixed with the well-crafted, complex interpersonal relationships that were fully integrated into the larger stories about the politics—and the science.

These stories built a world replete with social, political, and scientific machinations. Story arcs were based on what was accessible—and doable—to survive the constant threats found in space. They were grounded in a reality recognizable to twenty-first century audiences: the reality of unforeseen chaos. Watching that number on the whiteboard decrease in each set of opening credits, wondering what technical devastation would bring about the inevitable downward shift in the population count, made me invest in wanting to know how they were simply going to survive. Suddenly, I was having conversations with friends about the viability and the speculative reality of some of the story lines—but not just based on what made a good story. I was making connections with other works of science fiction, but more importantly, I was asking questions about the feasibility of faster-than-light travel, airlock expulsions, real life advancements in computing technology, transhumanism, singularity…questions that my scifi tv watching had never before engendered.

These questions needed answers. I wanted to know more about the science behind the stories. I wanted to know why I found the science such an integral part of the dramatic structure. That's when I discovered two revelatory things: 1) that there was a *real scientist* working with the show whose job it was to figure out the feasibility of the science; and 2) this science guy was going to be roughly in my neighborhood talking about the show and his job *with fans*. I was compelled. I made plans to go to the Phoenix ComicCon in May 2010 and inadvertently launched a multi-year, multi-Convention ethnographic study of how fans of science fiction interact with the science of scifi television at these intermedial, experiential Convention sites. With *Galactica* as my entry point, I started following the public appearances of the show's science advisor.

When *x* Meets Science Guy

In "Subcultural Celebrity," Matt Hills examines how fans "amass considerable knowledge of those behind, and in front of, the camera" to the point at which "the names and faces of production staff are often just as well-known as those of star actors: celebrity is not at all restricted to those receiving wider recognition (*8*)." Hills argues that these behind the scenes celebrities function as "objects of knowledge" through their accessibility and that they are likely to have "embodied, social interactions with their fans, meeting them at conventions, doing planned 'signing sessions' to promote niche products (*9*)." However, this interaction often extends beyond products, and embraces broader scientific and theoretical concepts.

I don't know what I expected of my first Con experience. That's a lie: I expected all of the stereotypes one could imagine. Some of those expectations were met in the best possible way, and some were completely dismantled. Ironically, the primacy of science in my longitudinal study began with a panel called "Bad Design in Science Fiction Universes" which featured John Scalzi, Seth Shostak, Michael Stackpole, and one of this anthology's editors, Kevin R. Grazier. While the panel ostensibly focused on when scifi gets the science wrong, my notes from the event are rich with anecdotes of when and how to get the science right. The panelists naturally pointed out the perilous consequences when incorrect science gets popularized by science fiction. For instance, Grazier used the example of the season three episode of *Galactica*, "A Day in the Life" as a cautionary tale of how previous scifi film and television have led fans to make certain false assumptions of what would happen to a body when expelled from an airlock.

This kind of example shows that a good story, a really compelling story, can sometimes get away with as much bad science as good science. But whether its Hollywood's responsibility to science or to its audience—or rather an opportunity for science to effectively entertain and educate an audience—there has been a steady increase in the integrity and veracity of the science embedded in science fiction television. Enter the National Academy of Sciences, and the creation of the Science and Entertainment Exchange (*10*). As fans have become more sophisticated and more media savvy, they demand a greater narrative integrity in science fiction. Those demands rely heavily on how story and science integrate. The science advisor spans boundaries between story and science, as I'm sure is clear not only in David Kirby's description of "boundary spanners" in *Lab Coats in Hollywood: Science, Scientists, and Cinema* (*11*) but within this anthology. Kirby builds his initial premise of the boundary spanner in "Hollywood Knowledge: Communication Between Scientific and Entertainment Cultures," where he defines the role as having the "ability to facilitate communication between these two unique social groups rest on their claims to membership in both (*12*)." While the two social groups to which Kirby refers are the scientific community and the entertainment industry, I would argue that to be most successful as a boundary spanner, the science advisor also has to negotiate membership with a third social group: the fans. Kirby's definition of boundary spanners doesn't preclude the addition of this social group; in fact, if a boundary spanner's effectiveness is based on his or her ability to "effectively inhabit multiple social identities," then including the fan community in this equation can only ground the argument more (*13*).

The distance between scientist and fan narrows in part at Cons, when fans experience the subcultural celebrity of the science advisor and realize what Sidney Perkowitz acknowledges as the preponderance of anecdotes from scientists and researchers about "how science fiction generated a sense of wonder that enhanced their youthful interest in science. For them, it didn't much matter whether the fictional science was exactly right. In fact, these protoscientists were stimulated and challenged by imagined science and technology that wasn't reality—yet. That forward-looking aspect is science fiction's most valuable property (*14*)."

I gained a wealth of foundational knowledge from my first Con experience, including a few key concepts which were reinforced not only during the initial panel I attended, but further still on a panel titled "Consulting on Science Fiction TV & Film." Looking back at my notes, I had circled three key points from this panel:

1. Consistency within the universes is the key to keeping fans closer to moments of "oh, wow"—and farther from moments of "oh, please!"
2. There is a difference between what is practically possible and what is practically plausible.
3. Sometimes the real science is even cooler than what the writers have imagined.

In the end, that fine line between possibility and plausibility is what captivates science fiction audiences, moving them to find ways to engage with the science beyond the structured narrative.

When *x* = "The Smart Ass Fan Boy"

While I attended the Phoenix Comic-Con as a fan and scholar of scifi tv, I did not embrace my fannishness towards science until later that summer, when I attended the first SETIcon in 2010. Created in part to honor the 50th anniversary of Drake's equation, this gathering drew fewer cosplaying fans fetishizing characters and storylines, and more profound examination—from fans—of the scientific underpinnings of both our culture and its Hollywood representations. I was thoroughly engrossed listening to Dr. Drake as he reminisced about the early days searching for radio signals from beyond our planet, and, looking around that ballroom, realized that I was surrounded by a very specific fan demographic.

At SETIcon, *x* equaled a specific type of fan: highly educated, and specifically knowledgeable in the hard sciences. These fans, numbering close to 1000, were primarily fans of science; any affinity for its fictionalization or creative representation was secondary. The few featured guests who weren't practicing scientists were dedicated science enthusiasts, and the balance of panels focused more on speculations of science than the speculative fiction of Hollywood.

A researcher is always looking for that "Ah-ha!" moment—that moment when they may not necessarily solve *x*, but come closer to understanding its nature and parameters. For me, that moment came at a SETIcon panel that was slated to give a "behind the scenes" look at *Battlestar Galactica* from Dr. Grazier's perspective as a science advisor. For me, however, it provided much more than that. As part of the panel, Dr. Grazier invited Phil Plait, author of the popular blog "Bad Astronomy," to join him in order to read a chapter from Grazier's newly published book, *The Science of Battlestar Galactica*, co-authored with Patrick DiJusto (*15*). The chapter in question was called "A Dialogue between a Smartass Fanboy and a Real Scientist, viz: The 'Silica Pathways' into the Cylon Head." It was as if this experiment I was so tentatively launching had crystallized in front of me: a performance of science, modeled after the dialogues of Plato and Aristotle. Of course, as a dramaturg, I had much more grounding in the Greeks' theories on drama; but this chapter, and the nature of it being read as a performance, presented evidence that made it simpler for me to begin to solve for *x* when *x* is a fan of both science and science fiction.

While Grazier contextualized the chapter for his live audience, noting that it had in fact developed through email conversations between himself and his coauthor, the chapter as printed in the book contains no such preamble. Instead, it launches *in medias res* with the Smartass Fanboy's complaint echoing a question I have heard asked at more than one *Battlestar* convention panel: "It's so frustrating that Cylons are supposed to be indistinguishable from Colonials, yet they can shove fiber-optic cabling into their arms and interface with a computer (*16*)!" This immediately positions the speaker within a specific subset of fans who may not necessarily find their agency and subjectivity through the creative outlet of fan fiction or fan art, but rather through the collection and cultivation of ancillary knowledge. The object of their fannishness is the science.

Whether a fan's snarky demeanor is intentional or not, those who focus on science's role in their favorite television shows have a tendency to adopt a position of certainty based on their assumed knowledge of science—often a blend of their experience of science through entertainment, or memories of science classes like those I have shared. In this instance, the fan performs his knowledge of the show in tandem with a basic knowledge of popular science, and how science operates in the fictive world of the series. It may not be the "real" science, but that's the point of intervention for the science advisor. The chapter goes on to debunk many of the misconceptions that fans developed about the science in *Battlestar Galactica*, offering practical scientific parallels to the technology and science employed within the fictive world of the series. Baltar's Cylon detector is the equivalent of a mass spectrometer. The Cylons' silica pathways are essentially fiberoptics. The final five go undetected despite numerous military and athletic physicals because routine medical exams would not necessarily pick up the anomalies.

Throughout *The Science of Battlestar Galactica*, the reader is reminded repeatedly of Adama's maxim, "Context Matters (*17*)." Hearing this chapter read aloud for the first time surrounded by scientists was a very different experience from the second staged reading of it I witnessed at Dragon*Con in 2011. At Dragon*Con, the performance of the chapter was much more metatheatrical. Fans move from a passive position as viewer or consumer to a more active position at conventions when they interact with both the people and the ideas at the heart of their fandom. At Dragon*Con, despite the specific Science, Space and Skeptic programming tracks that help guide fans to defined practices, the sheer number of attendees (46,000 in 2011) the fan variables mean that their knowledge of science is generally more diffuse (*18*). At this venue, when Stephen Cass, a fellow contributor to this anthology, stepped into the role of the Smartass Fanboy, the audience responded less to the specifics of the science, and more to the attitude of the Fanboy. Performance aspects aside, the audience attending the reading at Dragon*Con responded less to the content and more to the context, nodding and laughing in particular when the Fanboy's knowledge base fails him as he says "Dipole moment? I used to know what a dipole moment was (*19*)…"

In the end, while the Cylons' silica pathways may not have a dipole moment, perhaps the fans of *Battlestar Galactica* do. Perhaps they experience both the magnitude of their fannishness and their distance from the object of their fandom through interactions with those that work so diligently to make the science serve the story.

Some Possible Solutions for *x*—or Not

I'm not convinced there is a single, static solution for *x* when it comes to fans of science or science fiction. Perhaps the *x* that is representative of the Fan is more akin to π, in that fans are simultaneously an irrational, transcendental constant that both scientists and Hollywood writers acknowledge as integral to the entertainment industry. Perhaps fans are more variable than that—perhaps the infinite individual iterations of the fan create a kind of feedback loop between the science, the story, and those that engage intertextually, intermedially, and, most of all, experientially with the subject of their own study. Perhaps as fans fetishize the science in science fiction entertainment, and engage with the science and the story in substantively equal measure, all parties have the potential to evolve based on the interaction of the variable components.

References

1. Napoli, P. *Audience Economics: Media Institutions and the Audience Marketplace*; Columbia University Press: New York, 2003; p 5.
2. Hartwell, D. G. The Golden Age of Science Fiction is Twelve. Reprinted in *Speculations on Speculation: Theories of Science Fiction*; Gunn, J., Candelaria, M., Eds.; Scarecrow Press: Lanham, MD, 2005: pp 271–288.
3. Gaudiosi, J. Actress Ashley Eckstein Geeks Out for Girl Gamers at New York Comic Con. *Forbes*, 2011. http://www.forbes.com/sites/johngaudiosi/2011/10/15/actress-ashley-eckstein-geeks-out-for-girl-gamers-at-new-york-comic-con/ (accessed November 1, 2012).

4. Ross, A. Why We Don't Count: The Commodity Audience. In *Logics of Television: Essays in Cultural Criticism*; Mellencamp, P., Ed.; Indiana University Press: Bloomington, IN, 1990: pp 117–132.
5. Brunner, J. *The Dramaturges of Yan*; Ace Books: New York, 1972.
6. Booth, P. *Digital Fandom: New Media Audiences*; P. Lang: New York, 2010; p 11.
7. Jayemanne, D. Microstatecraft: Belonging and Difference in Imagined Communities. *Refractory: A Journal of Entertainment Media*, June 26, 2003. http://refractory.unimelb.edu.au/2003/06/26/microstatecraft-belonging-and-difference-in-imagined-communities-darshana-jayemanne/ (accessed November 12, 2012).
8. Hills, M. Subcultural Celebrity. In *The Cult TV Book: From Star Trek to Dexter, New Approaches to TV Outside the Box*; Abbott, S., Ed.; Soft Skull Press: New York, 2010; pp 233–238.
9. Hills, M. Subcultural Celebrity. In *The Cult TV Book: From Star Trek to Dexter, New Approaches to TV Outside the Box*; Abbott, S., Ed.; Soft Skull Press: New York, 2010; p 236.
10. The Science and Entertainment Exchange. http://www.scienceandentertainmentexchange.org/ (accessed November 12, 2012).
11. Kirby, D. A. *Lab Coats in Hollywood: Science, Scientists, and Cinema*; MIT Press: Cambridge, MA; p 45.
12. Kirby, D. A. Hollywood Knowledge: Communication Between Scientific and Entertainment Culture In *Communication Science in Social Contexts: Strategies for the Future*; Cheng, D., Claessans, M., Gascoigne, T., Metcalfe, J., Schiele, B., Shi, S., Eds.; Springer: New York, 2008; p 185.
13. Kirby, D. A. Hollywood Knowledge: Communication Between Scientific and Entertainment Culture In *Communication Science in Social Contexts: Strategies for the Future*; Cheng, D., Claessans, M., Gascoigne, T., Metcalfe, J., Schiele, B., Shi, S., Eds.; Springer: New York, 2008; p 185.
14. Perkowitz, S. *Hollywood Science: Movies, Science, and the End of the World*; Columbia University Press; New York, 2007; p 214.
15. Di Justo, P.; Grazier, K. *The Science of Battlestar Galactica*; John Wiley & Sons, Inc.: Hoboken, NJ, 2010.
16. Di Justo, P.; Grazier, K. *The Science of Battlestar Galactica*; John Wiley & Sons, Inc.: Hoboken, NJ, 2010; p 211.
17. Di Justo, P.; Grazier, K. *The Science of Battlestar Galactica*; John Wiley & Sons, Inc.: Hoboken, NJ, 2010; p 211.
18. Carroll, D. Dragon*Con 2012 Fact Sheet. http://mediarelations.dragoncon.org/wp-content/uploads/2012/04/DragonConFactSheet2012.pdf (accessed 12 November 2012).
19. Di Justo, P.; Grazier, K. *The Science of Battlestar Galactica*; John Wiley & Sons, Inc.: Hoboken, NJ, 2010; p 45.

Chapter 15

The Chemist as Anti-Hero: Walter White and Sherlock Holmes as Case Studies

Declan Fahy*

School of Communication, American University, 4400 Massachusetts Avenue N.W., Washington, DC 20016-8017
***E-mail: fahy@american.edu**

Compared to chemists in film, chemists in modern television drama are underexamined by scholars, even though the genre is an important processor of images and ideas about culture and society. This critical essay draws on ideas from science communication, media studies and literary studies to examine the representation of chemists and chemistry in the acclaimed television dramas *Breaking Bad* and *Sherlock*. A textual analysis of these shows, chosen as critical case studies, demonstrates that they both portray their chemist protagonists as anti-heroes, who are morally ambivalent. Both shows portray chemistry as uncommon knowledge, which is conducted largely in isolation or in secret. Although the shows represent chemistry as an empirical and experimental science, they demonstrate that the craft of chemistry is not ethically neutral. In *Breaking Bad*, Walter White chooses to stop using his chemistry skills to teach, and subsequently slides into an immoral world of drugs, death, destruction and destabilization. In *Sherlock*, Sherlock Holmes is an amoral, but benign, figure who uses his forensic knowledge to save lives. These representations demonstrate that ethical choices are entwined with the practice of chemistry, and these choices have social consequences.

Chemistry and Contemporary Television Drama

The comprehensive study of chemistry on celluloid, *Reaction!* (2009), identified *Dr. Jekyll and Mr. Hyde* (1931) as the film that established the prototype for subsequent characterizations of chemists in movies. The film codified the image of the chemist as at once good and evil, capable of care and harm, a dualistic portrayal that remains central to how chemistry is presented, and perceived, in Western culture (*1*). This ambivalence is enhanced by the recurrent representation in film of chemists, and other scientists, as ambiguous—even when they are benevolent characters. The chemists are idealistic figures, who become gradually corrupted. They are ambitious, but overlook the social consequences of their science. They are driven to gain new knowledge, but in doing so become willing to violate ethical principles (*2*). Their uncertain social stature is reflected also in their presentation as unusual in dress and behavior—and their scientific wisdom, like that of other scientists, is portrayed as "uncommon knowledge" (*3*).

Yet compared to their counterparts in movies, chemists in television drama have not received a similar amount of scholarly scrutiny. This is a significant shortfall because, as the scholar of television drama Helena Sheehan argues, the genre is a "processor of the collective images and ideas through which we as a society represent ourselves to ourselves and to others" (*4*). The genre reflects and refracts the values and experiences that exist in a particular culture at a specific time. This presentation of the world in dramatic television stories, Sheehan argues, is neither simple nor straightforward, but it does convey particular premises about how the world is organized, about how society is structured and about how culture is shaped by historical forces. The examination of the portrayal of chemistry and chemists in television drama, therefore, can reveal much about the image of chemistry in wider culture.

This critical essay aims to partially redress this overlooked portrayal of chemists, by investigating the patterns of representation of chemists and chemistry in contemporary television drama. Representation is used here as a concept from media studies to examine how the world is portrayed in television drama. Analyzing representations involves the close interrogation of media texts and their social contexts (*5*). This approach informs the central questions of this essay: What images of chemists are presented in television drama? What does television drama reveal about the position of chemistry in society? What contribution does television drama make to the public understanding of chemistry?

The essay offers answers to these questions by analyzing two television dramas, *Breaking Bad* and *Sherlock*, which have been purposefully chosen as critical case studies (*6*). Both shows have protagonists—Walter White and Sherlock Holmes—who are chemists or have expertise in chemistry. Both shows have received popular and critical acclaim, highlighting their value as influential cultural products that warrant critical analysis. The selection of *Breaking Bad*, which was produced in the U.S. by cable network AMC, and *Sherlock*, which was made by the U.K. public service broadcaster, the BBC, allows for the analysis of cross-cultural portrayals of chemists. Additionally, the shows present complex characters and stories that run across multiple episodes, which provide a rich body of material to analyze and allow for multiple patterns of representation to be examined.

Chemists on screen have been classed into categories, such as the evil alchemist, the noble scientist, the foolish scientist, the inhuman researcher, the scientist as adventurer, the mad, bad, dangerous scientist and the helpless scientist (*7*), or as eccentrics or anti-social geeks (*8*). But such categories, even in compound form, provide only simplified shorthand for scientist types. Placing White and Holmes within these broad categories risks draining them of their psychological complexity. Instead, this essay explores these characters using the idea from literary studies of the anti-hero, an approach that allows these complicated characters to be explored, with all their individual contradictions, tensions and quirks, across several episodes and series of the selected shows.

The anti-hero is a central character in a drama "who lacks the qualities of nobility and magnanimity expected of traditional heroes and heroines in romances and epics" (*9*). The anti-hero exhibits amoral and selfish tendencies, in contrast to the hero who emerges victorious after a significant struggle with the ability to bestow benefits on humankind (*10*). The anti-hero is essentially ambiguous and ambivalent in that he or she is neither heroic nor villainous (*11*). Critics have labeled White (*12*) and Holmes (*13*) as anti-heroic, but have not developed this idea to explore what it means for the wider representation of chemistry.

Yet this idea of the anti-hero is useful, because it resonates with chemistry's broad social and cultural position. Examining the field's status in society, the editors of *The Public Image of Chemistry* note that the popular associations of the field range from "poisons, hazards, chemical warfare and environmental pollution to alchemical pseudo-science, sorcery and mad scientists" (*14*). The chemist Luciano Caglioti writes that chemical products, like penicillin, dynamite, insecticides and petrochemicals, are characterized by ambiguity in that they can, at once, improve life and make living more hazardous (*15*). For chemist and popular science writer Pierre Laszlo, these associations contribute to the social impact of the field, as the public suffers from "chemophobia" (*16*). As context, it is important to note that *Breaking Bad* and *Sherlock* are produced, and circulate, in this social and cultural environment, where attitudes to chemistry as a science, and as an industry, are, at best, ambivalent.

Breaking Bad and Chemistry as Uncommon Knowledge

At the beginning of *Breaking Bad*, Walter "Walt" White, played by Bryan Cranston, is a self-described overqualified high school chemistry teacher. After contributing to the work of a Nobel Prize-winning research team early in his career, he has failed to live up to his initial academic promise. He earns $43,700 in his job in Albuquerque, New Mexico, a salary he supplements working in a local car wash. Married to Skyler, with a son, Walter Jr.—joined in season three by daughter, Holly—Walt has watched his former best friend at Cal Tech create a fortune as an industrial chemist and marry his ex-girlfriend. Diagnosed with inoperable lung cancer, Walt decides to provide for his family after his death by turning his prodigious talent as a chemist to something darker and more dangerous: the illicit production of crystal methamphetamine.

He teams up with a former student and small-time drug dealer, Jesse Pinkman, to manufacture a potent brand of meth identified by its distinctive blue color and its extraordinary purity. Walt progresses from "cooking" meth in the back of a dilapidated Winnebago using equipment stolen from his school, to industrial drug production in a secret laboratory with weekly quotas, run by meth kingpin Gustavo "Gus" Fring. Walt's immersion into the gruesome and dehumanizing drug trade provides him with what one critic called "a sort of existential rejuvenation" (*17*). Walt's motivation to provide for his family is gradually surpassed, as the show develops, by his desire to make his mark on the world through his chemistry.

The series features several recurring patterns about the nature of chemistry as a science. Chemistry is portrayed as a form of "uncommon knowledge," which must be earned. For example, sitting on a desk in front of his class, with a poster of the periodic table hanging in front of the chalkboard behind him, Walt discusses the arcane wonders of chemistry that only extended study reveals:

> Mono-alkenes. Di-olefins. Tri-enes. Poly-enes. I mean the nomenclature alone is enough to make your head spin. But when you start to feel overwhelmed—and you will—just keep in mind that one element: carbon. Carbon is at the center of it all. There is no life without carbon. Nowhere that we know of in the universe. Everything that lives…lived…will live…carbon (*18*).

Walt uses his rare knowledge to produce meth. His brand of the drug is so pure because he synthesized it himself using his advanced understanding. Walt trains Jesse—whose first forays into the chemistry of drug production were characterized by his addition of chili powder as a special ingredient—in the advanced chemical skills needed to cook high-grade meth. But that knowledge is not easily acquired. For example, when Jesse first manufactures a batch of blue meth on his own and shows the results to Walt, their conversation shows how technically accomplished Jesse has become, but how much more he needs to learn.

- Jesse: In the end I just went with two reflux condensers. I didn't want to loose track of my pH levels. But I did everything else just like you taught me…super-careful in my amounts and watched the numbers every step of the way. So, what do ya think? It's good, right?
- Walt: What in the hell is this?
- What?
- What? This, this, this is my product, this is my formula, this is mine…I mean look at the diameters here. What did you use for reduction? Don't tell me. Platinum dioxide, right?
- No, mercury aluminum amalgam. The dioxide's too hard to keep wet.
- Alright, well you must have done it wrong, then. Your color is all cloudy so you were struggling with distillation, too.

> No, this is…this is very shoddy work, Pinkman. I'm actually embarrassed for you (*19*).

The show dramatizes Jesse's progressive maturation as a scientist, as he learns hands-on in the lab, under Walt's tutelage (*20*). His training is complete in season four, when he travels to Mexico to show a drug cartel's collection of chemists how to make blue meth. He is initially dismissed as an amateur by the cartel's head chemist. But after telling the cartel to clean up its filthy laboratory, Jesse demonstrates his newfound "uncommon knowledge" by making a batch of the drug with a purity of 96.2 percent (*21*).

Yet, Walt's talent surpasses that of every other chemist. Gale Boetticher, the chemist originally picked to run the industrial meth production lab for Gus, admits that Walt's meth is the product of unique talent. Gale tells kingpin Gus:

> I can guarantee you a purity of 96 percent. I'm proud of that figure. It's a hard-earned figure, 96. However, this other product is 99. Maybe even a touch beyond that…But that last three percent, it may not sound like a lot, but it is. It's tremendous. It's a tremendous gulf (*22*).

Threatened with death by Gus, Walt argues that his specialist knowledge means that his brand of meth cannot be cooked by anyone who just follows a formula. After observing Walt over several weeks, the thug Victor claims to know the special process. He says: "It's called a cook because everything comes down to following a recipe." Walt responds:

> You're not flipping hamburgers here, pal. What happens when you get a bad barrel of precursor, huh? How would you even know it? And what happens in summer, when…when…when the humidity rises and your product goes cloudy (*22*)?

When Gus hints that Walt is proprietorial about his meth formula, Walt emphasizes that his skills are based on his deference to the intellectual integrity of his specialist field. Walt says: "I simply respect the chemistry. The chemistry must be respected" (*23*).

The Craft of Methamphetamine Production

While chemistry is portrayed as "uncommon knowledge," Walt also embodies the idea of the chemist as craftsman. Walt is talented, but he is also industrious and careful. He knows the importance of having the correct equipment. He is excited and astounded by the quality of the lab equipment that Gus procures for him. Viewing the resources for the first time, Walt says: "My God…thorium oxide for a catalyst bed. Look at the size of this reaction vessel. There's gotta be…There's gotta be 1,200 liters" (*23*).

The show represents the routine work of laboratory chemistry. It does not mystify scientific labor. It depicts the work of the field as experimental and empirical. It shows chemistry as a science of synthesis. Walt and Jesse turn up at the same time each morning and use glassware and specialized machinery to create compounds. They regularly take apart and laboriously clean their equipment. Jesse says that their work is art, but Walt corrects him by saying it is just basic chemistry.

The craft of chemistry is presented as a pure process, occurring in an enclosed world, cut off from politics and careerism. Gale tells Walt that he holds a master's in organic chemistry, with a specialty in X-ray crystallography, and that he had been studying for a doctorate at the University of Colorado, but left because he did not enjoy the politicking involved in academic chemistry. He enjoys the purity of laboratory life. He tells Walt:

- I love the lab because it's all still magic, you know, chemistry.
- Walt: It is. It is magic. It still is (*24*).

In one sequence, the two expert craftsmen make meth together and a whimsical song plays as background music. The chemists cook. They play chess in their breaks (*24*). They find beauty and peace and joy in the craft. They have turned their backs on establishment science and science education.

The portrayal of chemistry as socially or ethically problematic is signaled by the fact that the chemistry takes place in secret. Walt first produces his meth in an RV in the middle of the New Mexico desert. Then he makes it in an underground lab, hidden in an industrial plant tied to Los Pollos Hermanos, the fried chicken business run by Gus as a front for his drug network. The secret locations are another manifestation of the ambiguity of chemistry. Walt is engaged in what science communication scholar Peter Weingart and colleagues call a

> private science where the scientist has chosen to leave the community or was excommunicated by it because he or she transgressed the boundaries into forbidden research territory (*25*).

The Ambiguity of Walter White

The dramatic way Walt succumbs to circumstantial economic pressures means the show resonates with contemporary social uncertainties. For the *New York Times*, the show taps into the "sense of economic and social backsliding," as the middle-class White family engages in an "undignified struggle for dignity" (*26*). The same newspaper in another review—with the headline "Better Living Through Chemistry"—says that the dark mood of the series is so in tune with the post-bust economic times that its "extremist misery…feels virtually like reportage" (*27*).

Walt is presented as a deeply ambiguous figure. He faces a complex moral choice when he opts to use his chemistry talents for ill. The same passion he once used to teach class is used to manufacture meth. He breaks bad—the expression from the American southwest that describes a good person doing bad things—for his family, yet allows Skyler to become ever more involved in his illicit activities.

Walt remains loyal to Jesse, who is depicted as a surrogate son to him. But a key moment in Walt's moral slide is the death of Jane, Jesse's girlfriend, who is aware of Walt's secret life. She dies after choking on her own vomit while passed out in a heroin haze alongside Jesse. Walt does not intervene to save her. He watches her die. His motivation is selfish, as he knows Jesse will not break up their drug production partnership. And he knows Jane will never reveal his unlawful work.

But in a later episode, Walt, while unknowingly drugged, alludes to his suppressed guilt that has resulted from his corruption. He tells Jesse: "If I had just lived right up until that moment and not one second more. That would have been perfect." (The moment he is referring to is not explicitly stated.) In the same scene, Walt reveals that he is aware of the immorality of his work and that his continuation beyond the point when he had earned enough money meant he crossed an ethical boundary. He says:

> I missed it. It was some perfect moment and it passed me right by. I had to have enough to leave them. That was the whole point. None of this…None of this makes any sense if I didn't have enough. But it had to be before she found out, Skyler. It had to be before that (*28*).

Walt's ambiguity is linked to an ambivalent view of science. This is illustrated through his choice of pseudonym in the drug trade: Heisenberg. The German Nobel-winning physicist is best known for his Uncertainty Principle that says it is possible to know either the velocity or location of an electron, but not both. Heisenberg has remained controversial also for his atomic research in World War II. There are subtle references to the development of atomic weapons at Los Alamos in New Mexico. Walt meets Jesse's street dealer friends in the atomic museum in Albuquerque. The ambivalence surrounding the Manhattan Project and its scientific work is bound up subtly with Walt.

Through his scientific work, Walt dramatizes a fundamental philosophical tension about the social role of chemistry. Is the craft of the chemist morally neutral? Can purity of craft be separated from the social consequences that flow from laboratory work? At a time when they are crucial cogs in Gus's meth empire, Walt tells Jesse: "You are not a murderer. I am not and you are not. It's as simple as that" (*29*). Yet by this time, several victims had died as Walt and Jesse became more embedded in their dark trade. Walt killed two rival dealers, Emilio and Krazy-8, choking one of them to death with a bicycle lock. Jesse's friend Combo was shot while selling blue meth on a street corner in another gang's territory. By the end of season four, Walt was responsible for at least nine deaths (*30*)—fatalities that resulted from his complex moral choices.

The show makes clear the social consequences of illicit chemistry. After Jane dies, her father is so distracted by grief in his work as an air-traffic controller that he fails to prevent a mid-air plane collision over Albuquerque. Debris and body parts rain down on the city and on Walt's home, a metaphor for the social carnage wrought by his work.

The cold-blooded murder of Gale is similarly symbolic. After Gale learns to manufacture blue meth, Walt believes that Gus is planning to kill him and

Jesse. So to survive—and to continue to be the one holding "uncommon knowledge"—Walt tells Jesse to kill Gale. Jesse reluctantly does so. Gale—the herbal tea-drinking, karaoke-performing, vegan libertarian—loves the pristine isolation of the laboratory, but his execution shows that illicit chemistry has social costs.

Walt comes to realize the progressive corruption caused by his ambition. As his life is threatened by his associations with the drug trade, he is asked by his wife if he wants to go to the police to confess and get protection. But Walt tells her: "I am not *in* danger, Skyler. I *am* the danger" (*31*).

Holmes's Hidden Chemistry

Moving from Albuquerque to London, the portrayal of the eponymous detective in Sherlock is an illuminating point of comparison with White. Arthur Conan Doyle in his original novels and stories documents Holmes's chemistry credentials. In *A Study in Scarlet*, Holmes first meets Dr. Watson in a "chemical laboratory." Watson as narrator describes the scene:

> This was a lofty chamber, lined and littered with countless bottles. Broad, low tables were scattered about, which bristled with retorts, test-tubes, and little Bunsen lamps, with their blue flickering flames (*32*).

Holmes is presented in this initial encounter holding a test tube as he explains excitedly that he has developed a novel chemical test to identify human blood. Holmes says: "I've found it! I've found it…I have found a re-agent which is precipitated by hoemoglobin [sic], and by nothing else." Later in the same story, Watson evaluates Holmes's areas of specialized knowledge. Watson lists as "nil" Holmes's understanding of literature, philosophy and astronomy. Holmes's anatomical knowledge is accurate "but unsystematic." His knowledge of poisons is good and he can distinguish different types of soil. Crucially, Watson notes that Holmes's knowledge of chemistry is "profound" (*32*). With these and other depictions of Holmes as scientific investigator, Conan Doyle has done more than any other writer to present the social value of forensic science (*33*). Forensics and Holmes have become so aligned that Holmes is an archetype of a particular onscreen scientist—the forensic detective (*34*).

Sherlock updates Conan Doyle's Victorian setting to contemporary England. As played by Benedict Cumberbatch, the consulting detective is now a hyper-connected twenty-first century digital citizen. He continues to be fascinated by forensics. He is first shown in St. Bartholomew's Hospital beating a corpse with a riding whip, to evaluate the post-mortem formation of bruises. Critic Elizabeth Renzetti in *The Globe and Mail* writes: "Fascinated by forensics and chemistry, Holmes could have been a character on CSI Baker Street" (*35*).

But there is a telling omission in this representation of Holmes and his abilities as a forensic chemist. Nowhere is Holmes identified as a chemistry expert. Nor is his advanced knowledge of the science explicitly identified. His knowledge and love for chemistry is instead portrayed as implicit and allusive. Throughout the series, he is featured in repeated shots using a microscope in the laboratory. He is known for his ability to distinguish between more than 100 varieties of cigarette ash. His kitchen table is usually covered with laboratory glassware. (His science takes places in secret.) His love of chemistry is symbolized most clearly by the only decoration that appears on his bedroom wall: a colorful poster of the periodic table.

Sherlock as Sociopath

Sherlock presents Holmes as an ambiguous figure: impatient, anti-social, friendless, arrogant and cruel with a pronounced lack of empathy. Cumberbatch calls him "this character of the night, this sociopathic, slightly autistic, slightly anarchic, maverick, odd antihero" (*36*). Holmes is portrayed as amoral. For example, when asked by the police to help investigate a bizarre set of apparent suicides that have been linked by a mysterious note left by one of the victims, Holmes says: "Brilliant! Yes! Four serial suicides and now a note. Oh, it's Christmas." Holmes views the multiple killings as an exciting challenge for his mental capacities. When a clearly delighted Sherlock leaves his apartment to investigate the suicides, his landlady tells him his enthusiasm is not decent. Sherlock responds: "Who cares about decent? The game, Mrs. Hudson, is on" (*37*)!

This lack of empathy is noted by the police. When Dr. John Watson, recast in the series as a British army doctor who has returned traumatized from the war in Afghanistan, first accompanies Holmes to a crime scene, Sergeant Sally Donovan warns him to stay away from the detective. She says:

> You know why he's here? He's not paid or anything. He likes it. He gets off on it. The weirder the crime, the more he gets off. And you know what? One day just showing up won't be enough. One day we'll be standing around a body and Sherlock Holmes will be the one that put it there...he's a psychopath. Psychopaths get bored (*37*).

When he is once again called a psychopath by another police forensic expert, Holmes says: "I'm not a psychopath...I'm a high-functioning sociopath. Do your research" (*37*).

As Holmes tries to outwit his nemesis Jim Moriarty, who has threatened to leave corpses around London, Watson confronts his friend's amorality. He asks:

- There are lives at stake, Sherlock. Actual human lives. Just, just so I know: do you care about that at all?
- Will caring about them help save them?
- Nope.
- Then I'll continue not to make that mistake. Don't make people into heroes, John. Heroes don't exist and if they did I wouldn't be one of them (*38*).

But despite his ambivalence, Holmes is liked and respected by Watson. At the end of season two, after his climactic confrontation with Moriarty, Holmes is forced to jump from a rooftop to his apparent death. The closing scene of the series highlights Holmes's humanity. Watson stands over what he believes to be his friend's grave. He is watched, unknowingly, by Holmes, who has somehow manufactured his own death. The detective hears Watson say: "There's just one more thing, one more thing, one more miracle, Sherlock, for me. Don't...be...dead" (*39*).

Holmes's Uncommon Knowledge

Holmes's method of reasoning is obscured in Conan Doyle's stories. As the literary critic Steven Knight notes:

> The contexts of medical science, the chemistry and the exhaustive knowledge of crime are only gestured at, and we are actually shown no more than a special rational process (*40*).

This methodology continues to be largely hidden in "Sherlock." Examining physical evidence, Holmes rapidly forms conclusions that dazzle Watson (himself, a trained physician) with their insight. Elaborating on this point, Cumberbatch tells *The Times*: "You can have scientists on the ground who analyze forensics, but they won't be able to take that leap which takes them to a conclusion...It takes [Holmes's] leap of imagination as well as his knowledge to connect the dots" (*41*). For Holmes, the physical evidence is just the starting point. His "uncommon knowledge" lies in his powers of interpretation. With this portrayal, Sherlock presents scientific insight as the result of a process that is closer to artistic creation than experimental science. This is a recurring means of representing the work of the scientist, especially to audiences who may be unfamiliar with the process of scientific creativity. Its depiction of the imaginative process of science is informed by ideas from the Romantic movement of the late eighteenth and nineteenth century that saw imaginative creativity as the preserve of visionary artists (*42*).

But advances in production technology have allowed the creators of *Sherlock* to bring to the surface more elements of Holmes's uncommon reasoning process. As Holmes examines physical evidence, words flash up on screen that correspond to pieces of data that he identifies as significant. For example, when Holmes examines the corpse of a woman who apparently took her own life, labels are projected onto her clothing to reveal her history and circumstances of her death. Holmes identifies from her body, clothing and jewelry that she was left-handed and married unhappily for more than 10 years. When Holmes meets a man in Buckingham Palace, he reads the man's life history in an instant from his clothing and appearance. The viewer sees the evidence, sees what Holmes sees—but the conclusion can only be provided by Holmes. The conclusions remain the result of flashes of individual genius.

Holmes's "uncommon knowledge" is symbolized also by what reviewers interpret as the character's unusual appearance. *The Times* says this Holmes looks "as odd as you'd expect The Cleverest Man in the World to look. Eyes white, skin like china clay, and a voice like someone smoking a cigar inside a grand piano" (*43*). For *The Daily Telegraph*, Cumberbatch, with his "shock of blackened hair, his parchment-pale skin and liquid eyes [takes] on a translucent quality that made him appear both sickly and mesmerizingly other-worldly" (*44*).

The philosopher John Gray says this version of Holmes represents a conflicted modern attitude to science. He argues that the detective embodies reason in an age when systems of rationality—from security software to the mathematical formulae used by hedge funds—"have proved to be dangerously unreliable." For Gray, Holmes symbolizes the power of the mind, at a time when the "idea that intellect alone can be a guide in life is weaker than it has been for many years" (*45*).

Dramatizing the Dilemmas and Consequences of Chemistry

Although the patterns of representation in both shows are complex and sometimes contradictory, common themes about the public image of chemists and chemistry can be discerned. Chemistry is represented as an empirical and experimental field. Chemistry is portrayed as a type of "uncommon knowledge" held by particular experts. But the precise nature of this knowledge is depicted differently. The knowledge in *Breaking Bad* is gained through a process of experimentation and instruction, based on knowledge of the fundamentals of the field, although each chemist has particular talents and skills that distinguish their work. The special understanding in *Sherlock*, by contrast, is gained through a process of imaginative interpretations of physical evidence, a largely hidden process that is portrayed as unique to Holmes.

Chemists are represented in both shows as ambiguous figures. Neither White nor Holmes possesses the traditional heroic virtues. Instead, they are depicted as anti-heroes, who each exhibit various degrees of amorality, immorality and selfishness. White and Holmes see the practice of their work as ethically neutral, a value-free demonstration of their intellectual prowess. Their chemistry is largely conducted in secret, in private and in isolation. These common patterns mean the television characters of White and Holmes conform to, and reinforce, the wider cultural portrayal of chemists—and, by extension, chemistry—as socially and ethically problematic.

But the moral ambivalence of White and Holmes also allows for the dramatic portrayal of ethical issues in chemistry. White and Holmes face moral choices of varied complexity about how to apply their scientific skills. Although presented as amoral, Holmes's work improves society by causing the capture of criminals. White chooses to stop using his knowledge to teach. He then becomes progressively corrupted, as he slides into an immoral world of duplicity, deceit and death. By dramatizing the dilemmas of White and Holmes, the shows demonstrate that ethical choices are entwined with the craft of chemistry. And these choices, in turn, have social consequences. By circulating these ideas through culture, the shows can contribute to a deeper public understanding of chemistry—beyond the realistic depiction of laboratory chemistry, or the romantic portrayal of scientific creativity.

References

1. Griep, M.; Mikasen, M. *ReAction! Chemistry in the Movies*; Oxford University Press: Oxford, 2009.
2. Weingart, P.; Muhl, C.; Pansegrau, P. *Public Understanding Sci.* **2003**, *12*, 279–287.
3. Nisbet, M. C.; Scheufele, D.; Shanahan, J.; Moy, P.; Brossard, D.; Lewenstein, B. V. *Commun. Res.* **2002**, *29*, 584–608.
4. Sheehan, H. In *Mapping Irish Media: Critical Explorations*; Horgan, J., O'Connor, B., Sheehan, H., Eds.; University College Dublin Press: Dublin, Ireland, 2007; p 142.
5. Horgan, J.; O'Connor, B.; Sheehan, H. In *Mapping Irish Media: Critical Explorations*; Horgan, J., O'Connor, B., Sheehan, H., Eds.; University College Dublin Press: Dublin, Ireland, 2007; pp 1−13.
6. Creswell, J. W. *Qualitative Inquiry and Research Design: Choosing Among Five Traditions*; Sage: Thousand Oaks, CA, 1998; pp 118−120.
7. Haynes, R. *Public Understanding Sci.* **2003**, *12*, 243–253.
8. Nisbet, M. C.; Scheufele, D.; Shanahan, J.; Moy, P.; Brossard, D.; Lewenstein, B. V. *Commun. Res.* **2002**, *29*, 584–608.
9. Baldick, C., Ed.; *The Oxford Dictionary of Literary Terms*, 3rd ed.; Oxford University Press: Oxford, 2008; p 16.
10. Campbell, J. *The Hero with a Thousand Faces*; Bollingen Series XVII; Pantheon Books: New York, 1961; pp 19−30.
11. (a) Furst, L. R. *Stud. Lit. Imagination* **1976**, *9* (1), 53−67. (b) Lamont, R. C. *Stud. Lit. Imagination* **1976**, *9* (1), 1−22.
12. Koepsell, D. R.; Arp, R. Popular Culture and Philosophy. In *Breaking Bad and Philosophy: Badder Living Through Chemistry*; Koepsell, D. R, Arp, R., Eds.; Open Court: Chicago, IL, 2012; Vol. 67, pp vii−ix.
13. Kelly, S. *Guardian.co.uk*, [online], January 21, 2012. http://www.guardian.co.uk/commentisfree/2012/jan/21/sherlock-holmes-tv-antihero, (accessed December 1, 2012).
14. Schummer, J.; Bensaude-Vincent, B; Van Tiggelen, B. In *The Public Image of Chemistry*; Schummer, J., Bensaude-Vincent, B., Van Tiggelen, B., Eds.; World Scientific Publishing: London, 2007; pp 1−6.
15. Caglioti, L. *The Two Faces of Chemistry*; The MIT Press: Boston, MA, 1985.
16. Schummer, J.; Bensaude-Vincent, B; Van Tiggelen, B. In *The Public Image of Chemistry*; Schummer, J., Bensaude-Vincent, B., Van Tiggelen, B., Eds.; World Scientific Publishing: London, 2007; pp 329−365.
17. Bellafante, G. *New York Times*, March 6, 2009, p C1.
18. *Breaking Bad. Peekaboo*; Season 2, Episode 6; TV program, AMC, April 12, 2009.

19. *Breaking Bad. Green Light*; Season 3, Episode 4; TV program, AMC, April 11, 2010.
20. Kadonaga, L. Popular Culture and Philosophy. In *Breaking Bad and Philosophy: Badder Living Through Chemistry*; Koepsell, D. R, Arp, R., Eds.; Open Court: Chicago, IL, 2012; Vol. 67, pp 181−189.
21. *Breaking Bad. Salud*; Season 4, Episode 10; TV program, AMC, September 18, 2011.
22. *Breaking Bad. Box Cutter*; Season 4, Episode 1; TV program, AMC, July 17, 2011.
23. *Breaking Bad. Mas*; Season 3, Episode 5; TV program, AMC, April 18, 2010.
24. *Breaking Bad. Sunset*; Season 3, Episode 6; TV program, AMC, April 25, 2010.
25. Weingart, P.; Muhl, C.; Pansegrau, P. *Public Understanding Sci.* **2003**, *12*, 285.
26. Stanley, A. *New York Times*, January 18, 2008, p E29.
27. Bellafante, G. *New York Times*, March 6, 2009, p C1.
28. *Breaking Bad. Fly*; Season 3, Episode 10; TV program, AMC, May 23, 2010.
29. *Breaking Bad. Half Measures*; Season 3, Episode 12; TV program, AMC, June 6, 2010.
30. Koepsell, D. R; Gonzalez, V. Popular Culture and Philosophy. In *Breaking Bad and Philosophy: Badder Living Through Chemistry*; Koepsell, D. R, Arp, R., Eds.; Open Court: Chicago, IL, 2012; Vol. 67, pp 3−14.
31. *Breaking Bad. Cornered*; Season 4, Episode 6; TV program, AMC, August 21, 2011.
32. Conan Doyle, A. *A Study in Scarlet*; Bantam Classic: New York, 2003; pp1−120.
33. Snyder, L. J. *Endeavour* **2004**, *28*, 104–108.
34. Griep, M.; Mikasen, M. *ReAction! Chemistry in the Movies*; Oxford University Press: Oxford, 2009.
35. Renzetti, E. *The Globe and Mail*, July 31, 2010, p R3.
36. Wells, D. *The Times*, July 24, 2010, pp 6−7.
37. *Sherlock. A Study in Pink*; Season 1, Episode 1; TV program, BBC, October 24, 2010.
38. *Sherlock. The Great Game*; Season 1, Episode 3; TV program, BBC, November 7, 2010.
39. *Sherlock. The Reichenbach Fall*; Season 2, Episode 3; TV program, BBC, January 15, 2012.
40. Knight, S. *Form and Ideology in Crime Fiction*; Indiana University Press: Bloomington, IN, 1985, p 86.
41. Derakhshani, T. *The Philadelphia Inquirer* [online], May 6, 2012. http://articles.philly.com/2012-05-07/news/31598277_1_holmes-and-moriarty-irene-adler-mark-gatiss (accessed October 25, 2012).
42. Jones, R. A. *Public Understanding Sci.* **1998**, *7*, 135–147.
43. Moran, C. *The Times*, July 31, 2010, p 14, Review.
44. Davies, S. *The Daily Telegraph*, July 26, 2010, p 26.

45. BBC. A Point of View. Sherlock Holmes and the Romance of Reason by John Gray; Podcast, August 19, 2012. http://www.bbc.co.uk/programmes/b01lv7z1 (accessed October 17, 2012).

Chapter 16

Using the Space Program from Mercury to Apollo as Portrayed in the Movies *The Right Stuff* and *Apollo 13* and in the Mini-Series *From the Earth to the Moon* as a Teaching Tool

James G. Goll*

Department of Chemistry, Geoscience and Physics Edgewood College, Madison, Wisconsin 53711
***E-mail: jgoll@edgewood.edu**

This chapter will examine how popular media related to the space program can be used to demonstrate the nature and motivation of scientific inquiry and science concepts. For over fifty years, the space program has inspired students of science and engineering. The United States manned space program from project Mercury to Apollo is the subject of two movies, *The Right Stuff* and *Apollo 13*, and the mini-series *From the Earth to the Moon*. Many documentary style television productions are available to supplement these movies and the miniseries. These documentaries provide recollections from astronauts, flight controllers, and flight directors. *Moonshot* and *To the Moon* chronicle the manned space program during the Mercury, Gemini, and Apollo programs. The documentary *To the Edge and Back* inspired the movie, *Apollo 13*. The Science Channel's *Moon Machines* shows many behind-the-scenes people who made the trips to the Moon possible. The History Channel used the manned space program as a subject for several of its series: *Man, Moment, and Machine*, *Modern Marvels*, *20th Century with Mike Wallace*, and *Failure is Not an Option*.

Introduction

The movie *Apollo 13* (1995) and the follow-up miniseries *From the Earth to the Moon* (1998), have been used to teach many chemistry concepts (*1–5*). *Apollo 13* has been shown to produce both high impact upon audiences and high utility for instructors when used as a teaching tool (*6*). The movie *The Right Stuff* (1983) provides insight into the Mercury program (*7*). Many popular media documentaries about the space program, such as public television's *To the Edge and Back* (1994) and *To the Moon* (1999), Turner Television's *Moonshot* (1994), and the Science Channel's *Moon Machines* (2010) show behind-the-scenes people who made the trips to the Moon possible (*8–11*). The History Channel has several programs *Man, Moment, and Machine* (2006), *Modern Marvels* (2001), *20th Century with Mike Wallace Crisis in Space, The Real Story of Apollo 13* (1998), and *Failure is Not an Option* (2003) that contain recollections of the astronauts, flight controllers, and flight directors from Mercury to Apollo (*12–15*). Examples of teaching tools for scientific inquiry and chemistry concepts from these sources are provided.

Sending Astronauts to the Moon and Returning Safely to the Earth

How were people sent to the Moon? *From The Earth to the Moon* has insight into developing a problem solving approach to sending humans to the Moon. The method for going to the Moon needed to be decided between the early front runners, direct ascent and earth orbit rendezvous, and the controversial lunar orbit rendezvous that required two spacecraft to link up in lunar orbit. John Houbolt favored lunar orbital rendezvous and was able to convince the managers and engineers that this method was the most feasible way to send humans to the surface of the Moon. He successfully argued that only a small spacecraft could be landed safely on the Moon. The risk of lunar orbit rendezvous was much less than landing a much larger spacecraft required by either direct ascent or earth orbit rendezvous methods.

Once the method was decided, new materials for rocket engines, spacecraft, and spacesuits were developed, software and computers were designed, and a communication system devised. The challenges to be overcome started with getting an astronaut in orbit, developing extra vehicular activities (EVA), rendezvous, docking, and long duration flight.

A series of spectacular rocket failures made the challenge of sending a human into space seem too risky. Careful analysis of the causes of failures and motivation from the cold war competition with the Soviet Union were required to overcome this problem.

The first spacewalk by astronaut Ed White indicated that working in space would be readily managed. Rendezvous, docking, and long duration flight were then demonstrated in turn. Longer extra vehicular activities requiring astronaut to work in space was the next challenge. The next three astronauts to do EVAs worked to the point of exhaustion. Newton's Third Law of Motion, for every action there is an equal and opposite reaction, caused astronauts to be unable to maintain their position when working in space. Improved footholds and handholds to hold the astronaut in place when working were added to the spacecraft. Better training in water tanks to simulate the conditions in space was implemented. This allowed astronaut Buzz Aldrin on Gemini 12 to work effectively.

Practice in Problem Solving

One of the great lessons that science students should glean from *Apollo 13* is that practicing problem solving is useful. Practice hones skills both for problems that can be anticipated and quickly solved and for problems that were not anticipated. While practice is important, integrating knowledge and critical thinking are often required to answer the complex problems.

The astronauts and the ground crew underwent extensive training in problem solving. Problems developed by the simulation staff were possibilities that mission control members and the astronauts would face. The astronauts and flight controllers would give their best effort to solve the problems. The astronauts and mission control team would then analyze the solution that was devised and implemented or the failure to develop a solution to see what improvements could be made. Data was evaluated to be sure correct action could be taken. This was an arduous process that continued until nothing more could be done and all possible strategies were analyzed.

Gene Kranz described the debriefings as ruthlessly honest. These debriefings challenged the confidence of flight controllers and their ability to think critically and quickly. The simulations also required the astronauts and the controllers to reflect on their actions. This process developed trust, teamwork, toughness, and confidence. These attributes were instilled as part of the culture of mission control and were those valued by the flight directors. Gene Kranz clearly states this in many documentary-style programs such as *Failure is not an Option.*

Teaching Scientific Methodology and Content

The flight of Apollo 13 started off well. Liftoff has been described like shooting a bullet from a gun. The rocket goes 25000 mi/h and the astronauts feel four times the force of earth's normal gravity as they travel through the atmosphere. This is the amount of gravitational force one may experience during a roller coaster ride. When the rocket undergoes staging, the crew flies forward in their seats because of inertia. This is called the "little jolt" in the movie. The crew is securely fastened in their seats so they move around as little as possible. The problem of an engine cut-off on the second stage rocket then appeared. The problem was quickly evaluated as the other engines compensated and the mission continued.

During the mission, a routine procedure to stir the oxygen and hydrogen tanks used to produce electricity and water resulted in an explosion that placed the astronauts' lives in danger. The tanks required periodic stirring since the supercritical fluids striated due to the motion of the spacecraft, just as the earth's atmosphere is striated due to gravity. Homogenizing the tanks' contents was required to get an accurate reading on the amount of material remaining. One of the hydrogen tanks was reading low and stirring would allow for a better reading. The movie shows the switch turned on followed by the rupture of the oxygen tank and the subsequent explosion that blew away the side of the spacecraft. The question may be asked during a class about why such a depiction is in the movie. How is it known that this is what actually happened? The answer can start with a discussion about the observation of the loss of electrical power and the cause of this apparent power loss.

At Mission Control in Houston, the first hypothesis was instrument failure, since the tanks had been stirred earlier in flight and so many independent systems should not show errors at the same time. Subsequent reports from the astronauts eliminated this possibility.

Aboard the spacecraft, initial hypotheses were also formed. The mission commander, Jim Lovell, suspected a prank by fellow astronaut Fred Haise, who would press the cabin repressurization valve, resulting in a loud bang. A quick glance at Haise dispelled this possibility. Haise then added his observation of the bending of the tunnel connecting the command module and the lunar lander and a new hypothesis was formed: the thin-skinned lunar module may have been hit by a meteor. The attempt to seal the tunnel connecting the lunar and command module to prevent the loss of atmosphere was unsuccessful. The continuation of life aboard the spacecraft proved that this explanation was not possible. Instruments indicated dropping levels of oxygen and fuel cells becoming inoperable. The computer restarted and communication was disrupted. The spacecraft was buffeting and difficult to control.

Lovell then made an observation that a gas was venting from the ship into space. This could be correlated with the dropping levels of oxygen and could also explain the buffeting of the spacecraft. At this time, the working hypothesis was that a meteor hit the ship, causing an oxygen tank to leak, which in turn caused the damage to the fuel cells. Near the end of the flight, the service module that contained the oxygen tanks could be seen. The observation indicated more extensive damage, which had to be explained. At the end of the movie, the cause of the explosion is revealed when Tom Hanks, in a voice-over, tells about a damaged coil.

How did they reach this conclusion? An investigation of the history of the oxygen tank revealed that the tank had been dropped about two inches. This caused damage to a metal tube used to drain the tank. The tank was filled during a test to determine if the liquid oxygen would not leak. When the tank was drained after the test, it did not drain quickly. Heaters inside the tank were turned on to aid in draining the tank. It was discovered from the specifications of the oxygen tank that the heater had incorrect wiring, so the temperature in the tank was eventually raised to several hundreds of degrees, conditions that allowed some of the insulation to crack and burn exposing the wiring. The exposed wires then created a spark that set the remaining insulation on fire and resulted in the sudden conversion of the oxygen in the tank to a gas. The pressure caused by the oxygen gas caused the tank to rupture and blow the side off of the spacecraft. This scenario was recreated in a laboratory and the result was the rupture of the oxygen tank.

Throughout *Apollo 13*, scenes depict the ongoing problems to be solved and the management of resources. The crew needed to change the course of the spacecraft to return to earth. The challenge of piloting the connected spacecraft with a different center of gravity caused Jim Lovell to comment that he needed to learn to fly all over again. Fortunately some testing was done earlier during Apollo 9 and Lovell was a quick study and was able to adjust the course of the spacecraft.

Another problem the movie shows is rising carbon dioxide inside the spacecraft. If the amount of carbon dioxide becomes too great in the blood, a person goes into a state of narcosis. This is a condition of confusion accompanied by possible tremors, convulsions, and, ultimately, a coma. Eventually, it is possible to asphyxiate. To solve this problem they needed to sequester much of the carbon dioxide. Apollo 13 used lithium hydroxide to remove carbon dioxide from the air in the spacecraft. However, the number of round lithium hydroxide containers was inadequate for the three astronauts. Scientist and engineers on earth needed to devise a method to adapt the square lithium hydroxide cartridges used in command module for use in the lunar module.

The amount of available energy on Apollo 13 once the fuel cells were no longer operating was limited to that stored in batteries. To conserve energy, almost everything in the spacecraft was turned off. This caused the spacecraft to become cold and damp. The amount of water available was also reduced when the fuel cells no longer operated. The lack of water and the cold conditions were causes of an infection contracted by Fred Haise.

Chemistry comes into play in *From the Earth to the Moon.* During one of the final tests of Apollo 1, a spark in the spacecraft started a fire that killed astronauts Gus Grissom, Ed White, and Roger Chaffee. The atmosphere inside the spacecraft was composed of oxygen at a pressure of about 850 torr, well above normal atmospheric pressure and the pressure of oxygen in the atmosphere. A pure oxygen atmosphere was used to minimize the amount of atmosphere needed to be taken into space. It also removed the complication of taking nitrogen out of the atmosphere in space. This was needed to prevent nitrogen from forming gas bubbles in astronaut's blood that cause the bends. The pressure was elevated to ensure that the spacecraft was not leaking any of the internal atmosphere.

The increased oxygen concentration led to the rapid rate of the combustion of Velcro. Velcro was on the walls and was used to keep items from floating freely in the microgravity environment of space. In air, Velcro does not burn easily. In the oxygen rich environment of the spacecraft, Velcro burned very rapidly, almost explosively. In *From the Earth to the Moon*, the effect of oxygen concentration on the rate of Velcro combustion is dramatically demonstrated. This is a harsh but effective lesson on conditions causing changes in reactivity. It also serves as a reminder that safety must anticipate what might happen. Failure to anticipate the possibility of a fire during a high- pressure leak check led to disaster. Investigation of the fire revealed other issues such as careless work and a high-pressure schedule that reduced safety. A culture of accountability and safety was reinforced afterward.

The first flight to the Moon, Apollo 8, is chronicled in *From the Earth to the Moon.* A scene shows Sue Borman, the mission commander's wife, telling Chris

Kraft, a senior NASA administrator, about her concerns about the mission to put her husband in lunar orbit. Kraft assures her that the simplest engine system is used.

A hypergolic mixture of hydrazine and dinitrogen tetroxide that required no pumps and would combust without any outside ignition source was used in the engine. The chemistry of this reaction is interesting. Dinitrogen tetroxide is in equilibrium with nitrogen dioxide and a small amount of nitrosyl nitrate. The Lewis structure of nitrogen dioxide does not obey the octet rule and this is correlated with its reactivity. This shows a structure–reactivity relationship that is so important in explaining reactivity.

The relative weakness of a nitrogen-nitrogen single bond caused by the lone pair repulsion on the nitrogen atoms of molecules like hydrazine is typically taught during an inorganic chemistry course. The weak nitrogen-nitrogen bond is the reason for the low activation energy; thus the reaction occurs when the reactants are mixed. Both the fuel and the oxidant are highly toxic and great care must be taken when handling them. They are both very corrosive, and therefore no engine using them, including those used to land on and take off from the moon, can ever be tested before they are used.

Comparisons can be made between the nature of the rocket fuel and oxidant used for lift-off, with the engines used in space and for the small maneuvering thrusters. The first stage used a kerosene-liquid oxygen system for maximum thrust, and the second and third stages used hydrogen with oxygen. Since the spacecraft is still near the Earth, a more complex system requiring an ignition source was usable since the astronauts would be able to return to the surface of the Earth.

Apollo 12 has several scenes depicted in *From the Earth to the Moon* that can be used as teaching tools. It was raining on the day of the launch of Apollo 12, the second mission to the Moon. Thirty-seven seconds after liftoff, the rocket was struck by lightning. What would cause the rocket to be struck by lightning since no lightning was observed in the area? The rocket, as it ascended, left a trail of gaseous ions created by the high temperature combustion of the kerosene rocket fuel. These gaseous ions, left behind in the contrail, made a good conductive pathway for the lightning to travel from the rocket to the ground. The lightning strike disrupted the telemetry data required to guide the spacecraft safely to the Moon. The telemetry data needed to be restored or the mission would be aborted. Fortunately, the curiosity of flight controller John Aaron paid off from an incident that occurred the previous year. A series of unusual readings occurred after a rocket had been struck by lightning. John Aaron remembered the pattern of readings and remembered a switch that could be changed to restore the required telemetry data. An instruction was given to the crew of Apollo 12 to change the Signal Conditioning Equipment to an auxiliary setting (set SCE to AUX), as this instruction provided an alternative pathway that restored telemetry data about the spacecraft to mission control.

A major goal of the Apollo 12 mission was to demonstrate that an extremely accurate landing could be accomplished. This goal was required both for crew safety and planning scientific investigations on the Moon. Apollo 11, the first mission to land on the Moon, missed its landing site by about four miles. Accurate landings would be required if the Moon was to be explored relatively safely and systematically. Apollo 12 had a landing site near the unmanned lunar lander, Surveyor 3. An objective of the mission was to retrieve parts from Surveyor 3 to determine the condition of the spacecraft after years on the Moon. The lunar lander needed to be within walking distance of Surveyor 3 for the mission to be totally successful. The Apollo 12 lunar lander touched down 535 feet or approximately one-tenth of a mile from Surveyor 3. In a classroom, missing a target by 535 feet is a large error, but if the target is a landing site on the Moon, approximately 1.25 billion feet away from the Earth, it is very close. This point can be used to talk about the percent error and its difference from absolute error. The absolute error is the same but the percent error going to the Moon is very small.

A safety point can be made based on Apollo 12. Astronaut Alan Bean was painfully reminded that instructions on the checklist must be followed. He did not remove a camera from a window mount during reentry and it hit him in the head when it broke free from its mount on the descent to the ocean.

The Apollo 13 mission had a nuclear problem that is not shown in the movie but is brought out in *From the Earth to the Moon.* A small nuclear power source that contained 3.9 kg of plutonium 238, an alpha emitter with a half-life of 87.7 years, was in the lunar module, intended to be placed on the Moon to power experiments. Fortunately, the casing for the nuclear fuel was designed to survive an explosion of the rocket. The casing containing the plutonium, like the astronauts, apparently survived reentry since increases in radioactivity were not detected when it came to rest with what remained of the lunar module in the Tonga Trench south of Fiji, approximately 6-9 kilometers underwater.

Astronauts as Scientists

After the goal of sending a man to the Moon and returning him safely to Earth was met, why should any other mission be undertaken? The major political and technological objectives were achieved. Why should humans be sent to explore the Moon, a dangerous and costly endeavor, rather than unmanned probes? Could robots in space collect enough information to find out everything we want to know about the Moon? NASA asked these questions in deciding if it was worth the risk of sending humans to the Moon for further scientific exploration.

Few astronauts had any significant scientific background, since most of them were selected based on their ability as test pilots used to living with the hazards of space flight. In *To the Moon,* Walt Cummingham stated the position of several astronauts as being anti-doctor and anti-scientist. Would a random sampling of rocks and dust from the lunar surface done by robots be better than relying on untrained observations of astronauts? The decision was made to educate the astronauts to become scientific observers for geological surveying and sampling.

From the Earth to the Moon shows details of how the astronauts were educated to be scientific observers. In *To the Moon*, astronaut Jack Schmitt, a geologist with his doctorate, stated that the test pilot astronauts were bored by traditional geology education. He realized that a more engaging, hands-on approach was needed which emphasized active learning fieldwork rather than classroom lectures. The astronauts were taught to seek out unusual lunar samples that could provide clues to the history of the Moon and the Solar System by learning to collect samples and describe their context during field trips. Intellectually stimulating teachers, represented by Lee Silver and Farouk El-Baz, were able to get the astronauts motivated by simulating the observations that would be made during their missions.

The training of astronauts to become better scientists paid off during the mission of Apollo 15 when an interesting sample of crystalline rock was collected and taken to the earth. This sample was later estimated to be 4.5 billion years old and was part of the primordial lunar crust. As mission commander Dave Scott said, "We went to the Moon as trained observers not just to gather data with our instruments onboard, but also with our minds." That mission ended with "A little science on the Moon," as Dave Scott stated. Scott paid tribute to Galileo by demonstrating that a falcon feather and a hammer indeed fell at the same rate in the vacuum on the Moon.

The last mission to the Moon, Apollo17, had Jack Schmidt on its crew. He would be able to use his education and training as a geologist on the Moon. Schmidt was instrumental in the preparation of the astronauts and selection of the landing sites for the previous missions. As he and mission commander Gene Cernan were collecting samples, Schmitt noticed an unusual color on the lunar surface. As he examined the new location, he made careful observations, eliminated the possibility of light reflection, and concluded that it was the lunar dust itself that was orange. This discovery would have been less likely with unmanned probes on the Moon.

References

1. Goll, J. G.; Woods, B. J. Teaching chemistry using the movie *Apollo 13*. *J. Chem. Educ.* **1999**, *76*, 506–508.
2. *Apollo 13*; Universal Pictures: Universal City, CA, 1995, DVD, ISBN 1-4170-4294-X .
3. Internet Movie Database. http//www.imdb.com (accessed December 2012).
4. Goll, J. G.; Mundinger, S. L. Teaching chemistry using *From the Earth to the Moon*. *J. Chem. Educ.* **2003**, *81*, 292–293.
5. *From the Earth to the Moon*; HBO Home Video, 2005, DVD, ISBN 0-7831-2842-8.
6. Frey, C. A.; Mikasen, M. L.; Griep, M. A. Put some movie wow! in your chemistry teaching. *J. Chem. Educ.* **2012**, *89*, 1138–1143.
7. *The Right Stuff*; Warner Brothers: Burbank, CA, 1983, DVD, ISBN 0-7907-7769-X.
8. *Apollo 13: To the Edge and Back*; WGBH Boston Video: Boston, MA, 1994, videotape, ISBN 0-7832-1709-0.
9. *To the Moon*; WGBH Boston Video: Boston, MA, 1999, videotape, ISBN 1-57807-162-3.
10. *Moonshot: The Inside Story of the Apollo Project*; Turner Home Entertainment, 1994, videotape, ISBN 0-7806-0518-7.

11. *Moon Machines*; DOX Productions, London, 2010, DVD.
12. *Man, Moment, Machine Apollo 13: Triumph on the Dark Side of the Moon*; A&E Televisions Networks, 2006, DVD, ISBN 0-7670-9365-8.
13. *Modern Marvels: Apollo 13*; A&E Televisions Networks, 2001, DVD, ISBN 0-7670-7128-X.
14. *20th Century with Mike Wallace – Crisis in Space: The Real Story of Applo 13*; A&E Televisions Networks, 1998, videotape, ISBN 0-7470-0348-9.
15. *Failure Is Not an Option*; A&E Televisions Networks, 2003, DVD, ISBN 0-7670-5943-3.

Chapter 17

Using Movie Clips To Teach Chemistry Formally and Informally

Mark A. Griep* and Marjorie L. Mikasen

Department of Chemistry, University of Nebraska-Lincoln, Lincoln, Nebraska 68588-0304
***E-mail: mgriep1@unl.edu**

The goal of the Movie Wow! project is to develop chemistry educational materials targeted to underserved youth, including non-monolithic groups such as African Americans, Hispanics, and females. First, we identified over 50 movie clips that could be used to teach chemistry, some of which might prove especially engaging for our target groups because they feature certain performers. In our current phase, we seek to discover the movie clips associated with highest impact based on the responses from high school and middle school teachers, senior citizens, college students, science summer camp students, and middle school students. After only one year showing the same 10 clips to various groups, we discovered a strong correlation between movie "Wow!" (clips with famous actors, great dialog, incredible sets, etc.) and how much students report learning from the chemistry explanations that follow each clip. The correlation did not depend on the student's connection to a target group although special needs students gave higher Wow! scores compared to others. Instead, movie clips with the greatest impact make the strongest connections between movie chemistry, daily life, and classroom chemistry.

Introduction

Reasons To Use Movie Clips To Teach Chemistry Formally and Informally

Movie clips are an excellent addition to the chemical instructor's repertoire because they serve all the same purposes as demonstrations (*1*, *2*) plus they show "real people" (actors) interacting with chemicals or talking about them. The social dimension is the key to engagement because moviemakers work hard to create characters that people want to watch.

One of the hurdles in teaching chemistry is that it is the least popularized of the sciences (*3*), making it less likely that people will find engaging chemistry materials informally. This is significant because a much larger percentage of an individual's lifespan is available for informal learning than is actually devoted to formal learning (*4*, *5*). Since most people have heightened interest in movies and popular culture, we will use movie clips to engage them to want to know whether the movie chemistry is real or fake.

Chemistry in Movies Is Either Correct or, If Fictional, Based on Something Real

The most prominent example of a fictional molecule based on real chemistry is flubber, or flying rubber. According to a survey by the editor of the *Journal of Materials* (*6*), flubber is the most famous fictional material. It was introduced in *The Absent-Minded Professor* (1961) and updated considerably in *Flubber* (1997), starring Robin Williams. Realizing a mistake in his mathematical formula, Williams' character Phil Brainard says changing the temperature from "HOT" to "COLD" will allow the Cooper pairs to form a conductive polymer and complete the metastable sphere. He then uses his computer cursor to move an ethylene to complete a C60 fullerene. That is an awful lot of technobabble for the film's demographic although this film was made by Disney Studios, which has a long history of promoting science and technology, even when they're targeting 8-year-olds.

How did the *Flubber* moviemakers get so much right? Jeff Cruzan, University of California Berkeley chemistry graduate student at the time, served as their "science advisor" and "technical advisor." It was his job to ensure that the screen chemistry was scientifically and technically correct. The equation Brainard uses is the major prediction of superconducting theory — a material will superconduct below a critical temperature because the non-conducting electrons pair up, called Cooper pairs, to allow free flow of the conducting electrons. Perhaps the most remarkable property of superconductors is that they float when placed above a permanent magnet. Whereas fictional flubber was created long before superconductors were known to exist, the 1997 update made the connection that flubber doesn't fly, it floats.

Flubber is an example in which moviemakers use real science to anchor a fictional narrative. H. G. Wells initiated this device when he wrote his scientific romances in the 1890s Victorian England, a time and place when a large portion of the literate population was also scientifically literate. In the case of movies, it means being able to write or speak quality technobabble. Most of the audience doesn't know whether the real part of it is real but it has to LOOK real. Even more to the point, they know this is where they should suspend disbelief.

With so much engaging chemistry in this movie for young people, we imagined it must be possible to find many examples useful for teaching chemistry in and out of the classroom. After analyzing over 110 movies in our book *ReAction! Chemistry in the Movies* (*7*), we found most onscreen chemistry is described correctly. Although there are hardly any trained chemists who've become directors, screenwriters, set designers, or producers, they get the chemistry right. One possible reason for this accuracy is they don't know enough about chemistry so they ask someone knowledgable in the field to create incidental design material that looks good and relates to the action.

The quinone analog of the ATP molecule in Eddie Murphy's *The Nutty Professor* (1996) is an example of a molecule that made it to the screen based on a much less deliberate chemical choice than in *Flubber* (*8*). UCLA Pathology's Wayne Grody supplied the moviemakers with several examples from which they selected the one that ended up on the screen. Murphy's character is an overweight genetics professor who develops a molecular cure for obesity. ATP is one of the DNA building blocks. It is reasonable to expect ATP analogs to have physiological effects. After all, the successful drugs acyclovir for treating herpes and AZT for treating AIDS are both based on the DNA building blocks. It is an example of creative onscreen chemistry resulting from a conceptually uncoordinated collaboration containing a kernel of chemical truth.

Movies as Mediators of Public Understanding of Chemistry

The best movies arouse interest with themes, images, and actors that moviegoers want to see and are resolved with emotionally satisfying conclusions (*7, 9*). Moviemakers have to balance their use of the familiar with the non-familiar. At one extreme lie cardboard stereotypes and at the other extreme are stories so unfamiliar they don't connect. Every moviegoer brings his or her past experiences when he or she watches a movie and every moviegoer takes away something different from having watched it.

Occasionally, moviemakers will incorporate chemist characters or chemical imagery into their movies. A chemist watching such movies experiences the movie in a different way from the non-chemist. When the movie has a chemist character, movie-viewing chemists will scrutinize the nuances in the way the chemist is being portrayed, the level of discussion about the chemistry, the type of chemistry, and how other characters react to the chemist and chemistry. Non-chemist viewers will see the stereotype of a white lab coat, safety goggles (sometimes), and some type of glassware (often). More importantly, they will accept the on-screen chemical discussion as being true as stated. Their acceptance and suspension of disbelief allow them to enjoy the story.

Tortilla Soup (2001) is a movie that demonstrates these principles. It uses chemical imagery to tell part of the story even though it isn't really a movie about chemistry. The story begins at a transition point for a Mexican-American family living in Los Angeles. A widowed father and his three adult daughters gather every week for a shared meal but now each daughter's life is about to take her away from the home in a different way. The eldest daughter is a high school chemistry teacher played by Elizabeth Peña. She is an intelligent, competent teacher who falls in love with the high school baseball coach through the trickery of her students. Three or four scenes take place in her classroom. We discussed the lessons on her blackboard and their masterful connection to Peña's growing love in *ReAction! Chemistry in the Movies* (*7*).

What Qualifies as Chemistry in the Movies?

By the time our list of movies reached about 30 in number, we had identified a few recurrent themes and acquired a few borderline examples. This prompted a discussion as to how much chemistry a movie must have to be included on the list. We boiled it down to two rules. (1) A character is identified as a chemist, or more rarely a biochemist, geochemist, or chemical engineer. (2) A character mentions an element, isotope, compound, or simple mixure. The first rule has proven to be a solid rule because chemist characters tend to be the lead characters. The second rule is very broad so we developed a list of overly common things to exclude, such as gold, diamonds, and water. Sometimes these things are interesting, however, so we created exceptions to the rules of exclusion, such as Iron Man is actually Gold Titanium Alloy Man, synthetic diamonds imply their synthesis and supposed easy wealth, and heavy water implies its isolation and use in nuclear fission reactors.

With our rules in hand, we created a spreadsheet to begin tracking the movies. Each entry contains columns for the movie's title, year of release, brief plot description, and a tag for its chemistry. The tags fall into three major categories, element names (e.g. radium), compound class (e.g. local anesthetics), and chemist character's profession (e.g. pharmacist). We use the Internet Movie Database (imdb.com) to add the movie's viewer rating, two or three prominent actors, and the director.

To expand our list, we searched databases and encyclopedias for other movies using the same themes. For instance, the Internet Movie Database (www.imdb.com) is a curated list of thousands of movies in all languages including the earliest movies. Most of its entries contain plot descriptions and keyword tags among much more material. In 2012, after 12 years of collecting, our list contains over 1400 sound movies and over 400 silents. As the list grew, we added another column to the spreadsheet to indicate whether we had watched the movie.

Surprising Facts about Chemistry in the Movies

To be honest, we thought there would only be a few hundred movies with chemistry in them. Tops. So, the first surprising thing about this project is how much chemistry there is in the movies. The second surprising observation emerged as the project developed and was already noted above, fictional movie molecules like flubber are based on real molecules.

The third most surprising notion to reveal itself was the large number of movies featuring women chemists (*7*). Our current list includes over 70 such movies, mostly clustered into two eras—the Golden Era of movies during the

Depression and the Blockbuster/Remakes/Series Era from 1990's to present. During the Golden Era, the women chemists are usually the lead character who is serious about her science but also capable of falling in love. While it is possible to describe male chemist characters in the same way, the difference is that the romantic sides of women chemist characters are presented as being in conflict with their scientific sides.

Since the 1990's, women chemists have been portrayed in a wider variety of ways on the screen. The difference with those of the Golden Era is that they are now a better reflection of the large number of women chemists being trained and in the work force. Women have been earning chemistry graduate degrees since 1886. That's the year Rachel Lloyd, an American, earned her chemistry PhD from the University of Zurich. Dr. Lloyd became a chemistry professor at the University of Nebraska, where she initiated its first real research program. From the 1920's to the early 1970's, women earned about 5% of chemistry graduate degrees in the United States (*10*). The percentage has risen rapidly since the 1970's. Women are now earning one-third of all chemistry graduate degrees and are still rising as a percentage.

Method To Find Engaging and Useful Movie Clips

To find movie clips that are most useful for teaching and learning chemistry, we asked a variety of audiences to rate selected clips according to two subjective responses: Wow! and usefulness for teaching or learning (*11*). The viewers were asked to rate for Wow! (1 is low, 5 is high) immediately after watching each 2-5 minute clip. On the rating sheets, the factors contributing to Wow! were "famous actors, amazing special effects, memorable dialog, great movie sets, or other such things". After hearing an explanation of the real chemistry related to the clip, viewers were asked to rate the explanation (1 is low, 5 is high) for "pedagogical utility" if they were teachers and "How much chemistry did you learn?" if they were students or outreach participants.

All of the movies clips and explanations were part of theme-centered presentations. The movie clips tested by many audiences were from a one-hour presentation titled "Everything I Know About Chemistry, I Learned At The Movies" (Table I). These ten clips were selected from five clips used in Dr. Griep's chemistry course for non-science majors and five complementary clips to flesh out the themes.

The MPAA rating system was created in November 1968 as a consumer guide to movies shown in the United States; it has no legal standing. PG means Parental Guidance Suggested; Some Material May Not Be Suitable For Children. PG-13 means Parents Strongly Cautioned; Some Material May Be Inappropriate For Children Under 13. R means Restricted; Children Under 17 Require Accompanying Parent or Adult Guardian. Between 1930 and 1968, the so-called "Hays Code" ensured that studio-produced movies adhered to a fairly strict set of pan-studio self-censorship guidelines.

Table I. Movie Clips Used in "Everything I Know About Chemistry, I Learned At The Movies" Presentation

Movie (Year)	*Genre*	*MPAA Rating*	*Clip Summary*
	Theme 1: Chemical Sources and Their Physiological Effects		
Harry Potter and the Chamber of Secrets (2002)	Adventure, Family, Fantasy, Mystery	PG for scary moments, some creature violence and mild language	Repotting mandrakes
Senseless (1998)	Comedy, Romance	R for language and sexual content	Human clinical drug trial side effects
	Theme 2: Atoms and Elements		
Fuller Brush Girl (1950)	Comedy	Unrated	Chemical set elements
	Theme 3: Organic Chemistry		
Clambake (1967)	Comedy, Musical	Unrated	Super-hard, super-fast-drying varnish
Fuelin' Around (1949)	Comedy, Short	Unrated	Special effects
	Theme 4: Redox Chemistry		
Apollo 13 (1995)	Drama, History	PG for language and emotional intensity	Lithium hydroxide CO_2 scrubbers
The Bone Collector (1999)	Crime, Drama, Mystery, Thriller	R for strong violent content including grisly images, and for language	Nitrates formed during manure oxidation
Monkey Business (1952)	Comedy	Unrated	Chimpanzee mixes solutions
	Theme 5: Extraterrestrial Biochemistry and Chemistry		
Men in Black II (2002)	Action, Comedy, Sci-Fi	PG-13 for sci-fi action violence and some provocative humor	Alienicide leaves phosphorus residue
Duck Dodgers in the 24 1/2th Century (1952)	Animated short	Unrated	Illudium phosdex, the shaving cream atom

Two of the complementary clips served as controls for the presentation—*Duck Dodgers in the 24 1/2th Century* and *Monkey Business. Duck Dodgers* is the only cartoon in the set. It has movie appeal but not as much teaching utility since the post-clip explanation uses nomenclature rules to identify the nature of fictional matter (is it an element, compound, or mixture?). *Monkey Business* was added to the presentation because the scene in which a chimpanzee combines solutions has great visual appeal but the explanation lacks pedagogical utility. The explanation notes the chimp's excellent technique and then describes the next scene. All groups rated its Wow! as below average and its utility or learning as the lowest out of all ten clips.

Two audiences were able to rate many movie clips becaue they met 4-5 times and rated 10-14 clips per sitting. The clips and explanations were based on five chapters from *ReAction! Chemistry in the Movies* (*7*) supplemented by selections from other themed presentations.

Teacher Experts

The largest number of movie clips were rated by middle and high school chemistry teachers who took a 3-credit professional development course over one week in early summer. These 17 teacher experts rated 94 movie clips and explanations plus 5 other movies that used only a still image and an explanation. They are experts because of their profession but also because they seriously considered the theoretical and practical issues of using movie clips as part of a larger discussion about student engagement. Non-movie-related discussions covered the relationship between high school and college chemistry performance, comparative studies of various pedagogical practices, and the merits of using examples that target specific student demographics in and out of the classroom. Practical examples of engagement were experienced throughout the rest of the day, such as inquiry-based experiments, classroom discussion of current regional issues, and the many ways to use movies and movie clips.

Since our hypothesis was that more engaging movies would correlate with higher learning, we plotted the teacher expert ratings for pedagogical utility versus movie Wow! (Figure 1, top panel). While there is an overall trend from low to high for the ratings, the distribution suggested a different interpretation. When the plot was broken into quadrants, it was possible to see that 39, or nearly half, of the movies were in the most desirable "High Wow/High Utility" quadrant. The boring and unuseful "Low Wow/Low Utiltity" quadrant contained 18 movies, the educational "Low Wow/High Utility" quadrant contained 14 movies, and the entertaining "High Wow/Low Utility" quadrant contained 28 movies. Considered yet another way, the majority of clips have sufficient Wow! but differ in their amounts of pedagogical utility.

Since the Wow! draws learners into the chemical story and the pedagogical utility determines how much chemistry can be taught using that clip, we created the impact score (Wow! + Utility) to identify the top nine most impactful movie clips according to the teacher experts (Table II). These movie clips made it to the list because they are equally strong in Wow! and Utility. They cover a wide range of General Chemistry topics and should join chemical demonstrations as a regular part of every lecturer's pedagogical toolbox.

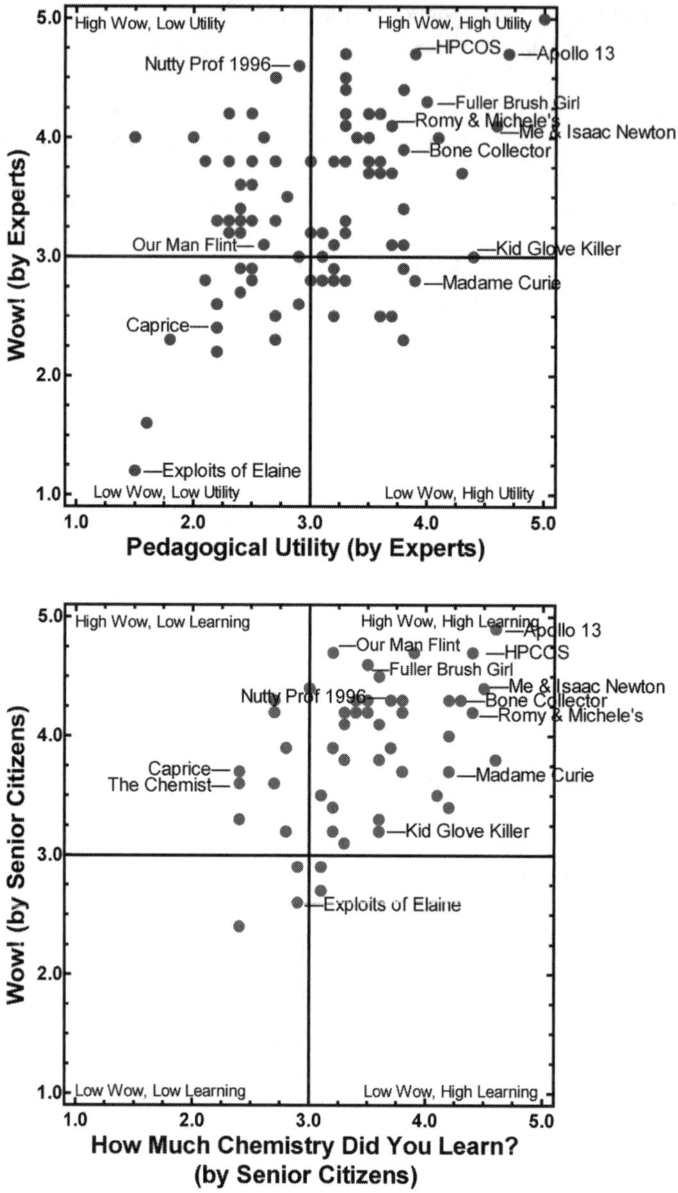

Figure 1. Quadrant analysis of the movie Wow! versus pedagogical utility or learning for the movie clips shown to 6-17 middle and high school chemistry teachers or 16-22 senior citizens. Both factors were rated from 1 to 5, where 5 is highest.

Table II. Highest Impact Movie Clips According to Teacher Experts

Title	*Impact (U+W)*	*Utility*	*Wow!*	*Chemical Theme*
Good Hair (2009)	10.0	5.0	5.0	NaOH hair relaxer
Apollo 13 (1995)	9.4	4.7	4.7	LiOH CO_2 scrubbers
Me & Isaac Newton (1999)	8.7	4.6	4.1	Nucleoside analog drugs
Harry Potter CoS (2002)	8.5	3.9	4.7	Natural products
Fuller Brush Girl (1950)	8.3	4.0	4.3	Chemistry set elements
Back to the Future (1985)	8.2	3.9	4.4	Fusion & fission
Erin Brockovich (2000)	8.1	4.1	4.0	Hexavalent chromium
Spider-Man 2 (2004)	8.0	3.3	4.7	Hydrogen fusion
Undying Monster (1942)	8.0	4.3	3.7	Line spectra

These are the top 9 clips out of 50 assessed.

The most recent movie clip in Table II was released in 2009, after our book *ReAction!* was published. It was rated by the teacher expert cohort of summer 2012 because it takes time to purchase each DVD, capture and edit each clip, and develop a mini-lecture about the chemistry in the clip before the teacher experts can rate it.

The teacher experts of summer 2012 rated *Good Hair* (2009) as having a 5.0 for both Wow! and Utility, making it the highest rated movie clip and explanation ever (Table II). It is a documentary about the use of sodium hydroxide as a hair straightener for nappy hair. The narrator Chris Rock is very engaging in the way he talks to the camera as though the audience was black with asides to the "white people in the audience" who may not know anything about black hair. In the clip, he visits a factory that produces sodium hydroxide gel for use in hair salons, discusses the safety issues, shows stylists being trained to use the product, shows it being applied to hair, and then has a white chemist perform a demonstration of sodium hydroxide strength while wearing goggles and standing in front of a period table.

Senior Citizen Informal Learners

The second largest number of movie clips were rated by senior citizens taking a lifelong learning course. These 22 informal learners rated 58 movie clips and explanations. The large class size shows there is a strong demand for science-oriented courses by this demographic. They chose this course from among a large selection of continuing education courses for senior citizens. While chatting with the students on the first day, several told us they didn't remember any of their high school chemistry but they always wanted to know more.

Since we wanted to know how their scores compared to the teacher experts, they were subjected to quadrant analysis (Figure 1, bottom panel). For most movies, the senior citizens gave much higher Wow! and learning scores than the experts. As a result, three-fourths (44 out of 58) of their ratings fell into the "High Wow/High Learning" quadrant. Only five movies were given a slightly low learning rating, indicating that every explanation was interesting to them.

Further analysis showed the relative placement of most clips within the quadrants were the similar to the teacher experts (Figure 1, compare locations of selected movie titles). This indicated both groups were using similar criteria for discrimination but that it was masked by the higher scores given by the senior citizens. They probably gave higher ratings because they enjoyed the experience and possibly because they wanted to please the teacher for going to the trouble of preparing and presenting the lectures.

When the senior citizen impact scores (Wow! + Learning) for the ten clips in the "Everything I Know" lecture were plotted versus those of the teacher experts, there was a very good correlation (Figure 2 open circles, $r^2 = 0.87$). For these ten clips, the senior citizens gave consistently higher Wow! but similar learning scores to those of the teacher experts. There were three modest outliers. They reported especially low learning for *Duck Dodgers* (nomenclature) and especially low Wow! for *Senseless*. The senior citizens gave higher than average impact to *Fuller Brush Girl* because they reported higher learning (elements in the periodic table).

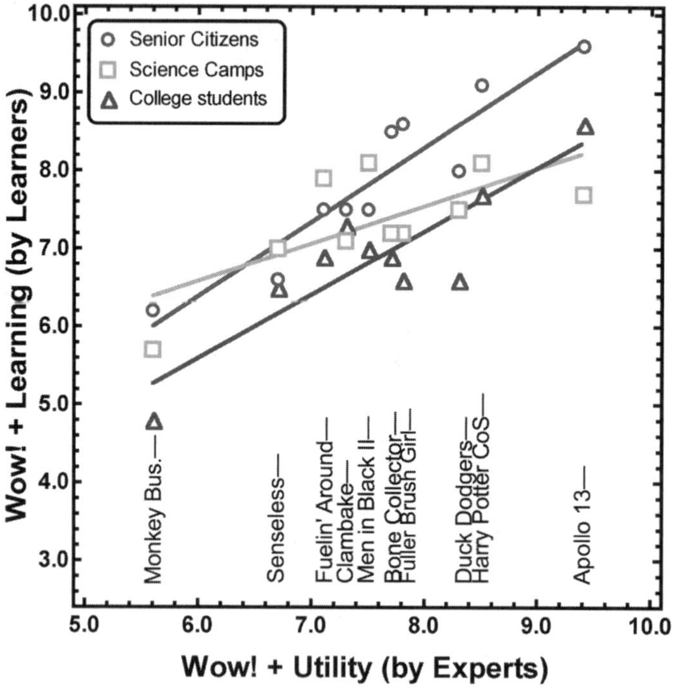

Figure 2. Ratings correlations between the Experts and three other groups for 10 movie clips and explanations. The symbol, slope, and correlation coefficient are: Senior Citizens (open circles, 0.96 ± 0.13, and 0.87); Science Camp Students (open squares, 0.48 ± 0.17, and 0.50); and College Students (open triangles, 0.81 ± 0.16, and 0.76).

Science Camp Students

The "Everything I Know" presentation was given to five science camps with a total of 62 underrepresented students from middle and high schools. Each camp lasts two and a half days and they receive the movie chemistry presentation on the second evening. When their impact scores (Wow! + Learning) for the ten clips were plotted versus those of the teacher experts, there was a poor correlation (Figure 2 open squares, $r^2 = 0.50$). This suggested the teacher experts were not good predictors of science camp students ratings. In one sense, these learners gave all clips a similar impact scores, indicating a lack of discrimination. Another view is to suggest they gave a very low impact score to *Apollo 13* compared to the teacher experts. When this movie clip is removed from the plot, the slope and correlation coefficient rise significantly (slope = 0.66; $r^2 = 0.60$) confirming that *Apollo 13* was the most significant outlier even though it has the fourth highest impact. Whereas adults see the drama of three men working to save their lives under extreme circumstances, the science camp youth may see slow drama. The stoichiometry explanation after the clip may also be too advanced for middle school students. In contrast, *Fuelin' Around* and *Men in Black II* had high impact because these informal learners gave them such high Wow!

College Students

The "Everything I Know" presentation was given to 37 students at a small private college. When their impact scores (Wow! + Learning) for the ten clips were plotted versus those of the teacher experts, there was a moderate correlation (Figure 2 open triangles, $r^2 = 0.76$). The college students gave lower scores than any group with impact scores that were an average 0.7 lower than the teacher experts. According to the correlation, they gave especially low impact to *Fuller Brush Girl* (due to a very low Wow!) and *Duck Dodgers* (due to very low learning). They gave especially high impact to *Clambake*, featuring Elvis as a chemical engineer who develops a varnish and then sings about it. *Clambake*'s impact was due to both high Wow! and high learning, showing the college students valued learning about varnish perhaps because it is material not normally emphasized in textbooks.

Middle School Students

Each of the five sections of 113 middle school students saw only nine out of the ten clips in the "Everything I Know" presentation. The need to fit the talk into a 50-minute period while allowing for ample interaction time with the well-behaved but excited students meant the 5-minute *Apollo 13* clip had to be eliminated. Their impact scores were so poorly correlated with the teacher experts ($r^2 = 0.33$) that they are not plotted on Figure 2. Instead, the middle school student scores were found to be most highly correlated with the summer science camp scores (Figure 3; $r^2 = 0.94$). The most significant outlier was *The Bone Collector* because the middle school students gave it such a low Wow! score. Compared to the teacher experts, middle school students were like the summer science camp students in giving especially high Wow! to *Fuelin' Around* and *Men in Black II*.

During the presentations, the two Special Needs sections had been significantly more boisterous than the one Advanced or two Regular sections and we wondered whether this would be reflected in their scores. When their average Wow! and Learning scores were compared (Figure 4), we saw that they differed in their Wow! scores but not according to their "How much chemistry did you learn?" scores. Like the teacher experts and senior citizens, the three middle school groups differ in Wow! but not Utility or Learning. It seems that everyone from middle school to senior citizens have the same high feelings about chemical learning. This underscores the power of movie clips to reach even the most challenging learners while creating impactful learning opportunities for everyone.

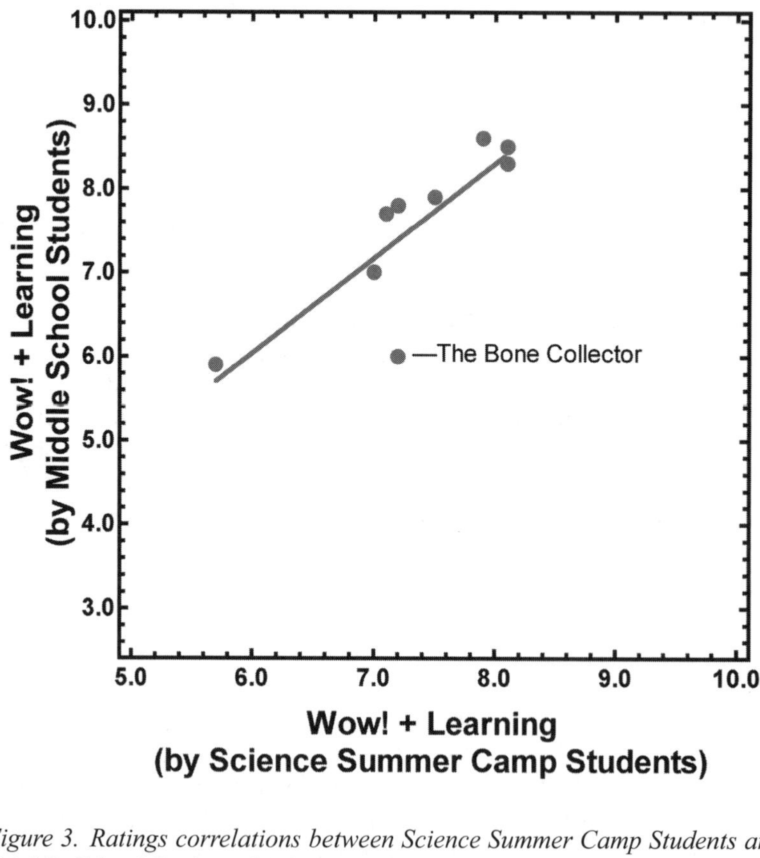

Figure 3. Ratings correlations between Science Summer Camp Students and Middle School Students for 9 movie clips and explanations. The slope and correlation coefficient for the plotted line are 1.13 ± 0.29 and 0.69. When the outlier point for The Bone Collector is omitted, the slope and correlation coefficient are 1.09 ± 0.12 and 0.94.

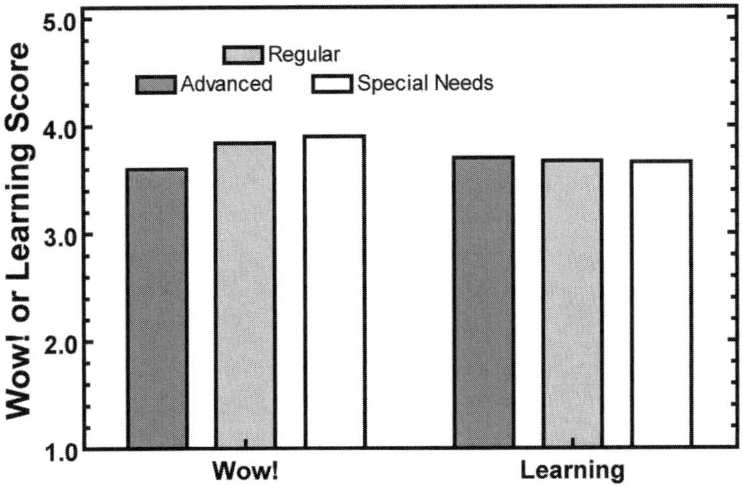

Figure 4. Bar chart of average Wow! and "How much chemistry did you learn?" as determined by Middle School students in advanced chemistry (27 students in one section), regular chemistry (40 students in two sections), and special needs regular chemistry (46 students in two sections).

Teachers Are Able To Predict the Most Useful Material

Despite the nuances of each group, we wanted to know whether teacher experts were good predictors of which movie clips and explanations the learners would find most impactful. To test this idea, we compared the rank order of clips for these two groups (Table III). The bottom two movies for teachers and learners were *Duck Dodgers* and *Monkey Business*. Both these movies make simple points that add spice to a lecture (naming elements using a fictional example, and watching a chimpanzee pour liquids together) but are only tangentially related to classroom material. The presence of *Monkey Business* at the bottom was expected since it was chosen as a control for its low pedagogical utility.

The top three clips for teachers and all learners were *Apollo 13*, *Harry Potter and the Chamber of Secrets*, and *Fuller Brush Girl*. Each of these clips have direct connections to the material often taught in chemistry courses (stoichiometry, the plant origins of molecules and drugs, and the periodic table, respectively). Nearly all groups rated them above average for Wow! and utility or learning.

There were several exceptions to the expert's ability to predict clip impact. *Fuelin' Around* and *Men in Black II* received especially high ratings from the middle school and science camp students who make up more than half the sample population. They undoubtedly enjoy cartoonish humor more than other audiences. The especially low rank for *The Bone Collector* was driven solely by the middle school students' low ratings. The dialogue moves quickly as the characters discuss a dung-covered cow bone. The forensic analyst playfully refers to the sample as *osso bucco*, veal shank, and finally a cow bone, which is probably more than your average 12-year-old can process.

Table III. Impact (Wow! + Learning) Ranks for the "Everything I Know" Presentation

Title	*Experts (N=17)*	*All Learners (N=297)*
Apollo 13 (1995)	1	1
Harry Potter CoS (2001)	2	2
Fuller Brush Girl (1950)	3	3
***Fuelin' Around* (1949)**	**7**	**4**
***Men in Black II* (2002)**	**8**	**5**
Senseless (1998)	5	6
Clambake (1967)	6	7
***The Bone Collector* (1999)**	**4**	**8**
Duck Dodgers (1952)	10	9
Monkey Business (1952)	9	10

The three clips in bold differ the most between experts and non-experts.

Conclusion

Students of all ages enjoy the wonder associated with learning about the chemistry found in movie clips. Learning happens when people can make connections between new knowledge and old knowledge. Instructors can take advantage of the moviemaker expertise in engaging audiences by using it as a prelude to an intellectually satisfying explanation. The audience will stay with the instructor even when digging quite deeply into chemical details because they want to know whether the movie got it right.

We find that all groups show similar utility or learning. It seems that students and teachers have the same perception as to what a real chemical explanation should look like. The two groups differ most in their perception of Wow! Students enjoy goofy humor much more than teachers whereas teachers enjoy dramas more than students.

References

1. Alyea, H. N. Tested demonstrations in general chemistry. *J. Chem. Educ.* **1955**, *32*, 28–29.
2. Bodner, G. M. Why lecture demonstrations are 'exocharmic' for both students and their instructors. *Univ. Chem. Educ.* **2001**, *5*, 31–35.
3. Masciangioli, T. *Chemistry in Primetime and Online: Communicating Chemistry in Informal Environments*; National Academies Press: Washington, DC, 2011.
4. Banks, J. A.; Au, K. H.; Ball, A. F.; Bell, P.; Gordon, E. W.; Gutierrez, K. D.; Heath, S. B.; Lee, C. D.; Lee, Y.; Mahiri, J.; Nasir, N. S.; Valdes, G.; Zhou,

M. *Learning In and Out of School in Diverse Environments*; University of Washington Press: Seattle, WA, 2007.

5. Bell, P.; Lewenstein, B.; Shouse, A. W.; Feder, M. A. *Learning Science in Informal Environments: People, Places, and Pursuits*; National Academies Press: Washington, DC, 2009.
6. Robinson, J. J. The reel thing: One editor's list of great material moments in the movies. *J. Mater.* **2004**, *56*, 9–13.
7. Griep, M. A.; Mikasen, M. L. *ReAction! Chemistry in the Movies*; Oxford University Press: New York, 2009.
8. Griep, M. A. Big screen, big influence. *Nat. Chem.* **2011**, *3*, 496.
9. Griep, M. A.; Mikasen, M. L. Based on a true story: Using movies as source material for general chemistry reports. *J. Chem. Educ.* **2005**, *82*, 1501–1503.
10. *Doctorates in the 20th Century*; NSF Report 06-319; National Science Foundation: Washington, DC, 2006.
11. Frey, C. A.; Mikasen, M. L.; Griep, M. A. Put some movie wow! in your chemistry teaching. *J. Chem. Educ.* **2012**, *89*, 1138–1143.

Chapter 18

The Materials Science of Marvel's *The Avengers*—Some Assembly Required

James Kakalios*

School of Physics and Astronomy, The University of Minnesota, Minneapolis, Minnesota 55455
***E-mail: kakalios@umn.edu**

The Avengers, Earth's mightiest heroes of the Marvel comics universe, saved our planet from an army of hostile aliens and the Norse god of mischief Loki in 2012's blockbuster hit, Marvel's *The Avengers*. We consider here the previously underappreciated role that chemistry and materials science played in keeping Earth safe from this extraterrestial invasion. The chemical composition of Hawkeye's bow, Iron Man's suit of armor and Captain America's shield are discussed, and a scientific mechanism is proposed to account for the "enchanted" aspects of Thor's hammer, Mjolnir.

The roots of the 2012 Marvel Entertainment film Marvel's *The Avengers* stretch back to a golf game in 1961. In the late 1950s, comic book publisher National Periodical Publications, the home of Superman, Batman and Wonder Woman, had reintroduced superheroes to the newsstand and corner candy store, reinventing many of their classic characters from the 1940s, such as The Flash, Green Lantern, the Atom and Hawkman. Following the premise of a popular comic from the 1940s, the *Justice Society of America*, where these heroes would team up and fight supervillains, in 1961 National created the analogous *Justice League of America*. This comic was a huge sales success. Jack Liebowitz, National's publisher, casually remarked on how popular their *Justice League of America* was during a round of golf with Martin Goodman, a comic book publisher at a different company that what would eventually become known as

Marvel Comics. At the time Goodman's company was putting out a series of giant monster comics, western comic books, comics featuring teen humor as well as stories of soldiers at war. Goodman was always happy to jump on a bandwagon (it was said that he always wanted to be second on any new thing—not unlike many in Hollywood!), and, the story goes, upon returning to the office he instructed Stan Lee, his nephew-in-law and sole permanent employee, that they should immediately begin publication of a comic book featuring a team of superheroes. Thus, in November 1961, was born: *The Fantastic Four*!

As Marvel was not publishing superhero comics in 1961, Lee had to create a team of super-powered do-gooders from scratch in order to satisfy Goodman's demand that they put out a comic featuring a team of superheroes. The Fantastic Four proved to be very popular with the comic buying public, and their success at Marvel Comics was followed up by the creation of other offbeat heroes, including the Astonishing Ant-Man (soon joined by his girlfriend, the Winsome Wasp), the Incredible Hulk, the Amazing Spider-Man, the Invincible Iron Man, the Mighty Thor, the Uncanny X-Men and others. By 1963 Marvel comics had a sufficient roster of their own characters that Lee could now publish a comic where various characters from other books joined together to form a team of superheroes, and in March 1963 the first issue of *The Avengers* hit the stands. In a plot that will seem familiar to filmgoers in 2012, the evil Norse god Loki tricks the Hulk into causing mayhem, leading Thor, Iron Man, Ant-Man and the Wasp to join forces in order to try to contain the Hulk. Once it is clear that the Hulk is being manipulated by Thor's half-brother, they turn their attention to their real foe, and working together as a team, manage to defeat Loki.

Fast forward nearly fifty years, and a similar narrative played out on the silver screen, where Thor, Iron Man, Captain America, the Hulk, Hawkeye and the Black Widow come together to face a threat to Earth that no single hero could handle (The film was known in the U.K. as *Marvel's The Avengers Assemble*, to alert audiences that they should not expect a film featuring secret agents John Steed and Mrs. Peel.) There are many reasons why the feature film debut of the *Avengers* was such a crowd-pleasing summer blockbuster. Certainly, the direction and screenwriting; the interactions between the actors reprising their roles from other Marvel films; and the eye-popping special effects and stunts, all contributed to making the film a success. Little appreciated until now, Fearless Reader, was the key role that Materials Science played in making the Avengers the box-office smash of 2012.

Who could imagine Captain America without his mighty shield, composed of a unique alloy of steel and vibranium? Or Hawkeye without his special bow and trick arrows? Or the Norse demigod Thor without his mystic hammer Mjolnir? Take away Tony Stark's suit of armor, and what do you have? (You'd have a genius, billionaire, playboy, philanthropist - which is still pretty awesome, but not quite as useful when facing an army of alien warriors.) Materials science has played a central role in superhero comic books for years. Wonder Woman's bracelets, with which she is able to deflect bullets, are composed of "Amazonian metal" (which apparently has the strength of cold-rolled steel). In *Green Lantern* #8 (Sept.-Oct. 1961) the Emerald Ace needs to dissolve a gold object, and prepares a mixture of hydrochloric and nitric acids, which he informs us is known as Aqua Regia (and a caption box in the panel helpfully explains that it is so known for it alone could affect gold—"the king of metals") (Figure 1). Chemistry and materials science have always been crucial elements of any hero's ability to fight for truth, justice and the American way. So it is no surprise that when these heroes make the transition to the multiplex, materials science would come along.

Figure 1. Green Lantern, in issue no. 8 (Sept.-Oct., 1961) of his self-titled comic book, employs some real world chemistry, creating Aqua Regia, a mixture of hydrochloric and nitric acids in order to dissolve gold. ***GREEN LANTERN™*** *and © DC Comics. All Rights Reserved. Used with Permission.*

Let's consider Captain America and his amazing shield. When Cap debuted in *Captain America* No. 1 in March 1941 (a good eight months before the attack on Pearl Harbor), he carried a triangular shield whose composition and properties were not particularly noteworthy. By the second issue he had taken up a circular shield with which he is associated to this day. This change in shield geometry was motivated not by chemical or physical considerations, but legal ones. A rival publisher was already selling a popular comic featuring the star-spangled hero The Shield, who had a triangular chest shield very similar to Cap's original three-pointed one. Captain America, wanting to save his fighting for the Red Skull in Europe, and not with lawyers in United States courts, switched to the circular disc. This new circular shield turned out to be practically indestructible, and would, time and again, protect Captain America from bullets, grenades, laser beams, and death rays. Which raises the question: What is the chemical composition of Captain America's shield?

In the comic books, Cap's shield was described as consisting of a unique alloy of steel and vibranium. The former was employed for its strength and rigidity. The shield would have to be fairly stiff, as Cap would often throw it with sufficient force that it would ricochet multiple times before finally returning back into his hands. Only a material with a very high Young's modulus, such as steel, could serve as such an effective offensive weapon. The ratio of uniaxial stress (pressure applied in a given direction) to the uniaxial strain (the deformation that results, in the same direction as the applied force) is defined as the Young's modulus. The higher its value, the more pressure has to be applied in order to generate a given deformation, and thus the Young's modulus provides a quantitative measure of a material's stiffness and rigidity. In units of pounds per square inch, tooth enamel has a Young's modulus of 12 million; titanium is 16 million, and steel is 29 million. There are materials with higher Young's modulus than steel, but they are either very rare and heavy, such as osmium with a value of nearly 80 million, or very expensive (diamond has a Young's modulus of over 150 million) or had not yet been discovered in 1941, when Captain America was first given this shield by President Franklin Delano Roosevelt (single walled carbon nanotubes or graphene, with values over 145 million).

However, Captain America's shield could not be composed solely of steel, for the more rigid an object, the less useful it is as a shock absorber, the main defensive property of a shield. Ideally a shield should be easily deformable, in order to better absorb the energy of impact from a hostile force. When stunt men and women in a superhero action film leap from the top of tall buildings, they tend to avoid landing on the hard, unyielding pavement, and rather aim for large, deformable mattresses. These mattresses yield under pressure, and the kinetic energy of the falling stunt person is converted into the work of deformation of the mattress. Another way to think about the benefits of a soft, squishy material as a shock absorber is that it increases the time necessary to bring a rapidly moving object to rest. That is, for a given change of momentum, the longer the time used to arrest the motion, the lower the necessary force that must be applied to effect the deceleration. As Newton's Third law of Motion informs us that 'forces come in pairs,' the lower force applied to the projectile by the shield corresponds to a lower force transmitted to the shield by the projectile, and in turn, to the person holding the shield.

Captain America is the first Avenger (at least according to the movies), in addition to being the ultimate super soldier, and thus one would expect his shield to be the ultimate shock absorber. Fortunately for Cap (and us), a crucial component of the shield is "vibranium," an imaginary extraterrestrial mineral found in the equally fictitious African nation of Wakanda. Vibranium has the unique property that it absorbs all vibrations, and thus protects Captain America as long as he keeps the shield between himself and any blow or explosion.

Vibrations in a solid are oscillations of the atoms about their equilibrium positions. For low energy, long wavelength oscillations, the atoms can move collectively as in sound waves propagating through the material. For very high-energy, short wavelength vibrations, such as arise when the shield deflects a bullet or an explosion, each atom may act as an independent harmonic oscillator, shaking violently back and forth. If the amplitudes of the vibrations in a real solid become too large, then the atoms leave their equilibrium lattice positions, and the solid undergoes a melting phase transition. But what about Vibranium? This material can apparently absorb a near infinite amount of vibrational energy without melting, or even experiencing any sort of atomic scale disruption, such as atoms moving out of their equilibrium lattice locations into interstitial positions. If vibranium can absorb any and all vibrations, then regardless of the source of the oscillations, whether from an explosion or an alien blaster or Thor's hammer, the atoms in the shield will never leave their crystalline positions, and in this sense the material may be considered to be indestructible.

The principle of conservation of energy requires that the energy of the atomic vibrations cannot be destroyed but may only be transformed into another form.

In order for the shield to not grow too hot to handle when absorbing concussive energy, it is likely that the vibrational energy is transformed, via the vibranium, into photons of light. It would take about 120 quantized lattice vibrations in a solid, each with an energy comparable to room temperature, to equal the energy in a single photon of light in the visible portion of the spectrum. Converting sound to light would thus be a very effective way to dissipate away the energy absorbed by a blow or blast. Vibranium is therefore the Marvel universe's most efficient sonoluminescent material. Indeed, the *Avengers* film provides support for this hypothesis.

In the first half of the film, Iron Man and Captain America have managed to capture Loki and are bringing him to S.H.I.E.L.D. (Comic version: Supreme Headquarters, International Espionage, Law-Enforcement Division; movie version: Strategic Homeland Intervention, Enforcement and Logistics Division.) headquarters in a quinjet, when Thor appears for the first time in the film and snatches his half-brother from the ship. Thor intends to bring Loki back to Asgard, but (Spoiler Alert!) S.H.I.E.L.D. needs Loki to provide them with the location of the stolen Tessaract. As in nearly every comic book story featuring the first meeting of two different super-powered heroes, they initially fight until they inevitably discover that they are both working on the side of good. In the *Avengers* film, Iron Man and Thor have an extended battle until Captain America arrives, intending to break up the fight. Telling Thor to "put the hammer down," the God of Thunder does just that, and strikes a powerful blow onto Cap's shield. The shock wave knocks all three heroes off their feet, but the shield is, of course, undamaged. At the moment that the hammer strikes the shield, a bright flash of light, mostly in the blue/ultraviolet portion of the spectrum, is indeed observed. Which only goes to show, they couldn't put it in the movie if it wasn't true!

What can materials science tell us about the S.H.I.E.L.D. agent, and fellow Avenger, Hawkeye, archer extraordinaire? In the film Hawkeye uses a variety of specialty or 'trick' arrows to combat an invading alien horde. The tips of these arrows contain either a timed explosive, corrosive acid or a grappling hook, though no doubt there are many other "trick arrows" that Hawkeye, also known as Clint Barton, could employ as circumstances dictate. Of course, arrows themselves are not very effective without a bow to propel them toward a target, and here again materials science plays a key role. Barton's bow in the film is a "recurve bow," for which, when strung, the tips curve away from the archer. This style of bow stores more potential energy when stretched, and when released is able to deliver more kinetic energy to the arrow than a bow with straight limbs. While the design of the bow is important, just as vital is the material of which it is comprised.

Not unlike the discussion above concerning Captain America's shield, the ideal bow material should couple great strength with high elasticity. An archer needs to be able to bend the bow, but it must also be fairly stiff, that is, it should have a high 'spring constant.' The bending of the bow will be proportional to the force exerted by the archer, and the constant of proportionality connecting force and displacement is termed the 'spring constant' or "Hooke's constant," after Robert Hooke, who first proposed this relationship. The greater the spring constant, the more elastic potential energy can be stored in the bow, which, once the bowstring is released, is transferred into kinetic energy of the arrow. However, the higher the spring constant, the more force is needed to achieve a given bending of the bow. Prior to the twentieth century and the development of light-weight, high strength materials, most bows were made of wood. No doubt early bow makers experimented with the wood of various trees for their bows, but by the middle-ages it was recognized that for the best bows, that shot an arrow further for a given deflection, one should employ the wood of the yew tree.

Yew trees, such as that shown in Figure 2, have a very hard inner core, surrounded by a shell of softer wood. If one cuts along the length of the tree, one can excise a cylinder whereby one side, along the length of the rod, is composed of the harder wood with the softer, more bendable wood on the other side. Pound for pound, a bow made from such a choice cut will store more elastic energy when flexed than one from any other tree. Using such a bow, one could stand farther away from one's foes, beyond their range if they used bows made from other trees, and still be able to reach them with an arrow. In this way one can use the yew wood's superior spring constant as a form of shielding! Of course, the wood for this ideal bow should not have any knots or other imperfections, for these "weak links" will lead to fracture. One might therefore have to cut down several yew trees before finding a log that is defect free. The desirability of yew trees in order to fabricate high performance bows lead to the deforestation of entire regions of Europe, as the yew tree became the "weapon of mass destruction" of the middle ages.

Figure 2. A thousand-year-old yew tree in Britain that has managed to escape the archery maker's attention. (photo by Lairich Rig, under a Creative Commons Attribution-ShareAlike license (1)).

Nowadays, the yew tree is still highly prized, not for its wood but rather for its bark, which contains a key component of the anti-cancer drug 'taxol,' or 'paclitaxel.' In modern times, bows still rely on carbon for their strength and elasticity, only now one forgoes the processing by nature through the growth of trees, and relies on graphitic fibers synthesized by materials scientists.

Carbon's special role in chemistry and materials science derives in part from its ability to rehybridize its atomic electronic orbitals in multiple ways. Quantum mechanics provides an understanding of atoms, and enables us to map out the regions in space where the electrons are most likely to be. A spatial plot of the probability of finding electrons in the atom is termed an "orbital," and depending on the electron's energy, these plots may look like spheres (called s orbitals) or 'figure eights' (termed p orbitals), see Figure 3. In certain elements such as carbon, these s and p orbitals can remix (a process termed 'rehybridization') and form new orbitals that have very different spatial patterns and properties. With only six electrons, the wide variety of bonding configurations that carbon can achieve is impressive. The first two core electrons typically remain tightly bound in the first s orbital, but the next two s-orbital and two p-orbital electrons are like contortionists, combining in various arrangements to form new orbitals. All four of these second and p orbitals may rehybridize to form four sp^3 orbitals, each one pointing to the vertex of a tetrahedron with the carbon nucleus at its center (Figure 4).

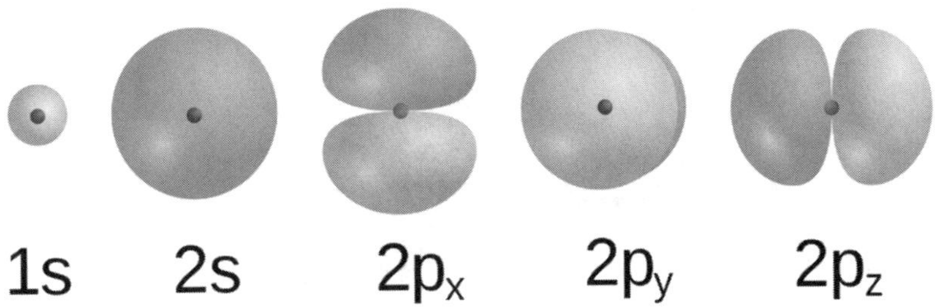

Figure 3. Normal configurations for s and p orbitals. This file is licensed under the Creative Commons Attribution-Share Alike 3.0 Unported (//creativecommons.org/licenses/by-sa/3.0/deed.en) license.

Two adjacent sp^3 orbitals on the same atom, which map out the probability of finding the electron in the carbon atom at a given point in space, form an angle of 109.5° with each other. When many carbon atoms form covalent bonds with these sp^3 orbitals, the resulting array of carbon atoms generates what is known as a 'diamond lattice.' These covalent bonds are very rigid and strong, and diamond is famous for its hardness. However, there is a form of carbon bonding that is even stronger than diamond, and it is what is used in pencil lead.

In this form, three of the four second s and p electrons in carbon can rehybridize to form sp^2 orbitals that can accommodate three Electrons (Figure 5). These sp^2 orbitals lie in a plane, and form an angle of 120° between each other. The fourth p-orbital is unchanged by this process, and protrudes perpendicular to the plane formed by the three sp^2 orbitals. When many such carbon atoms are brought together, the resulting structure, termed graphite, consists of stacks of sheets of carbon atoms, like the multiple layers of phylo-dough in a pastry, where covalent bonds between the atoms within each plane lead to a hexagon pattern of carbon (Figure 6). The p-orbitals sticking out of the planes hold the planes together, but each carbon atom has only one such p-orbital electron and needs to bond two planes, one above and one below. Consequently, the strength of the bonds between planes is weaker than between the atoms within a plane. The between-plane bonds are so weak in fact that that they can be easily pulled apart, by grabbing onto the top surface using scotch tape, or by simply pushing the graphite across a sheet of paper. In the latter case, multiple layers of the graphitic solid will peel off, and adhere to a piece of paper due to electrostatic forces. In the former case, if one repeats the process with the tape many times, one eventually can obtain a single hexagonal plane of carbon atoms, only one atom thick, which is termed 'graphene.' Such a low-tech approach to sample preparation was utilized by Andre Geim and Konstantin Novoselov at the University of Manchester in 2004. Their pioneering studies of graphene were recognized with the Nobel Prize in Physics in 2010.

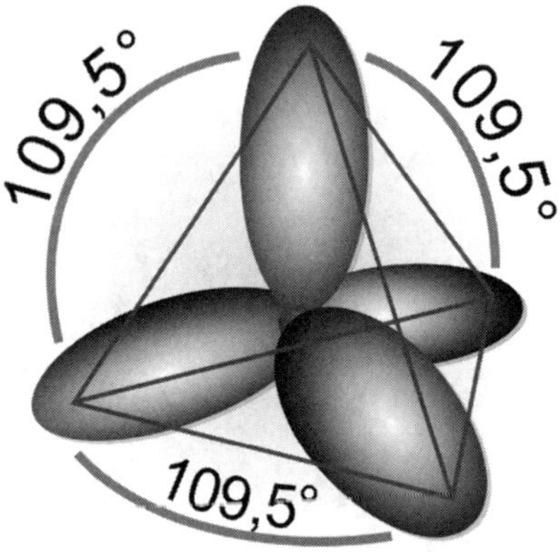

Figure 4. The geometry of s and p orbitals after sp³ hybridization. This file is licensed under the Creative Commons Attribution-Share Alike 3.0 Unported (//creativecommons.org/licenses/by-sa/3.0/deed.en) license.

While the bonds between planes of carbon atoms in graphite are very weak—those within the plane are stronger than diamond. Filaments of graphite, suitably aligned can be braided into filaments, and can be epoxied into whatever shape one desires, including that of an archery bow. The strength-to-weight ratio of this material is very high, much greater than that of yew tree wood, and is almost certainly used in Hawkeye's bow in the *Avengers* film.

Though given the resources of S.H.I.E.L.D., I would not be surprised if Barton's weapon is composed of graphite sheets that have been rolled into cylinders, termed 'carbon nanotubes' (Figure 7). These structures are hollow tubes whose walls can be only one atom thick with a diameter of three atoms, yet are extremely strong, thanks to the sp2 bonds between the carbon atoms, with a strength-to-weight ratio two hundred times greater than steel. While consumer goods composed of carbon nanotubes are not yet available to the general public, efforts to manufacture this material are likely underway at Stark Enterprises, using various proprietary processes developed by the firm's CEO.

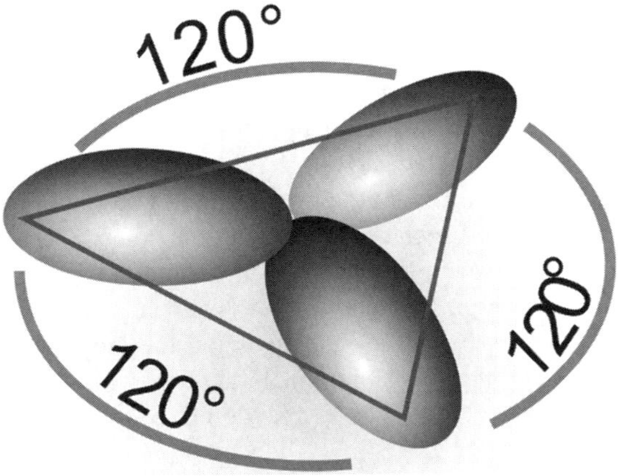

Figure 5. The geometry of s and p orbitals after sp^2 hybridization. This file is licensed under the Creative Commons Attribution-Share Alike 3.0 Unported (//creativecommons.org/licenses/by-sa/3.0/deed.en) license.

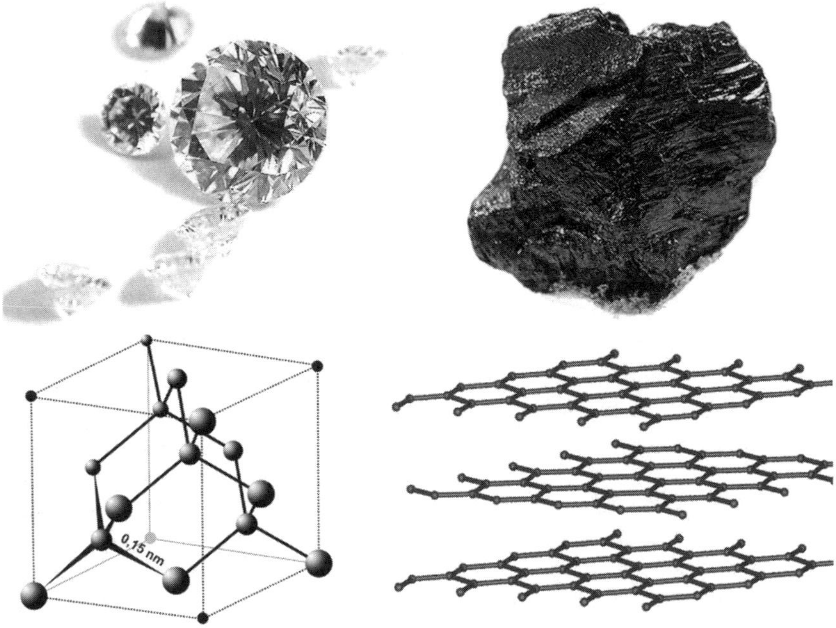

Figure 6. The atomic structure of carbon in its diamond and graphite forms under the Creative Commons Attribution-Share Alike 3.0 Unported (//creativecommons.org/licenses/by-sa/3.0/deed.en) license.

Speaking of Tony Stark, let's consider his armored alter-ego, the invincible Iron Man. The origin of the golden avenger in his 2008 feature film debut essentially followed that of his comic book counterpart, described in Marvel Comics' *Tales of Suspense* no. 39 (March, 1963). Captured during wartime and held in a prisoner camp behind enemy lines (Afghanistan in the movies, Viet-Nam in the comics), brilliant inventor, industrialist Tony Stark is ordered to create weapons of mass destruction by his captors. Unbeknownst to his jailers, Stark uses his access to a machine shop and lab to design and construct a suit of armor, containing a variety of offensive and defensive weapons. Originally, his suit is a bulky dull-grey cylinder of iron, but upon his escape and return to the States and his own lab, he constructs a replacement suit. In the comics, the next iteration of the suit looked identical to the first with the important exception that rather than dull grey it was now a bright golden yellow. This change was instituted for one very important reason—a female bystander found the grey suit frightening and off-putting. If you are going to fight criminals and supervillains in a high tech suit of armor, you might as well look good while doing so!

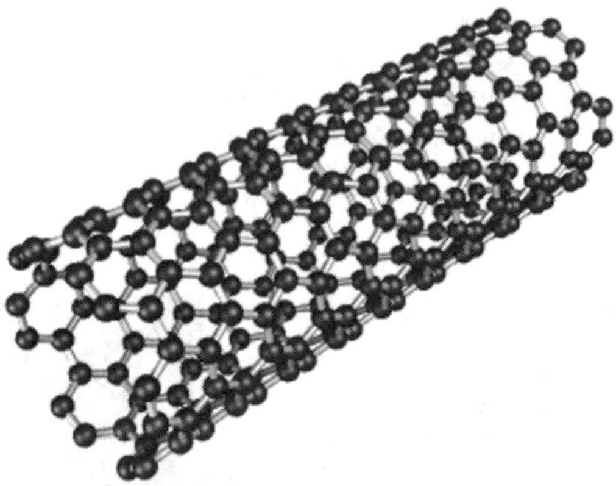

Figure 7. Carbon nanotube structure (image by Eric Weiser under the Creative Commons Attribution-Share Alike 3.0 Unported (//creativecommons.org/licenses/by-sa/3.0/deed.en) license.

In the film version, the next generation of the suit is a more radical redesign, even down to the composition of the armor. The newer model is slimmer and more form-fitting than the original incarnation (which was constructed in a cave with a bunch of scraps), but in an initial test flight displayed an unfortunate tendency to ice up at elevated altitudes. To solve this problem, Tony Stark constructs a new suit not out of iron, but composed of a gold-titanium alloy. Is this insertion of materials science realistic? Sadly, not really.

Gold and titanium are indeed miscible, and form a homogenous alloy of 90% gold and 10% titanium. The resulting material is hard but also very brittle, and is most commonly used in dental applications, rather than superheroics. Iron would make for a strong suit of armor, providing reasonable protection from small arms fire and would be a good choice when escaping from an overseas prisoner of war camp. However, iron is a fairly dense metal, and the suit itself would thus be rather heavy, as illustrated in the origin recap story in *Iron Man* #144, when Tony Stark, having bested his captors, now has to slog through the jungles of Viet-Nam wearing the Mark I suit. Ignoring the weight of the weaponry hidden within the suit, a full body suit of iron would weigh approximately 150 pounds. Titanium is a much lighter metal, and an equivalent titanium suit would weigh only 86 pounds. However, gold is heavier than iron, and if Tony Stark truly were the "golden avenger," his gold-titanium alloy suit would weigh 340 pounds. Maybe the Hulk or Thor might not notice wearing such a heavy raiment, but normal human Anthony Stark would most likely seriously question the superhero business if he were to wear such a suit.

Weight is a consideration, not only for the strain a heavy suit would put on the human inside, but also for the energy required to lift Iron Man off the ground using boot jets. The greater the mass, the greater the potential energy when it is raised a given distance above the ground, requiring more fuel. Tony Stark's flight time would be restricted to a few minutes at most if he relied upon chemical reactions for his boot thrusters. Fortunately for Tony, as explained in his 2008 feature film debut, his suit is powered by an "arc reactor." While the exact mechanism by which this small reactor, about the size of a hockey puck, produces energy is never explicitly stated, it is possible that Stark has managed to manufacture a small, hand-held nuclear fusion reactor. Originally constructed to power the electromagnet that prevents the metal shrapnel in his chest from moving towards his heart and killing him (the location of the shrapnel makes surgery too risky), the arc reactor is also the energy source for the suit's offensive capabilities. This small device can provide a vast amount of energy—equivalent to three nuclear power plants—and if such a power supply existed in the real world, we wouldn't need superheroes to solve the world's problems!

In his second movie *Iron Man 2*, it is revealed that the palladium used in the arc reactor, in such close proximity to his body, is slowly but definitely killing Tony Stark. Presumably this metal is employed in the arc reactor as a storage platform for the isotopes of hydrogen that fuel the fusion reactor. While there are alternative metals, such as platinum, that could serve a similar function, the delicate and intricate construction of the arc reactor precludes such a straightforward solution. Nevertheless, in true entrepreneurial spirit, Tony Stark uses a homemade particle accelerator to artificially synthesize a new element that not only replaces the palladium in his arc reactor, but actually improves upon its performance. This is fairly accurate after all, as colliding large nuclei to create larger still, albeit often unstable, nuclei of atoms that are not normally found in nature, is a standard technique for the manufacture of new elements with atomic numbers above the Actinides (elements such as Thorium, Uranium, Plutonium and heavier atoms).

The next superhero in the *Avengers* film we'll consider is Thor, a Norse god turned Marvel comics superhero. Introduced in 1962 in *Journey into Mystery* #83, Thor is portrayed in the comic books as a hero whose powers are magic-based (thus relieving writer Stan Lee from having to concoct a rationale for his powers and abilities involving radioactivity or cosmic rays). However, in his big screen debut, the 'magical' source of his powers was deliberately changed. In the film, Thor explains to the mortal Jane Foster that while the denizens of Asgard may appear to us as supernatural gods, they in fact are beings from another world who have mastered a branch of science that is so far advanced beyond the capabilities of twentieth century Earth, that it appears to be like magic. This explicit invocation of what is sometimes referred to as "Clarke's Principle," coined by famed science fiction author Arthur C. Clarke—that any sufficiently advanced science will appear to a less technologically evolved people as indistinguishable from magic—as made with the *Avengers* film in mind. When the *Thor* film was made, it was intended that, at some point, the cinematic versions of Iron Man and Thor would inhabit the same universe in a big screen version of the Avengers. Consequently, it was deemed important that there should be some justification for Thor's abilities to be able to co-exist in Tony Stark's world. Iron Man's technology may seem like magic to us, but scientist/engineer Tony Stark would not be able to accept "magic" as an explanation for a super-powered being.

A being from another world could very well exhibit great strength on Earth, especially if his or her muscles and skeleton structure were adapted to a planet with a much higher gravity than on Earth. After all, on the Moon, with a gravitational attraction only one-sixth that of Earth, we would seem super-powered to the Moon people, astonishing them with our ability to leap Moon buildings in a single bound. What, then, could be the scientific justification for Thor's mystical hammer, Mjolnir?

In the 2011 film *The Mighty Thor*, and in the Norse god's comic book adventures, the hammer has an enchantment placed on it by Thor's father, Odin, such that "whosoever holds this hammer, if he be worthy, shall possess the power of Thor." In this day of voice recognition software in smart phones, it is not such a stretch to suppose that nanotechnology embedded within Mjolnir could be easily reprogrammed by a simple command by Asgard's All-Father. In his big screen debut, when Thor learns the lesson of humility after being banished, powerless, to Earth; he regains his special bond to his hammer. What does it mean to "possess the power of Thor?" In both the comics and the film, only Thor can lift Mjlonir, and when he throws it at an enemy, the hammer returns to his hand, as if by some strange form of personal magnetism. The reason for the hammer's unique properties is simple, and can be readily explained.

In his Marvel comic book stories, Thor's hammer is said to be composed of "uru metal," forged in fiery pits by Dwarven blacksmiths. Based upon its observed properties, that it is nigh indestructible, cannot be lifted by anyone except if they be as worthy as Thor, and always returns to his hand, there can be only one explanation—"uru metal" must actually be an exotic form of matter that can be induced to emit gravitons, most likely in response to an external stimulus provided by the nanotechnology within the hammer's head. Gravitons are particles (theoretically predicted, but, unlike the Higgs boson, still not experimentally confirmed) that mediate the force of gravity, just as photons transmit the force of electromagnetism. While we are unable to forge uru metal on Earth, the Dwarven blacksmiths, being as advanced compared to us and we are to our early ancestors, could craft a hammer whose properties seem like magic to us.

Being able to change the rate of emission and absorption of gravitons is equivalent to being able to change an object's mass. If the nanotechnology within the hammer determines the person attempting to lift it to be unworthy, such as the

Hulk in his battle with Thor in the aircraft hanger on the S.H.I.E.L.D. helicarrier in the *Avengers* film, the uru metal will dramatically increase the rate of graviton emission. (In the comics, though not yet on the silver screen, Captain America; a noble alien named Beta-Ray Bill; and Superman, in a special Justice League of America/Avengers cross-over, have all been worthy enough to lift mighty Mjolnir.) This will result in an exponential increase in the gravitational attraction between the Earth and the hammer, such that it cannot be budged, even by the Jade Giant. When Thor grips Mjolnir's handle, the identity recognition programming causes the graviton emission to cease, and the hammer resumes its normal weight.

Changing an object's mass via graviton emission or absorption will alter not only its gravitational attraction, but will also affect its inertia, that is, its resistance to a change in its motion. (A similar mechanism to account for the Man of Steel's super strength and ability to fly, through independent control of inertia, was first proposed by Ben Tippett (*2*)). When Thor throws Mjolnir at the alien Chitauri warriors or some Frost Giants, his hammer always inevitably returns to Thor's hand. One can only presume that the hammer is programmed so that when it is a given distance away from the thunder god, it will alter its mass, and hence its inertia. If the hammer can also change its rest frame, from that relative to the "fixed stars" to some other reference frame, then it could alter its trajectory and reverse its motion, back to Thor. This would also account for how the hammer, if initially stationary (relative to the Earth) would fly to Thor's hand when "summoned." Thor's hammer, one of the most amazing of the materials presented in *The Avengers*, shows the power of real world science and fictional technology!

The final two members of the Avengers, Bruce Banner, who turns into the incredible Hulk when angry (and apparently, he is always angry), and Natasha Romanov, the Black Widow, make use of advanced materials science in their "costumes." Both the Black Widow's form fitting body suit, and the Hulk's stretchy purple pants, share one crucial property that makes them more important than Captain America's shield, Hawkeye's bow, Iron Man's armor or Thor's hammer. That is to say, the Hulk's and Black Widow's clothing ensure that *The Avengers* avoids an R-rating from the Motion Picture Association of America!

We thus see, Fearless Reader, how chemistry and materials science play crucial roles in the adventures of Earth's mightiest heroes in 2012's *Marvel's The Avengers*. While we often must grant a "one time miracle exception to the laws of nature" when dealing with super-powered adventurers, they nevertheless provide an entertaining way to discuss real world science. Advances in the lab today will lead to amazing new materials tomorrow, which in turn will aid the next generation of superheroes in their battles with the forces of evil. And so, on behalf of all materials scientists, let me say to those who have thrilled to the Avengers on the big screen: 'you're welcome!'

References

1. http://s0.geograph.org.uk/geophotos/01/01/10/1011057_c04d13d3.jpg.
2. Tippett, B. *A Unified Theory of Superman's Powers*, 2009. http://www.qwantz.com/fanart/superman.pdf.

Chapter 19

Censoring Science in 1930s and 1940s Hollywood Cinema

David A. Kirby*

Senior Lecturer in Science Communication Studies, Centre for the History of Science, Technology, and Medicine, University of Manchester, Manchester M13 9PL, England
***E-mail: david.kirby@manchester.ac.uk**

From 1930 to 1968 movie studios sent their screenplays to Hollywood's official censorship organizations to make sure these scripts met the standards of the Motion Picture Production Code or "Hays Code." The Motion Picture Producers and Distributors of America, also known as the "Hays Office," administered the Hays Code from 1930 to 1934. From 1934 a newly formed organization within the Hays Office, the Production Code Administration, oversaw enforcement of the Hays Code. In this chapter I explore how these censorship organizations applied their interpretation of the Hays Code to issues involving movie science in the early years of its administration in the 1930s and early-1940s. I examine the censorship files of three science-heavy movies from these years, Warner Brothers' *Dr. Ehrlich's Magic Bullet*, Paramount Studios' *Island of Lost Souls* and Universal Pictures' *Captive Wild Woman*. These films serve as case studies showing how film censors' concerns about scientific research often dovetailed with their primary concerns over sex and violence in movies. These case studies also demonstrate how censors considered the potential moral consequences of science and scientific ways of thinking including the theological implications of scientific research, the blasphemy of scientism, and science's usurping of religion's role.

Columbia Pictures released the 1942 horror-comedy *The Boogie Man Will Get You* at the end of the 1930s' horror cycle that began with *Dracula* (1931) and *Frankenstein* (1931). Most successful film cycles eventually run their course and reach an end with spoofs or parodies of the genre. Such was the case with *The Boogie Man Will Get You.* The film mimicked the classic Universal Pictures' horror films of the 1930s with its old dark house setting and over-the-top mad scientist character who experiments on travelling salesmen to create a race of super humans. Horror icon Boris Karloff plays the "boogie man" of the title alongside another horror legend, Peter Lorre. The film also featured a basement laboratory that the script wanted filled with exotic scientific equipment including a "transparent woman" model. Full body transparent figures were relatively new at this time having first been developed by Franz Tschakert in 1930 for display in the German Hygiene Museum (*1*). These scientific objects consisted of life-sized human statues, either female or male, with a clear plastic skin that housed models of specific anatomical features such as the internal organs, the cardiovascular system, or skeletal structures. The makers of *The Boogie Man Will Get You* wanted a transparent woman figure showing the nervous system with its nerve centers lit up.

Transparent models were popular attractions in the 1930s and by 1940 they had become fixtures in several museums in Europe and in the U.S (*1*). There was even a transparent woman, famously known as "Miss Science," who toured the U.S. in 1936. Filmmakers have historically been quick to capitalize on emerging trends and novelties. So, it would seem natural for the filmmakers to want one of these objects as a prop in their mad scientist's lab. When you watch the film, however, you will not find a transparent woman model in the laboratory. Why did the filmmakers choose not to include the figure in their laboratory set? Was the prop too expensive? Did they ultimately not like the way the transparent figure looked next to the other props?

The transparent woman is not in *The Boogie Man Will Get You* because the choice of whether to use the prop was effectively taken out of filmmakers' hands by Hollywood's censorship body at the time, the Production Code Administration (PCA). The PCA was created in 1934 by the film industry to curtail calls by religious groups, primarily by the Catholic Church and Protestant groups like the National Council of Churches of Christ in America, for a governmental censorship organization that had been growing since the 1910s. Hollywood's initial response to threat of governmental censorship had been to bring in Postmaster General Will H. Hays to head a new self-censorship organization called the Motion Picture Producers and Distributors of America, which became popularly known as the "Hays Office." In 1930 studio heads agreed to abide by a code of standards called the Motion Picture Production Code. Since the Hays Office administered the Code it became known as the "Hays Code." Continued pressure from religious groups on the Hays Office to strictly enforce the Code led Hays to create the PCA, which was run by the tough-minded Catholic Joseph Breen (*2*). From 1934 until 1968 studios sent their scripts to the PCA for review, approval, and recommendations on how to get past the various local, state and international censorship boards. Failure to get PCA approval, or to get past censorship boards in the largest markets, could be a financial kiss of death for a film.

The PCA's recommendation to the producer of *The Boogie Man Will Get You* reveals that the censors were concerned that a scientific object—the transparent woman—was potentially indecent and immoral:

> We do not know exactly what the glass figure of a woman, "with lights denoting various nerve-centers," on page 12, will look like in the finished article, but there must be nothing sex suggestive about it (*3*).

The censors did not know what a transparent woman was, but it was obvious to them that in order to be "transparent" the woman was not going to be wearing any clothing. The fact that this was a movie aimed at young adults made it even more imperative that the filmmakers took steps to insure that their prop was not indecent. The filmmakers could have appealed the PCA's judgment by arguing that the transparent woman was a scientific object that hundreds of thousands of people had already seen in museums. But, the filmmakers would still have to worry about any subsequent difficulties with local, state and international censor boards that might also find a statue of a naked woman problematic, no matter how scientific it might be claimed to be. As was often the case, scientific integrity was not the primary concern for the producers of this film; getting through the censorship process was. This meant that regardless of how visually interesting they found the transparent woman to be, it was far easier to remove the object from the script than it was to challenge the PCA's recommendation.

It is always important to remember that scientific depictions in movies are the result of specific decisions made by filmmakers during production. What *The Boogie Man Will Get You* highlights, however, is that decisions about these scientific depictions are not always left to the filmmakers. Sometimes, organizations and individuals external to the production process make determinations about what aspects of science can and cannot be included in a film. My concern in this chapter is how the PCA applied its understanding of the Code to issues involving movie science in the early years of its administration in the 1930s and early-1940s. I will examine the PCA's recommendations for three science-heavy movies in this time period, Warner Brothers' *Dr. Ehrlich's Magic Bullet*, Paramount Studios' *Island of Lost Souls* and Universal Pictures' *Captive Wild Woman*, in order to explore how the PCA responded to movie science.

These films were chosen as case studies for two reasons. First, they were all produced by major studios. The numerous independent studios that produced B-movies during this time also had to send their scripts to the PCA. However, the major studios were initially responsible for sanctioning the PCA and, thus, they were the most willing to work with the organization on its recommendations during this period. Second, the PCA's judgment on these three films covers many of the issues the organization had with scientific content in movies during this early time period. Although there were 12 categories of the Hays Code, issues related to science primarily violated only five of these categories: Sex, Repellant Subjects, Crimes Against the Law, Vulgarity, and Religion. The other seven categories, such as Profanity, National Feelings and Dances, were mostly irrelevant for scientific content. With these three cases studies I can show how the PCA's concerns about scientific issues dovetailed with the censors' primary concerns over sex and violence in movies. I will also show how censors viewed science from a moral standpoint including the theological implications of scientific research, the blasphemy of scientism, and science's usurping of religion's role.

Mixing Science and Sex in *Dr. Ehrlich's Magic Bullet*

The issue that censor boards were most concerned with were scenes that involved, or even hinted at, sexuality. "Sex" is one of the twelve primary categories of concern listed in the Hays Code. As the anecdote about *The Boogie Man Will Get You* opening this chapter indicates, the definition of what constituted an immoral depiction of sexuality was broad enough that many aspects of science and medicine were included. As medical historians Susan Lederer and John Parascandola note, one aspect of medical science was explicitly listed within the strictures of the code: "sex hygiene and venereal diseases are not subjects for motion pictures" (*4*). This provision meant that any mention of STDs or "social diseases" in a motion picture was forbidden regardless of how seriously or scientifically the film treated the issue. Susan Lederer points out that several dramatic films based on novels or plays were forced to change a character's disease from syphilis to tuberculosis in order to gain approval from the PCA such as with *Dead End* (1937) or they were released without PCA certification which killed their box office earnings as was the case for *Marriage Forbidden* (1936) (*5*).

The PCA's restriction on mentioning venereal disease made things difficult for Warner Brothers' producer Hal Wallis in 1938 who was interested in making a film about German scientist Paul Ehrlich's discovery of Salvarsan, the first effective medical treatment for syphilis. *Dr. Ehrlich's Magic Bullet* (1940) is a film about science, not sex, but such distinctions were irrelevant to the PCA. Even films warning about the dangers of venereal disease needed to acknowledge how a person contracted the disease. Venereal disease was associated with other medical issues involving procreation such as eugenics, abortion, and sterilization. All of these issues fell under the heading of "repellent subjects," which was another of the Code's twelve categories.

Warner Brothers Studio considered *Dr. Ehrlich's Magic Bullet* to be a "prestige picture" that was expected to be a box office hit and to boost the "respectability" of the studio (*6*). The film was produced at the tail end of the medical biopic cycle of the 1930s which included the Warner Brothers' films *The Story of Louis Pasteur* (1936) and White *Angel* (1936). *Dr. Ehrlich's Magic Bullet* hit a topical sweet spot for Harry and Jack Warner at the end of the 1930s. It was an anti-fascist medical picture featuring a Jewish scientist whose research addressed an important social problem. Producer Hal Wallis also claimed that he wanted to make this film in order to refute a widely quoted 1938 statement by Adolf Hitler that "a scientific discovery by a Jew is worthless" (*7*).

There was a growing American interest in syphilis in the late-1930s, so the topic also made sense from a box office standpoint. Surgeon General Thomas Parran had begun a high profile public health campaign on syphilis in 1936 and he published a popular book on the topic in 1937 called *Shadow on the Land.* Parran's goal for the public health campaign was to shift discussions of the disease away from issues of morality towards conversations based on science and medicine (*8*). Parran was excited about the idea of a major Hollywood film on the topic and he ultimately served as a script consultant during production. From Parran's perspective, a Hollywood movie was the natural extension of his existing PR campaign consisting of pamphlets, posters and educational films. Parran had, jointly with the author of *Microbe Hunters*, Paul de Kruif, even approached Will Hays about the likelihood of receiving PCA approval for a film based on the scientific development of a syphilis cure well before Warner Brothers was involved (*7*).

Producer Hal Wallis knew that this film was going to be a tough sell with Hays and the PCA and he spoke informally with Hays about the possibility of making a film about Ehrlich. When Wallis's plans for an Ehrlich picture were leaked to press, the PCA sent a letter to Warner Brothers telling them that this film had not yet been approved and that VD was not a "proper subject" for a movie (*9*). But, Hallis went on with his preparations for the film and the film title "Paul Ehrlich" was accepted by the PCA in March of 1939. However, when the script for the now re-titled film "Test 606" was submitted for PCA approval in August of 1939 the film was immediately rejected (*10*). Breen reminded the studio that, despite the film's scientific merit, the topic was considered inappropriate:

> The acceptability of this story, under the Code, is dubious, for, while the basic story may not be said to be a story of 'sex hygiene and venereal diseases', but rather the life story of a great bacteriologist, nevertheless it might well be argued that the several references to venereal diseases, which are basic in the story, come within the purview of the provision, which states that 'venereal diseases are not subjects for motion pictures'. True, the story is a dignified presentation of the life of a great scientist, but we wonder whether this is sufficient warrant for our approval of this picture (*10*).

The title "Paul Ehrlich" was approved in March based on an understanding that the film would be about Ehrlich's life as a whole, with his syphilis cure being a minor part of the story. By re-titling the picture after Ehrlich's successful syphilis experiment, "Test 606," the filmmakers had emphasized the syphilis cure as the primary focus of the film.

Breen was also angered that after promising the PCA that the film was not going to feature syphilis, he found that syphilis was already a major part of the film's initial publicity. He attached to his letter a news clip from the *Hollywood News* in which the title of the film was announced as "Test 606" and the film's subject was discussed:

> It will be seen from this news item that the emphasis in the story is to be placed upon Dr. Ehrlich's discovery of 'salvarsan', the specific for syphilis, and that it is in the minds of the studio to tie up this film 'with the

> anti-syphilis campaign now being conducted by the United States Public Health Service'. This, we feel, is an ugly angle to the whole thing (*10*).

In a subsequent meeting between Jack Warner, Wallis and Breen, Wallis questioned the basis of the Code if it meant that the "life story of a great bacteriologist" could not be produced (*11*). Hallis called the Code "old fashioned" and claimed that his film was no different from the public discussions already being promoted by Parran in his public health campaign. But for Breen movies *were* different. For him, cinema was a much more influential mass media than a pamphlet ever could be. It reached a much larger audience than any other medium and its visual qualities made it even more seductive than other fictional forms like novels.

Will Hays defended the PCA's decision by emphasizing that the problem from his perspective was not that syphilis was going to be in the film, but that syphilis was the main focus of the narrative writing "there is a difference between a picture in which <u>venereal disease is the subject</u> and a picture in which the <u>discovery of a cure for venereal disease is an incident</u>" (emphasis in original) (*12*). For Hays, the use of syphilis in this film was allowable under the Code because it was based on a historical figure. Removing syphilis from films that were entirely fictional like *Dead End* was acceptable because the presence of syphilis in the story was artificial to begin with. But, syphilis was an historical fact in Ehrlich's life and it would be intellectually dishonest not to allow mention of the venereal disease in a film about him. For Hays and the PCA, however, there was no need to make an entire film about his cure for syphilis.

Ultimately, Hallis and the PCA came to an agreement about how to make the film acceptable for exhibition. First, Warner Brothers had to change the name of the film from "Test 606" to *Dr. Ehrlich's Magic Bullet*. They were also to revise the script to minimize the number of references made to venereal disease and they would remove any scenes featuring syphilitic patients or treatment of the disease. Most importantly, they had to promise not to send out any advertising or publicity material mentioning syphilis. Indeed, the brief that Warner Brothers sent out to distributors about how to promote the picture included "some vital 'DON'TS'" emphasizing the importance of not mentioning syphilis (*13*). The day of the film's premiere the PCA's parent organization, The Motion Picture Producers & Distributors of America, sent a letter to its distributors explaining why they finally had decided to allow a film which mentions venereal disease directly:

> [This movie] transgresses none of the principles that together we have evolved as guides to production but presents, forthright and dramatically, the story of a great scientist who gave his life for the progress of medicine and human happiness (*14*).

The PCA wanted to celebrate a scientific hero without the need to talk about his science. In the end, the censors convinced themselves that the final film was not about Salvarsan the drug, but about Ehrlich the man. By putting the focus on the scientist and not the science, the story became morally acceptable to the censors.

The Blasphemy of Evolution in *Island of Lost Souls*

There were several instances where scientific ideas themselves were subject to censorship because they were found to be offensive to religious minded individuals. *Island of Lost Souls* is an interesting case study because it was initially released in 1932 in what is known as the "pre-Code era" which was the time between the adoption of the Code in 1930 and the establishment of the PCA in 1934, but it was also re-released in 1941 which meant that it had to obtain approval again from a much more rigorous PCA run by Joseph Breen. The movie, based on the H.G Wells' 1895 novel *The Island of Dr. Moreau*, was Paramount Pictures' attempt to get in on the early-1930s fad of horror "shockers" such as *Dracula, Frankenstein*, and *Dr. Jekyll and Mr. Hyde* (1932).

The film had little problem with the Hays Office during its production in 1932. Paramount sent its shooting script to the Hays Office for approval. At the time Colonel Jason Joy ran the Hays Office and his approach to the Code was extremely permissive (*2*). Joy found nothing in the script which he believed would violate the Code accept for one blasphemous line of dialogue uttered by Moreau which indicates that he believes he is equal to God as a creator of life. In the scene Dr. Moreau is explaining to the shipwreck victim Mr. Parker that he is getting close to succeeding in his goal of evolving a new human race from the "lower animals" when he utters the line: "Mr. Parker, do you know what it means to feel like God?" Joy indicated to one of the film's producers that they would have trouble with state censor boards over that line "since a similar line in a recent picture was eliminated by the majority of boards" (*15*).

The "similar picture" Joy was referring to was 1931's *Frankenstein*, which ran into significant censorship issues over a similar line with the Hays Office and state censor boards. The censors' conflict with both Frankenstein's and Moreau's claims are that they directly conflict with the Bible's account of Genesis. In the case of *Frankenstein* the censors ruled "it being a dogma of the Catholic Church that only God can create, it is not advisable to be shown on screen" (*16*). The line in *Island of Lost Souls* has an additional blasphemous implication that God himself also used the lower animals to create humanity. The film has numerous lines of dialogue explicitly indicating that Moreau's experiments are supported by evolutionary theory including the lines "Man is the present climax of a long process of organic revolution. All animal life is tending towards the human form." By using evolutionary processes to create beings that resemble humanity, the line gives credence to the notion that an evolutionary process could create humans.

In most cases, filmmakers would remove the line and count their blessings that this was the only recommendation. In this case, however, the studio was not willing to remove the line because they believed it was crucial in establishing Moreau's hubris (*17*). Ultimately, in 1931 the Hays Office suggested they leave the line in the script and "use the line honestly and sincerely, and let it take its chances with the censor Boards" (*18*). In anticipation of trouble with state censor boards, the filmmakers filmed the scene in such a way that the line could be edited out seamlessly without any loss of continuity. The official approval letter from the PCA states that outside the possibility that the God line would be removed by state censors: "In our opinion it is satisfactory from the standpoint of the code."

The letter concludes by telling the studio that the censors "enjoyed this picture thoroughly" and they believed others would enjoy it as well (*19*).

It turns out that the Hays Office was seriously mistaken. The film was "rejected *in toto* by fourteen state censor boards" as well as being banned in numerous countries (*20*). In addition to the usual censorship concerns over the revealing costumes of Lota the Panther Woman and the film's allusions to vivisection, many censor boards found the plot's overt reliance on evolutionary theory to be unacceptable. Even worse, Moreau's evolutionary experiments were depicted as successful in the context of the story. When the film was re-released in 1941, a much more stringent PCA rejected the film outright because of the evolutionary basis of Moreau's experiments. Joseph Breen explained to the studio that the underlying evolutionary aspects of the film meant that it could not be approved under the Code, writing in his judgment letter to Paramount "The general unacceptability of this picture is suggested by the blasphemous suggestion of the character, played by Charles Laughton, wherein he presumes to create human beings out of animals" (*21*). In order to obtain the PCA's approval Paramount had to eliminate every line of dialogue in the film which suggested that Moreau was "creating" humans by evolving them from animals. Paramount's response letter to the PCA outlines the cuts they made to the film claiming that, "we feel that these cuts eliminate from the picture the suggestions that Moreau considers himself on par with God as a creator, and reduces him to the status of a scientist conducting bio-anthropological experiments (*22*)" In the edited film there is no longer any indication that Moreau made the creatures on the island, he is now merely an anthropologist studying their behaviors. In this way, the beast-people merely become another of God's creations.

Science, Souls, and Gorillas in *Captive Wild Woman*

The PCA's censors did not always aim their comments at specific scientific aspects like research on venereal disease or evolution. The censors' recommendations often reflected general concerns over scientism or science's potential incursion into the spiritual realm. Despite being produced by a major studio, there was little to distinguish Universal Pictures' 1943 science fiction/horror film *Captive Wild Woman* from other low-rent horror movies of the time such as the films Val Lewton produced for RKO. The initial plot of *Captive Wild Woman* was a standard "mad science" scenario featuring horror legend John Carradine's first role as a mad scientist. In the script, Carradine's Dr. Walter combines glandular transfusions with the transplant of a woman's brain into a gorilla's body. After submitting their script for review, Universal received a letter from Joseph Breen explaining that the PCA found the brain transplant to be problematic and that the film could potentially meet "objections from religious groups" (*23*).

The PCA objected to every line of dialogue specifying that a woman's brain was going to be transplanted into a gorilla and asked that the lines be removed or modified. Breen provided specific recommendations for how this dialogue might be changed so that the movie could retain its main narrative element—the physical transformation of an ape into a beautiful woman—without the need for a brain transplant from human to animal. He suggested that the mad scientist's complaint that "my new creation would have to have a human brain" could be changed to read, "my new creation would not respond to hypnosis." While he felt that Dr. Walter's statement to the woman whose brain he wants to use in his experiment "but you are a woman with a clever brain," could be changed to "you are a woman with a well-developed memory, the kind I need for the experiment." In Breen's recommendations the brain could still be involved in the experiment, but the mad scientist would only need to transfer the "memory segment of the human brain" for it to be successful.

The film's mad science scenario was ludicrous and scientifically impossible. So, why did the PCA believe that a partial human brain transplant into a gorilla was acceptable, but that a full brain transplant was problematic? In his letter, Joseph Breen explained to the producers that the full brain transplant was unacceptable because it violated a directive from a 1927 precursor to the Hays Code (the "Don'ts and Be Carefuls") that cautioned filmmakers to "not cause willful offense to any nation, race or creed." According to the Catholic Church, non-human animals do not possess an immortal soul. Catholic doctrine also holds that the human soul only leaves a person's body upon death. In the film's original scenario, the woman's brain, and thus the woman, was kept alive before the transfer. Once the scientist put her brain into the gorilla, the animal possessed a human soul.

As a devout Catholic, the theological implications of Dr. Walter's experiment offended Breen and he believed they would also offend other Catholics. This is why Breen made his suggested dialogue changes to the filmmakers:

> We believe [these changes] will obviate any objections from religious groups, who would object to any idea of transferring a human soul into an animal body. The changed lines we suggest are merely hints for revision—where you will probably find a more clear explanation of the transfer of a memory segment of the human brain to the brain of the animal (*23*).

From Breen's perspective, if only a part of the woman's brain was transferred into the gorilla, it meant that the woman was dead and that her soul had left her body before the experiment. Breen and his censors were not literally afraid of scientists performing brain transplants between humans and gorillas in real life. Their concern was that the movie script sent by the studio implicitly promoted the theologically problematic perception that science could be used to transfer human souls into animals. Breen's recommendations for this film were also in line with concerns about movie depictions of scientific experiments involving human souls. For example, scriptwriters removed brain transplant storylines in *Captive Wild Woman* (1943) and *Mesa of Lost Women* (1953) because the PCA believed that any demonstration of a success with this medical procedure—even in a fictional setting—implied that scientists were capable of manipulating the human soul and, thus, that the soul had a materiality.

For their part, the filmmakers were more than willing to make the changes suggested by Breen. Their intention was not to make a theological point. They were just making a mad science B-movie aimed at teenagers. It did not matter to them what the mad science experiment entailed, as long as it resulted in an ape transforming into a beautiful woman. Ironically, the studio incorporated Breen's suggestion to use the "memory segment of the human brain" by having the mad scientist transplant the dead woman's cerebrum into the gorilla. The cerebrum, which is the Latin word for brain, is the largest part of the human brain. It is the part that makes us human and controls all the higher functions including speech, thought and memory. Breen showed an extraordinarily nuanced understanding of Catholic theology, but his approval of this substitution showed a complete lack of basic scientific understanding of the brain. It also demonstrated the absurdity of his initial concerns. Despite the fact that the mad scientist essentially performed the equivalent of a full brain transplant between a woman and a gorilla, there is no evidence that the film offended any religious minded people.

Conclusions

The created nature of movies—their "made-ness"—makes them useful in understanding society's relationship with science because movies reveal the kinds of stories people *want* to tell about science. Censorship is of interest in this context because it tells us the ways in which certain groups would like to control how, or if, these stories about science should be told. The threat of censorship forced filmmakers to make decisions about which science to include or remove based on reasons that had nothing to do with artistic merit as they anticipated censure. In the case of the Hays Office and the PCA, censorship decisions were founded on beliefs rooted in mid-twentieth century American Christianity. These organizations' sense of moral certainty did not require their censors to understand many of the topics upon which they were passing judgment.

This chapter's analysis of how the Hays Office and the PCA applied their code of standards to scientific depictions in 1930s and 1940s cinema provides us insights into what religious-minded people saw as morally offensive, indecent, horrific or threatening about science and scientific ways of thinking. In the case of *Dr. Ehrlich's Magic Bullet*, scientific research became an "unmentionable" as it was overtly tied to the morally repellant topic of venereal disease. Ehrlich was a scientific hero as long as nobody overtly discussed his greatest scientific achievement. In *Island of Lost Souls* it was the spectre of Darwin and evolutionary thought that the censors feared, as the film's mad scientist was blasphemously close to usurping God's role as creator. Even nonsensical mad science was threatening to morality, according to the censors, if it potentially tampered with the sanctity of the human soul as in *Captive Wild Woman*.

Ultimately, the PCA began to lose its influence in the late-1950s after a few films, including *The Moon is Blue* (1953) and *Some Like It Hot* (1959), attained financial success without having the PCA's approval. With this loss of influence the PCA became less worried about depictions of science in movies and more focused on retaining some influence over their primary concerns with sex and violence. The theological implications of human-gorilla brain transplants seemed far less significant to the PCA than the growing depiction of graphic sex and violence that had crept into mainstream cinema.

While official Hollywood censorship is no longer a threat for movie science, there are still ways for religious groups to influence cinematic stories told about science. Censorship is possible today through means other than directly changing a movie's script or by banning a film. Movies can face a *de facto* ban if theaters are unwilling to show the film or if the film is unable to find distribution. The 2009 BBC film *Creation*, for example, was unable to initially find a distributor in the United States because of its sympathetic portrait of Charles Darwin. This example shows that religious groups still perceive movies as a battleground over science's impact on morality. As long as audiences continue to support movies with good science, though, it is a battle we can win.

References

1. Vogel, K. The Transparent Man - Some Comments on the History of a Symbol. In *Manifesting Medicine*; Bud, R., Ed.; Science Museum: London, 1999; pp 31–61.
2. Black, G. D. *Hollywood Censored*; Cambridge University Press: Cambridge, 1996.
3. Letter from PCA to Harry Cohn, October 24, 1941. Production Code Administration files, Margaret Herrick Library, Academy of Motion Picture Arts and Sciences, Los Angeles, CA; (hereafter cited as PCA files).
4. Lederer, S. E.; Parascandola, J. Screening syphilis: *Dr. Ehrlich's Magic Bullet* meets the public health service. *J. Hist. Med. Allied Sci.* **1998**, *53* (4), 345–70.
5. Lederer, S. E. Repellent subjects: Hollywood censorship and surgical images in the 1930s. *Lit. Med.* **1998**, *17*, 91–113.
6. Pokorny, M.; Sedgwick, J. *Stardom and the Profitability of Film Making: Warner Brothers in the 1930s*; University of North London: London, 2001.
7. Wallis, H. B.; Higham, C. *Starmaker*; MacMillan Publishing Co.: New York, 1980.
8. Parascandola, J. Syphilis at the Cinema: Medicine and Morals in VD Films of the US Public Health Service in World War II. In *Medicine's Moving Pictures*; Reagan, L. J., Tomes, N., Treichler, P. A., Eds.; University of Rochester Press: Rochester, NY, 2007; pp 71–92.
9. Letter from Breen to MacEwen, October 24, 1938; PCA files.
10. Letter from Breen to Warner Brothers, August 16, 1939; PCA files.
11. Memorandum for the Files, Test 606, August 21, 1939; PCA files.
12. Letter from Hays to Breen, August 22, 1939; PCA files.
13. To all showmen (undated); PCA files.
14. Letter from Milliken to theater owners, February 8, 1940; PCA files.
15. Letter from Joy to Hurley, September 26, 1932; PCA files.
16. Telegram from Leduc to Fithian, January 12, 1932; PCA files.
17. Letter from Bailey to Joy, October 3, 1932; PCA files.

18. Memo of phone conversation between Shurlock and Bailey, October 5, 1932; PCA files.
19. Letter from Wingate to Hurley, December 8, 1932; PCA files.
20. Letter from Breen to Hammell, September 18, 1935; PCA files.
21. Letter from Breen to Luraschi, March 4, 1941; PCA files.
22. Letter from Luraschi to Breen, March 15, 1941; PCA files.
23. Letter from Breen to Pivar, December 4, 1942; PCA files.

Chapter 20

Entertainment Media Portrayals and Their Effects on the Public Understanding of Science

Matthew C. Nisbet*,[1] and Anthony Dudo[2]

[1]School of Communication, American University, 4400 Massachusetts Ave. NW Washington, DC 20016
[2]Department of Advertising & Public Relations, The University of Texas at Austin, Belo Center for New Media, 300 W. Dean Keeton (A1200), Austin, Texas 78712
***E-mail: nisbet@american.edu**

For decades members of the scientific community have lamented the state of entertainment media, which they often assume portrays science and scientists negatively and creates public animosity toward science. In this chapter, we review research that provides important context for these longstanding concerns. We first discuss research examining patterns in Hollywood portrayals of scientists and science, which suggest that over the past decade there has been a trend toward ever more positive "hero" portrayals of scientists. We then review research examining the contributions of entertainment media to perceptions of science, highlighting their potential to reinforce beliefs in the promise of science and support for controversial areas of research.

Since as early as the 1970s, many members of the scientific community have criticized entertainment, television, and film portrayals for promoting negative stereotypes about scientists, for featuring improbable or inaccurate scenarios and depictions, and as contributing to a perceived culture of "anti-science." In this chapter, we review research that offers important insight and context for these longstanding concerns. First, we discuss relevant research analyzing trends and patterns in the portrayal of scientists and science over time and across Hollywood genres. Contrary to the fears of many scientists, this research in fact suggests that over the past decade there has been a trend towards ever more positive "hero" portrayals of scientists. We then review research correlating individual forms of entertainment media use with perceptions of science generally and of specific controversial topics such as biotechnology. As these findings suggest, entertainment TV viewing tends to reinforce belief in the promise of science and support for controversial areas of research, though these effects vary by genre and by the background of the audience member.

Science on Screen

Previous studies suggest there is no one portrayal of scientists or theme about science that appears consistently across film and television programming. Instead, portrayals are marked by their diversity, with multiple images often appearing within the same film or program. These portrayals have also shifted across decades depending on social or historical context. Moreover, even those presentations or programs conventionally perceived as hostile to perceptions of science, such as *The X-Files*, offer important opportunities for social learning.

Archetypes of Scientists

Scholars have identified four main archetypes for scientists as characters. Depictions prior to the 1990s featured some of the most negative archetypes, yet over the past two decades, the most positive archetype—scientist as hero—appears with increasing frequency as a central character both in film and television. This trend suggests that somewhat contrary to scientists' impressions, their image is not as negatively slanted as they might presume. Yet the trend towards more positive archetypes does not mean that scientists are portrayed realistically. Whether a nerd, a villain, or a hero, each of these archetypes are not reflective of scientists generally as a profession or as citizens. Only in biopics depicting the lives of real-world scientists such as Charles Darwin, Alfred Kinsey, or John Nash is realism likely achieved (*1–4*).

The first archetype is that of Dr. Frankenstein, the sinister scientist who pursues socially irresponsible research only to be doomed by failure and often death. Noteworthy examples in films include Gregory Peck as *Dr. Mengele in Boys from Brazil*, Marlon Brando as the title character in The Island of Dr. Moreau, and Jeff Goldblum as a scientist ruined by self-experimentation in *The Fly*. On television, the long-running series *Dr. Who* provides numerous examples of scientists who fit this archetype, some of which include Professor Richard Lazarus and Davros, the creator of Dr. Who's nemesis, the Daleks.

A second enduring archetype is that of the scientist as a powerless pawn serving industry, the military, or a master evil figure. Examples include Robert Duvall as Dr. Griffin Weir in the cloning film *The 6th Day* and the scientists in *Jurassic Park* working under CEO John Hammond's dinosaur cloning company InGen.

The scientist as an eccentric, anti-social geek is the third archetype. This scientist is so personally committed to their research that they forego families, friends, or romantic relationships. Examples include Christopher Lloyd as the Doc character in the *Back to the Future* series, the teenage boys in John Hughes' *Weird Science* who create the perfect woman, or Val Kilmer and his fellow post-docs in *Real Genius* who work in the lab of a sinister professor seeking to control Star Wars-like anti-missile technology.

The fourth archetype, the aforementioned character increasingly portrayed in current entertainment, is that of the scientist in a lead role as the action hero and protagonist. These figures also often serve as the voice and force for ethical decisions and virtue. Examples include Dr. Alan Grant as the main protagonist in the *Jurassic Park* series, Spock in the new version of *Star Trek* who takes on leading man and action hero qualities to rival Captain Kirk, Jodie Foster's character in *Contact*, Sigourney Weaver's character in *Avatar*, Dennis Quaid as the climate scientist hero in *The Day After Tomorrow*, Chiwetel Ejiofor as the geologist hero in *2012*, William Peterson as Gil Grissom in the original *CSI* television series, Emily Deschanel as Dr. Temperance Brennan in the *Bones* television series, and Robert Downey Jr. as Tony Stark in the *Iron Man* films.

Closely linked to this archetype is the increasingly common role of the scientist as a trusted, loyal, and brainy "sidekick," a character who supports the main hero in a film as a friend or compatriot, and who often does the scientific "digging" and "uncovering" that leads to important revelations or discoveries that advance the cause of the hero. Examples include Leonard Nimoy's Spock in the original *Star Trek* series, Gillian Anderson as Dana Scully in *The X-Files*, Morgan Freeman as Lucius Fox, inventor and CEO of Wayne Industries in the recent Christopher Nolan directed *Batman* films, and the supporting casts of crime scientists in the popular television series' *CSI* and *Bones*.

There also seems to be an emerging archetype wherein scientists are depicted as ambiguous protagonists. This archetype depicts scientists prominently, but in ways that highlight the multi-dimensionalty of their personalities. Gaius Baltar from the reimagined *Battlestar Galactica* TV series is an excellent example. Baltar is often depicted as a deeply-troubled, narcissistic scientist who is capable of egregious ethical lapses and destructive behaviors. The Baltar character, however, also frequently demonstrates intense empathy, moral sensitivity, and contempt for his questionable decisions. In this regard, this particular archetype shows scientists as fundamentally human—capable of the good, the bad, and everything in between.

Analyzing Images of Scientists

Still, despite evidence to the contrary, a belief in a one-sided negative portrayal of scientists persists, and is promoted in recent commentaries and books, usually to reinforce a narrative about an alleged loss of standing for science in society. An example is the chapter discussing entertainment media in Chris Mooney and Sheril Kirshenbaum's *Unscientific America: How Scientific Uncertainty Threatens Our Future* (*5*).

The authors argue that the negative stereotype of a mad, dysfunctional scientist still dominants Hollywood, citing as evidence a quantitative study of portrayals from the mid-1980s by former University of Pennsylvania communication researcher George Gerbner and colleagues (*6*) and an analysis by Stanley Goldman (*1*) from the same time period. The Gerbner study showed that in comparison to other occupations, scientists featured in primetime television suffered a higher ratio of negative stereotypes and were more likely to be victims of violence.

Yet subsequent research documents a shift towards the positive for the image of scientists over the past two decades. In a 1999 report to the U.S. Department of Commerce, Gerbner and colleagues updated their analysis, concluding that based on data collected during the mid-1990s, "there is no basis to claim that any kind of systematic negative portrayal of scientists exists. Changes have occurred in Hollywood since the time of our initial study, which found scientists to be typically evil, disturbed, sexually dysfunctional villains…this is no longer the case" (*7*).

More recent analysis of TV content confirms this trend. A study of primetime content appearing between 2000 and 2008 replicates Gerbner's methodology and finds that scientists—in accord with their professional distribution among the general population—remain relatively rare characters in the TV world (just 1% of characters are scientists), but when they are shown, it is almost exclusively in a positive light. Of the scientist characters, 81% were characterized as good with some of the best known examples being Gil Grissom from *CSI* and Leonard Hofstadter from *The Big Bang Theory*, 26% as both good and bad such as Dr. Gaius Baltar from *Battlestar Galactica*, and just 3% as bad which includes examples such as villainous David Robert Jones.

Depictions of Science

Apart from the image of scientists, a number of historical and critical analyses conclude that science in general is often depicted as mysterious, magical, or dangerous with both positive and negative consequences for society. Depictions tend to break down along two lines. According to the first standard portrayal, scientists lose control of their research or their technology, to the detriment of society; as a consequence scientific achievement and technology are distrusted because of possible unforeseen ramifications. This is the *Jurassic Park* vision of genetic engineering, and the vision of science offered in *The X-Files*, the reimagined *Battlestar Galactica*, in horror movies, and sometimes in comic books turned into movies such as *The Incredible Hulk* or *Spider Man*.

In the second portrayal, science and technology are shown as truthful, sacred endeavors. This is the *Star Trek* vision of social progress through science, the *CSI* vision of science as glamorous crime solving and a force for justice, and the PBS *NOVA* and *An Inconvenient Truth* vision of glorified, overly certain science (*2*, *3*, *9–11*).

Not all scholars, however, view presentations such as those in *The X-Files* as negative, and instead see these portrayals as valuable complements to programming such as *NOVA* or *Star Trek* that might, in the words of the editors at *Nature*, dogmatically present scientists "as truth's ultimate custodian" (*12*). For example, Dhingra observes that *The X-Files* emphasizes several important realities about science, namely that it is uncertain, sometimes offers few clearly defined answers, and can often be interpreted in multiple ways (*13*).

Audience Perceptions

Analyses of how science and scientists are represented in film and television are valuable because they help us understand how these portrayals influence perceptions of science among the public. Entertainment media comprise the dominant source of information for the public about science and are an integral part of the social context by which the public judges and makes decisions relative to debates and controversies. For students and adults, entertainment media also likely pre-shape the impressions, views, knowledge, and orientations that they bring to school-based and informal learning settings such as museums or science centers or to reading about policy-debates or events in the news.

Despite concern about this topic within the scientific community, few studies have addressed directly how the image of scientists in film and television impact adult stereotypes about scientists. A study by Losh, however, does provide indirect evidence (*14*). As entertainment portrayals have shifted since the 1990s from more negative archetypes to more positive hero portrayals, so have the stereotypes held by adult audiences. The study concludes that in comparison to 1985, American adults in 2002 were far less likely to hold negative stereotypes about scientists and were much more likely to believe that a career in science was a desirable choice for their children or for themselves.

Though evidence on the direct connection to stereotypes is limited, a series of studies have considered how patterns of television viewing are connected to generalized perceptions of science, either in terms of beliefs relating to reservations about the impact of science on society or belief in the promise of science to improve life and society. Most Americans simultaneously hold both mental models. Depending, for example, on how issues such as stem cell research or nanotechnology are framed in fictionalized programming, in news coverage, by opinion-leaders, and in personal conversations, one or the other model—reservations or promise—can become activated, influencing public evaluations of the issue (*3*, *8*, *15*). As discussed in the previous section, these mental models also map closely to the dominant narratives told about science in entertainment, as either a force out-of-control to the detriment of society; or as a glamorous tool for societal improvement and justice.

Viewing Science As Portrayed on TV

Studies examining the connection between public beliefs about science and entertainment media use have been conducted using analysis of cross-sectional, nationally representative public opinion surveys. These efforts have been predominately guided by a body of work in the field of communication called cultivation theory. This research approach hypothesizes that heavy viewers of television will be more likely to hold conceptions of the world consistent with what is seen on television than individuals who view television less frequently (*16*).

In these studies, differences in perceptions are looked at across variations in frequency of general entertainment TV viewing along with the viewing of specific TV genres such as science fiction programming. Education, age, gender, values, race, science knowledge, and other background factors are statistically controlled in an attempt to identify the unique relationship between TV viewing and beliefs.

During the 1980s, studies done by Gerbner and colleagues (*6*, *17*, *18*) represent the first research on the relationship between television viewing and attitudes toward science. This work found that heavy television viewing was associated with more negative views of scientists and new technologies, more willingness to place restrictions on science, a tendency to think science makes life change too fast, an increased anxiety about science, and an erosion of appreciation for the benefits of science.

Yet consider that in two recent survey-based studies conducted during the 2000s, heavier viewers of entertainment television held stronger reservations about the impact of science on society but they were also more likely to score higher on belief in science as contributing to societal progress. These findings suggest that the general influence of TV viewing reflects the dual imagery of science in entertainment. Forms of TV use are also linked to public knowledge. Heavier entertainment viewers tend to be less knowledgeable about science, since entertainment use likely displaces other media behaviors such as newspaper reading, online news or blog reading, or documentary TV viewing (*3*, *8*). But not all entertainment programming has been found to have the same effects. Heavier viewers of science fiction programs, tend to be more positive in their views of biomedical research (*15*), agricultural biotechnology (*19*), and science more generally (*8*, *20*).

These studies demonstrate how exposure to unique genres of televised entertainment programming can cultivate different attitudes toward science. The studies also compel additional explanation and exploration relative to the relationship between entertainment television and science. Nisbet and Goidel, for example, raise two possible factors explaining the direct influence of science fiction television on evaluations of stem cell research and therapeutic cloning (*15*). First, the science fiction audience is by nature strongly enthusiastic about science, meaning that their viewing habits capture an underlying natural support for science. Repetitive viewing of science fiction simply strengthens this orientation, further cultivating an audience naturally receptive to new innovations in science. Yet as the authors describe, the analysis controlled for generalized views about science, which should account for at least some of this underlying predisposition.

A second possible factor suggested by the authors was that by familiarizing themselves with the moral dimensions of human genetic engineering through TV and film portrayals, audiences may assuage some of their reservations about the technology. In part, science fiction TV viewing may in fact desensitize an audiences' natural "yuck factor," shielding viewers from the influence of some of the more dramatic claims made by political opponents of stem cell research.

Not only do effects tend to differ by entertainment genre, but they also vary by the background of the audience member. In a 2011 survey study, Dudo and colleagues, for example, observe that heavy entertainment TV use among Americans *without* college science experience is linked to a *stronger belief in the promise of science* (*8*). Conversely, heavy entertainment use among Americans *with* college science experience is linked to *stronger reservations about science.* The reason for these differences, however, remains unclear. Similarly, when focus groups are used to prompt audiences to think more deliberatively about entertainment portrayals of issues such as genetics, personal experience and social background factors are observed to alter how respondents draw upon these portrayals to arrive at judgments about the social implications of genetics (*21*). Recent research also has shown that watching religious TV programming contributes to religious individuals' negative perceptions of science (*20*).

The CSI Effect

Perhaps one of the most widely presumed influences of entertainment is the so-called "CSI effect," the alleged influence that long running syndicated TV series *Crime Scene Investigation* (CSI) has on public perceptions about forensic science and its role in jurisprudence. For example, CSI has commonly been associated with increased demand for DNA evidence in the courtroom (*22*, *23*) and with burgeoning enrollment in collegiate forensic science departments (*24*, *25*).

In one of the few studies empirically investigating the CSI effect, Brewer and Ley find that overall TV viewing was correlated with stronger belief in the reliability of DNA evidence, greater weight attached to the absence of DNA evidence in a case, and support for a national DNA databank (*26*). The survey analysis also included an experimental design. In the half of the sample that were presented first with questions asking them to reflect about their media use and then to evaluate a series of DNA questions, this process of thinking about what programs they watch primed respondents to give greater weight to DNA evidence in court cases. In short, for these respondents, when thoughts about TV programming were made more salient, their answers about real-world court decisions were more likely to be in line with TV portrayals of crime solving.

Conclusion

Overall, the studies and findings reviewed offer important context for the longstanding concerns voiced by scientists about the image of their profession and work in Hollywood productions. Contrary to the fears of many scientists, this research indicates that over the past decade portrayals of scientists have become more complex. Scientists in 21st century entertainment programming are more frequently imbued with intricate, multi-faceted personalities and are more often being depicted as "hero" protagonists. Furthermore, as these findings reviewed in this chapter suggest, entertainment TV viewing often strengthens beliefs in the promise of science and support for quickly evolving fields like biotechnology or nanotechnology. Overall, this growing body of research does not support commonplace assumptions that entertainment media are hostile toward science. This is not to say that members of the scientific community should not strive to enhance the presence of their profession in Hollywood. Indeed, the Science and Entertainment Exchange, a program created by the National Academy of Sciences, was recently created to help bolster collaborative partnerships between entertainment producers and scientists. We believe that such efforts, however, would be best served to operate from an understanding of the Hollywood-Science relationship that is more sophisticated than the "hostile media" trope and that instead seek to capitalize on some of the encouraging trends reviewed in this chapter.

References

1. Goldman, S. L. Images of technology in popular films: Discussion and filmography. *Sci., Technol., Hum. Values* **1989**, *14* (3), 275–301.
2. Kirby, D. A. Hollywood Knowledge: Communication Between Scientific and Entertainment Cultures. In *Communicating Science in Social Contexts*; Cheng D., et al., Eds.; Springer: New York, 2008; pp 165−181.
3. Nisbet, M. C.; Scheufele, D. A.; Shanahan, J.; Moy, P.; Brossard, D.; Lewenstein, B. V. Knowledge, reservations, or promise? A media effects model for public perceptions of science and technology. *Commun. Res.* **2002**, *29* (5), 584–608.
4. Perkowitz, S. *Hollywood Science: Movies, Science, and the End of the World*; Columbia University Press: New York, 2007.
5. Mooney, C.; Kirshenbaum, S. *Unscientific America: How Scientific Illiteracy Threatens our Future*; Basic Books: New York, 2009.
6. Gerbner, G.; Gross, L.; Morgan, M.; Signorielli, N. *Science and Television*; Research Report; Annenberg School for Communication: Philadelphia, PA, 1985.
7. Gerbner, G.; Linson, B. Images of Scientists in Prime Time Television; U.S. Department of Commerce: Washington, DC, 1999
8. Dudo, A.; Brossard, D.; Shanahan, J.; Scheufele, D. A.; Morgan, M.; Signorielli, N. Science on television in the 21st century: Recent trends in portrayals and their contributions to public attitudes toward science. *Commun. Res.* **2011**, *48* (6), 754–777.
9. Collins, H. M. Certainty and the public understanding of science: Science on television. *Social Stud. Sci.* **1987**, *17* (4), 689–713.
10. Hornig, S. Television's Nova and the construction of scientific truth. *Crit. Stud. Mass Commun.* **1987**, *7* (1), 11–23.
11. Ley, B. L.; Jankowski, N.; Brewer, P. R. Investigating CSI: Portrayals of DNA testing on a forensic crime show and their potential effects. *Public Understanding Sci.* **2012**, *21*, 51–67.
12. Opinion. How not to respond to *The X-Files*. *Nature* **1998**, *394*, 815.
13. Dhingra, K. Thinking about television science: How students understand the nature of science from different program genres. *J. Res. Sci. Teach.* **2003**, *40* (2), 234–256.
14. Losh, S. C. Stereotypes about scientists over time among US adults: 1983 and 2001. *Public Understanding Sci.* **2012**, *19* (3), 372–382.
15. Nisbet, M. C.; Goidel, R. K. Understanding citizen perceptions of science controversy: Bridging the ethnographic-survey research divide. *Public Understanding Sci.* **2007**, *16* (4), 421–440.

16. Morgan, M.; Shanahan, J.; Signorielli, N. Growing Up with Television: Cultivation Processes. In *Media Effects: Advances in Theories and Research*, 3rd ed.; Bryant, J., Oliver, M. B., Eds.; Routledge: New York, 2009; pp 34–49.
17. Gerbner, G.; Gross, L.; Morgan, M.; Signorielli, N. Scientists on the TV Screen. *Society* **1981**, *18*, 41–44.
18. Gerbner, G. Science on television: How it affects public conceptions. *Issues Sci. Technol.* **1987**, *3*, 109–115.
19. Besley, J. C.; Shanahan, J. Media attention and exposure in relation to support for agricultural biotechnology. *Sci. Commun.* **2005**, *26* (4), 347–367.
20. Brossard, D.; Dudo, A. Cultivation of Science Attitudes. In *The Cultivation Differential: State of the Art Research in Cultivation Theory*; Shanahan, J., Morgan, M., Signorielli, N., Eds.; Peter Lang: Bern, Switzerland, 2012; pp 120−143.
21. Bates, B. R. Public culture and public understanding of genetics: A focus group study. *Public Understanding Sci.* **2005**, *14* (1), 47–65.
22. Houck, M. M. CSI: Reality. *Sci. Am.* **2006**, *295* (1), 84–89.
23. Pratt, T. C.; Gaffney, M. J.; Lovrich, N. P.; Johnson, C. L. This isn't *CSI:* Estimating the national backlog of forensic DNA cases and the barriers associated with case processing. *Crim. Justice Policy Rev.* **2006**, *17* (1), 32–47.
24. Cole, S. A.; Dioso-Villa, R. CSI and its effects: Media, juries, and the burden of proof. *New Engl. Law Rev.* **2007**, *41* (3), 435–469.
25. Lee, A. The *CSI* effect: TV's impact on the future of forensic science. *Triple Helix* **2007**, *7*, 22–23.
26. Brewer, P. R.; Ley, B. L. Media use and public perceptions of DNA evidence. *Sci. Commun.* **2010**, *32* (1), 93–117.

Chapter 21

The Character of Science on Television

Thomas Parham[*] and Jovana J. Grbic

Azusa Pacific University, 901 E. Alosta Avenue, Azusa, California 91702
[*]E-mail: tparham@apu.edu

An examination of the different ways contemporary television series have used science to evoke plot, character, and theme. This will be accomplished through a series of case studies.

Science on television series tends to serve three main purposes, which are often intertwined. First, it functions as a catalyst character. Script consultant Linda Seger defines catalyst characters as "the people who provide a piece of information or cause an event to happen that pushes the protagonist into action" (*1*). For example, a series can use a science project (or its results) as a way to prompt characters into motion for a prototypical episode. Another function of science on television is as a "MacGuffin." Alfred Hitchcock coined that term referring "to an object, document, or secret within a story that is of vital importance to the characters and thus motivates their actions and the conflict, but that turns out to be less significant to the overall narrative than we might at first imagine" (*2*). Finally, science often serves as a crucial component of setting for television series. In his seminal book, *Story*, screenwriting guru Robert McKee notes that setting is four-dimensional and encompasses period, duration, location, and level of conflict (*3*). This chapter will examine the different ways contemporary television series have used science as exemplified through a series of case studies.

Science as Punchline: *The Big Bang Theory*

On September 24, 2007, the first episode of the CBS sitcom *The Big Bang Theory* aired. Billboards throughout Los Angeles had proclaimed, "Smart is the new sexy." This advertising slogan seemed at odds with the show's academic-sounding title. Although the show premiered to modest ratings, they have grown robustly. Now in its sixth season, *The Big Bang Theory* has become the highest-rated situation comedy on television, both with total viewers and the coveted 18-49 demographics. On November 15, 2012 the show achieved a series high of 17.4 million viewers and 5.5 rating/17 share in the coveted 18-49 demographic group. At the core, *Big Bang* resembles *Friends*-type sitcoms that focus on a group of friends working and living in a large metropolitan area and grappling with universal themes of friendship, love, rejection, career success, and finding one's identity in the world. However, these friends are different than others previously portrayed on television: they are physicists employed by the elite California Institute of Technology, commonly known as CalTech. This setting allows the show to observe nuances of the scientific world while shattering stereotypes.

Often in television and film, the scientist is a geeky, one-dimensional sidekick without a life, opinions, or an emotional connection to the main characters. *Big Bang*'s scientists have healthy libidos and yearn for the perfect woman. The crux of the show is Dr. Leonard Hofstadter's (Johnny Galecki) on-again/off-again relationship with beautiful neighbor Penny (Kaley Cuoco), with Leonard's roommate Dr. Sheldon Cooper (Jim Parsons) serving as comic foil and verbal sparring partner. Mama's boy but self-styled lothario Howard Wolowicz (Simon Helberg) and notoriously shy around women Dr. Rajesh Koothrappali (Kunal Nayyar) rounded out the initial supporting cast. Over the years, co-creators/executive producers Chuck Lorre and Bill Prady have allowed Leonard and his egghead chums to evolve "lest they be dismissed as a bunch of socially inept dweebs who only gab about math" (*4*). The addition of two women to the cast in season three—neurobiologist Dr. Amy Farrah Fowler (Mayim Bialik) and Dr. Bernadette Rostenkowski (Melissa Rauch)—has allowed the show to soar creatively and ratings-wise. Yahoo contributor Mark Paul concurs:

> Amy and Bernadette have brought a whole new element to the show. Both characters started in small roles and that allowed fans to slowly get used to them. Now they fit perfectly into the show, and they've allowed Penny's character to expand beyond just her storylines with the guys. It's not easy to incorporate new characters while still keeping the heart of a show, but "The Big Bang Theory" has done it brilliantly (*5*).

Also, since *Big Bang*'s writers themselves are self-professed geeks, they draw on their own experiences to humanize the characters. Prady reveals, "One of the things that makes this writers' room work is the emotional bravery in the room. We talk about the things in our lives that have hurt, the things we wanted and we didn't get. That's allowed us to get some cool stuff" (*4*).

The writers do not shy away from use of extremely inside science jokes or knowledge. For example, a physicist colleague dumps Leonard after he sides with Sheldon to support string theory rather than loop quantum gravity, which she advocates. Episode titles often pay tribute to science with names like "The Higgs Boson Observation," an episode in which Penny quips to Sheldon, "I'm just a blond monkey to you." Recurring duels, feuds, and competitions between characters revolve around scientific work and the egos of academic researchers. A running gag is the schism between the hard and "soft" sciences (psychology, sociology, *etc.*), with jokes often targeting the latter. Although Sheldon is caught off-guard when platonic girlfriend Amy Farrah Fowler uses her neurobiology acumen attempting to speed up their relationship in "The Launch Acceleration." *The Big Bang Theory* writers enjoy a close working relationship with science advisor, UCLA physics professor David Saltzberg. As they shared with ScriptPhD.com during the 2010 Comic-Con International press junket, the writers often place open-ended dialogue in the scripts and entrust Dr. Saltzberg to complete it with technical jargon and details of scientific experiments (*6*). Perhaps the most refreshing aspect of the show's success is the lack of gimmicks. High-profile guest stars on the show have included astrophysicist and Nobel laureate George Smoot, astrophysicist Neil deGrasse Tyson, cosmologist Stephen Hawking, NASA astronauts Buzz Aldrin and Michael J. Massimino, as well as *Star Trek* actors Wil Wheaton (recurring as Sheldon's arch-nemesis), Leonard Nimoy, and LeVar Burton, and other sci-fi personalities like Summer Glau, Katee Sackhoff, and Stan Lee. Chuck Lorre notes, "These characters are the reason people watch. We don't have car chases. Helicopters don't come up over the horizon. It's just people talking. So they have to be great" (*4*). Yahoo's Mark Paul summarizes, "*The Big Bang Theory* has attracted and maintained its audience by keeping its own formula" (*5*).

Science as Gateway: *Fringe*

The title of *Fringe*, which ran five seasons on the Fox Broadcasting Company television network, epitomizes the importance of science to the series. On September 9, 2008 the series premiered. "*Fringe*! Words float on the screen, like 'teleportation' and 'dark matter' and 'nanotechnology.' These are the opening credits, not even long enough to accommodate a jaunty theme song," complained Daniel MacEachern, a reviewer for Television Without Pity (*7*). The show focused on fringe science, described in this exchange between lead characters FBI Agent Olivia Dunham (Anna Torv) and Peter Bishop (Joshua Jackson), estranged son of Dr. Walter Bishop (John Noble), a veteran of classified Army genetic experiments.

> Peter: When you say "fringe" science you mean pseudo-science.
>
> Olivia: I suppose. Things like mind control... teleportation... astral projection... invisibility... genetic mutation... reanimation... ("Pilot")

Agent Dunham, the Bishops, and FBI junior agent Astrid Farnsworth (Jasika Nicole) investigated "Fringe Events"—many of which traced back to technology developed by Massive Dynamic—a mysterious, "multi-faceted corporation working for the betterment of medical, communications, energy, transportation, and entertainment technology" (http://www.massivedynamic.com/our_history.php). William Bell (Leonard Nimoy), Walter's former research associate, founded Massive Dynamic. Their partnership dissolved after a lab accident that led to a worker's death and Walter's commitment to a mental institution.

Although MD's Executive Director Nina Sharpe (Blair Brown) claimed to have no knowledge of "The Pattern," a codename for various, connected fringe events, she seemed to know far more than she let on. FBI Special Agent in Charge Phillip Broyles (Lance Reddick) described The Pattern to Olivia, "Someone out there's experimenting, only using the whole world as their lab" ("Pilot").

In the roundtable discussion "The Culture of *Fringe*," Dr. Nicholas Warner, professor of Physics, Mathematics and Astronomy at the University of Southern California, distinguishes between sci-fi space opera and the science of *Fringe*.

> So [there's] science fiction where you change the law of universe, say let's run with that—*Star Trek*, whatever, warp drive, something. But then [there's] science where you say, "Okay, let's try and extrapolate what we currently know and push it beyond" (*8*).

That type of science Warner defines as fringe science. The series' premise coupled science as a discovery device—rather than just a "cool toy"—with aspects of a procedural drama. Though the first season unfolded like Fox's previous sci-fi series *The X-Files*, the primary difference was the nature of *Fringe*'s arc stories or mythology. Instead of alien abductions, government conspiracies, and cover-ups, Dunham's investigation of The Pattern led to the terrorist group ZFT, an acronym representing "*Zerstörung durch Fortschritte der Technologie*," German for "Destruction by Advances in Technology," and its mastermind David Robert Jones ("In Which We Meet Mr. Jones"). Throughout the season, clues unfolded: the presence of Observers—mysterious, bald men omnipresent throughout episodes; the disclosure Olivia as a child had been subjected to Walter and William Bell's genetic experiments; and the existence of parallel universes. The ultimate revelation to viewers was the season-ending shocker that Peter Bishop was not Walter's son who died as a child but a doppelgänger from another Earth. After repeated attempts to locate and interview William Bell (Leonard Nimoy), Olivia found herself face to face with him in his New York World Trade Center office—an unsettling scene for the post-9/11 audience ("There's More than One of Everything").

Differences between the prime Earth and the Earth "Over There" can be explained with basic precepts of Chaos Theory. Dr. Kevin R. Grazier notes, "The decisions we make on a daily basis can be viewed as perturbations to the paths of our lives—forks in the road known in Chaos Theory as bifurcations—with different, sometimes very different, possible outcomes" (*9*). Grazier continues, "[S]ome events simply have so much weight, so much gravity, that they cannot be changed ... even in societies that [show] pronounced divergence. The attacks of 9/11 were such events." Thus on the prime Earth, the Twin Towers were destroyed and Pentagon attacked, whereas Over There only the White House and Pentagon were attacked. Viewers eventually learned The Pattern originated at Reiden Lake when Walter, unable to save his own son, opened a portal into the alternate universe to save its Peter. This act had devastating consequences as foretold by Walter's colleague, Carla Warren.

> I may go to church every Sunday, Walter, but I also have three degrees in theoretical physics, and I am telling you, you cannot do this. We both know the amount of energy required to create a portal will forever ruin both universes. For the sake of one life, you will destroy the world. Some things are not ours to tamper with. ("Peter")

If one examines the series as a cohesive narrative text, Walter's abduction of Peter from the alternate universe served as Inciting Incident because that event destabilized both universes. During season two, Fringe Events escalated until The Pattern's architect is revealed as none other than "Walternate" ("Northwest Passage"). As U.S. Secretary of Defense, Walter Bishop from the alternate universe has declared war on the prime universe "motivated by personal fury over the kidnapping of his son and his personal view that only one of the two realities can survive in the long run," as explained by Paul Levinson in his essay "The Return of 1950s Science Fiction in *Fringe*" (*10*). He continues, "This Walter is more ruthless and less cracked than our Walter, and possibly more intelligent, since our Walter had his friend and colleague William Bell remove parts of his brain (which would account for at least part of our Walter's crackage)."

Fringe's archplot took several twists and turns during its run. Fox Broadcasting's website synopsizes seasons three and four:

> Ultimately, Peter realized his kidnapping was the cause for the alternate world's impending doom. So he chose to sacrifice himself by activating a device that saved the universe, bridged the two worlds together but erased him from existence.

Season Four found the Fringe Divisions working together to rebuild the alternate universe. Despite Peter's seemingly permanent sacrifice, he returned very much alive and helped stop William Bell's mad plans to collapse the two worlds together. Then, after Olivia learned she was pregnant with Peter's child and all seemed right with the world(s), Walter received an ominous warning that "they" were coming.

The writers set episode 419 "Letters of Transit" in the year 2036, two decades after the Observers seized control of the universe with the core Fringe Team having been awakened from suspended animation. This episode served as prelude for the final season as Olivia, Peter, Walter, and Astrid fought alongside the grown Etta Bishop (Georgina Haig) to liberate the world from the Observers.

Although some fans felt the change in time period blunted the series' effectiveness, others appreciated the emphasis of the series leads as a literal family unit, not just a figurative one. Episodes 502 and 503, "In Absentia" and "The Recordist," especially focused on Olivia's relationship with Etta. Haig describes her approach to playing the grown-up daughter of Olivia and Peter:

> It's bizarre to try to imagine what it would be like not having spent that time with your parents. I try to think of it as long-lost relatives that you feel a really deep connection to, but you have to get to know them all over again. There's that kind of unconditional love, but at the same time

there's an awkwardness sort of trying to get to know each other and how to be around each other. ("Etta" fox.com/fringe)

The first arc of season five ended with the reappearance of Philip Broyles—revealed to be working undercover to supply information to the Resistance—and the shocking death of Etta Bishop. As the final season unfolded, she became a martyr as her family coped with grief and continued to combat the Observer occupation. Throughout the series, whether the Fringe Team fought against ZFT, Walternate, or the Observers, scientific discovery was juxtaposed with its doppelganger—scientific destruction. The dialectical tension between the two is not unlike conceptions of Yin and Yang or order v. chaos—a resonant theme that will be discussed later in this chapter.

Science as Commodity: *Eureka*

The dichotomy between technology's almost limitless potential and its capacity to do harm was also a facet of the SyFy original series *Eureka*, which premiered on July 18, 2006. Season one's marketing tagline was simple yet effective: "Small town. Big secret." In the pilot episode, U.S. Marshall Jack Carter (Colin Ferguson) and his daughter Zoe (Jordan Hinson) get lost in the Pacific Northwest and stumble upon Eureka, a town "where the children are geniuses and the car mechanics are former NASA engineers" (SyFy website). While waiting for his car to be fixed, Carter learns of a missing child whose disappearance may be connected to a mysterious vortex. Despite warnings from Deputy Jo Lupo (Erica Cerra) and Department of Defense representative Dr. Allison Blake (Salli Richardson-Whitfield), Carter finds the boy. After the town's sheriff is sidelined by a vortex, Dr. Blake divulges the town secret to Carter: Eureka is a DoD think tank for the nation's leading scientists. The major player in Eureka is Global Dynamics, an advanced research facility. (Although *Fringe*'s Massive Dynamic has a similar name, *Eureka* premiered more than two years earlier. The writers of both series were probably inspired by General Dynamics, the real-life United States defense contractor founded in 1952.) *Discover* magazine's Eric Wolff described the *Eureka* writers' approach to science:

> So, sometimes the science leads directly to a show, but [co-creator Jaime] Paglia says that story and science have about equal weight in driving the arc of a given episode. Paglia and his team spend a lot of time with science magazines, blogs, and Web sites, and they track all their science and sci-fi ideas on the most important of all scientific tools: a white board.
>
> "In Season One, we put all of our characters on one board, with episodes across the top, and for each one we wrote what we want to have happen to these characters," Paglia said. "Meanwhile, we had a separate board with all the sci-fi ideas. We made a concerted effort, without being too on the nose, to tie what's happening with the science to what's happening to the characters" (*11*).

Science featured on *Eureka* ranged from tachyon accelerators and nanoids to time loops and an Einstein-Rosen Bridge device capable of creating wormholes.

The story arcs for seasons one and two dealt with "The Artifact," the last remaining object in the universe from before the Big Bang and that is housed in GD's mysterious Section 5. The third season focused on less technology-driven, more personal story arcs like the DoD sending in a corporate fixer to increase profits, an extended visit from Carter's very pregnant sister, as well as the introduction of Dr. Tess Fontana (Jaime Ray Newman), a new love interest for the sheriff. In the season four premiere "Founders Day," executive producer Paglia literally hit the reset button by sending Carter, Allison, Jo, engineer-turned-auto mechanic Dr. Henry Deacon (Joe Morton), and hapless scientist Dr. Douglas Fargo (Neil Grayston) back in time to 1947 when Eureka was a military base. The characters returned to a timeline slightly different than the one they left: Carter is still in a relationship with Tess; Allison's son, Kevin (Trevor Jackson), no longer has autism; Henry is married to Dr. Grace Monroe (Tembi Locke), whom he barely knows; Jo learns she and bad-boy scientist Zane Donovan (Niall Matter) were never in love; and Fargo is no longer a lackey at Global Dynamics but its director. Plus, Dr. Trevor Grant (James Callis), an associate of Albert Einstein they met in the past, has returned with them to contemporary Eureka. About.com reviewer Mark Wilson opined the following:

> *Eureka* started out in dangerous territory, with the outsider getting drawn into the life of a secret town, in a format that mines comedy from drama: the result could have been a "what explodes this week" kind of show. But the combination of exceptional writing and a truly compelling and always pleasant performance from Colin Ferguson has led *Eureka* in a deeper and more fulfilling direction (*12*).

That direction resulted in Season 4.0 focusing on the quintet dealing with repercussions of changes to the timeline—made trickier because not only are other Eurekans unaware of the shift but telling them would invoke DoD protocols and cause further problems. Perennial villain Beverly Barlowe (Debrah Farentino) also resurfaced to wreak havoc as season 4.5 centered on preparations for the manned Astraeus Mission to Titan, a moon of Jupiter. The season culminated with Allison finding herself trapped aboard the spacecraft as it launches, then disappears...

Alasdair Wilkins from the website io9 regards season 4.5 and the Astraeus project as "a bridge between *Eureka*'s contemporaneous scifi and the world of hard scifi space opera that lays [*sic.*] ahead." He continues his explanation.

> While this show's science has always been pretty out there—we've had time travel, human cloning, whatever the hell the Artifact was, and even some light space exploration before now—but it all happened in a world that was basically our own, just with one top-secret town in Oregon doing a bunch of crazy crap. Astraeus feels like a giant leap into a more broadly scifi world, and while I highly doubt we'll see the ramifications on *Eureka* itself, it's cool to think that what we've been watching over the last six

years is how we get from "our" world to one of starships and endless exploration (*13*).

Eureka's final season began with Astraeus landing on Earth four years after its disappearance. While the crew was missing and presumed dead, many changes transpired. Jo moved in with Carter and helped him take care of Allison's kids, while Deputy Andy headed GD's security division army of robots. Rather than another reset as the previous season began, however, this dystopian Eureka was an elaborate virtual reality simulation. Beverly Barlowe, Senator Wen (Ming Na), and their cronies in the Consortium, have hijacked the Astraeus Mission and networked the crew's brains as a human render farm to develop new technology. Barlowe boasted, "The greatest minds in Eureka are working for us now. And they don't even know it" ("Lost").

The initial triptych of episodes covered the return of the Astraeus crew to the "real" Eureka with unexpected assistance of Beverly, who reconsidered her alliance with Wen after the latter ordered the death of Fargo's girlfriend, Dr. Holly Marten (Felicia Day), lest the ruse be discovered. The remainder of the final season focused on the characters' romantic entanglements: Carter & Allison's rocky road to the altar; Henry's efforts to save Grace after her arrest for conspiring with the Consortium; Jo & Zane's long-awaited reconciliation in the new timeline; Fargo's attempt to implant Holly's neural imprint into a synthetic body. Perhaps the most insightful episode was "Smarter Carter," in which Kevin Blake (Trevor Jackson) spikes his stepfather's lattes with a cognitive enhancer so Jack could impress Allison's genius brother. However, as the sheriff's intelligence grows, his empathy wanes until she devises a cure. Throughout the series, Carter had been belittled because of his average IQ; yet he still managed to save the proverbial day in most episodes. *Eureka*'s science adviser Dr. Kevin R. Grazier explained.

> Carter has the ability to make connections between two seemingly independent events or two different kinds of technologies that may be interacting. It's the ability to make those types of connections that is a trait many successful scientists share.
>
> Summed up, we collectively call those abilities—the ability to let one's subconscious mull over our problems and the ability to make seemingly disparate connections—intuition. In that sense, what Carter does is *very* scientific (*14*).

Therefore, as the series moved toward its close, this episode confirmed what the writers had known all along: Jack Carter may have seemed a fish out of water in Eureka, but his intuitive gifts made him right at home there.

Science as Power: *Breaking Bad*

While *Eureka* focused on potential scientific discoveries that could be used to benefit humanity, AMC's drama *Breaking Bad* serves as a cautionary tale about how present-day science can be abused, contribute to a tragic flaw, and lead to a man's downfall. Protagonist Walter White seems like a contemporary Everyman when the series begins. A high school chemistry teacher who leads a devoted, albeit uninspired family life, Walter learns he has an advanced stage of lung cancer, nearly impossible to treat. Then after a chance encounter with former student turned meth cook Jesse Pinkman (Aaron Paul), Walter makes an impulsive decision. He decides to pay for his treatment and secure his family's financial future by using his knowledge of chemistry to manufacture and sell crystal meth. With warped altruism justifying the venture and science as his primary accomplice, Walt synthesizes a 99.1% chemically pure crystal methamphetamine. As he learns to navigate the violent crucible of the drug underworld, science often comes to his aid. He and Jesse melt the corpses of "collateral damage" using extremely corrosive hydrofluoric acid. In the episode "Crazy Handful of Nothin," Walt escapes a confrontation with dangerous drug kingpin Tuco Salamanca by substituting a crystalline nugget of highly volatile, mercury fulminate for meth. Stranded with Jesse in the middle of the desert ("4 Days Out"), Walt improvises a working galvanic cell battery out of everyday parts to repower the RV that houses their traveling meth lab.

Walt's rationalizations and good intentions inevitably pave a private road to hell. His needs to pay medical bills and provide for his family are supplanted by a lust for power, the craving to expand his production and distribution network for the sake of ego rather than need, and murder for pleasure and retribution rather than instinct and survival. He adopts the pseudonym Heisenberg (named after physicist Werner Heisenberg, creator of quantum mechanics) and eventually becomes feared and renowned as evidenced by the *narcocorrido* song "Negro Y Azul: The Ballad of Heisenberg," which served as the cold opening for the episode "Negro y Azul." Not coincidentally, the science of Walter White's meth production parallels his character arc. The loftier his ambitions for production, the more sophisticated his equipment and cook methods become. By season three, Walt and Jesse are operating out of a covert "superlab" financed by Gus Fring (Giancarlo Esposito) with industrial-grade materials and equipment that rival drug manufacturers like Pfizer and Merck. This lab eventually becomes a plot point as Gus schemes to replace Walt—who has turned into a liability—first with the ill-fated Gale Boetticher (David Costabile) then with Jesse. After outmaneuvering Gus and striking back at the end of season four, Walter devises a more sophisticated mobile manufacturing operation. He merges the equipment and cook chemistry methodology from the superlab with a new front, Vamonos Pest Control. By working under the guise of a fumigation company, Walt and Jesse can use a tented house as a temporary cook location, then after finishing, the real bug bombing can occur.

Breaking Bad creator/executive producer Vince Gilligan did not intend science to play a crucial role in the show despite his lifelong fascination with science, especially chemistry. In a 2010 interview with science blog ScriptPhD.com, he admitted structuring the show itself as an experiment—a character study of a man undergoing a rapid and radical transformation. In this case, that man just happens to have a profound and intimate knowledge of science, one that would serve as the primary means and centerpiece of his downward spiral. "An element I thought we could have fun with is the *MacGyver* aspect, and the idea of using [Walt's] knowledge to get him out of a jam every now and then," reveals Gilligan. His intense love of science also allowed him to overlay a framework of moral ambiguity atop *Breaking Bad* in direct contrast to the certainty of scientific principles, many of which Walt has used or taught. Gilligan further explains:

> I love the idea that there are real, concrete, black-and-white answers in science and math that we don't unfortunately get in the rest of our life. The world, and our lives, is full of gray areas and uncertainties and opinion versus fact. And yet in mathematics, 2 + 2 = 4 and always has and always will. In science, certain chemicals put together in a certain way always create the same compound (*15*).

People are more valuable, however, than the chemical compounds they are composed of—a lesson Walter White needs to learn.

In season two's penultimate episode "Phoenix" he stands idly by while Jesse's girlfriend Jane (Krysten Ritter) asphyxiates on her vomit. Walter perceives her as a threat to his relationship with Jesse and allows her to die. This act is unquestionably heinous, but the point of no return comes toward the end of season four when he poisons the young son of Jesse's girlfriend as a means to solidify an alliance against Gus Fring. Walter's willingness to manipulate and abuse Jesse as part of an oft-improvised master plan demonstrates an internal struggle between chaos and order. This is perhaps best exemplified by a quote from the season three episode "The Fly."

> My God, the universe is random. It's not inevitable; it's simple chaos. It's subatomic particles in endless, aimless collision. That's what science teaches us, but what does this say? What is it telling us that the very night that this man's daughter dies, it's me who is having a drink with him? I mean, how could that be random?

Despite Walter's internal conflict, his inability to temper science and reason with common sense and emotion may lead to his demise as the series completes its five-season run.

Conclusion

Ultimately, using science to tell a story involves a human story. Good stories center on people, the choices they make, and their relationship to good and evil. Like most effective science fiction, *Fringe* is about more than gadgets and technobabble. In Todd Aaron Jensen's interview with *Fringe* co-creator J.J. Abrams for *Written By*, the Writers Guild of America's magazine, the journalist opines *Fringe* is "a television cult hit about destiny, free will, and warring parallel universes" (*16*). Abrams, fellow co-creators Alex Kurtzmann and Roberto Orci, and showrunners J.H. Wyman and Jeff Pinkner deserve credit for organically integrating these themes and story elements. Abrams explains the original impetus for the series was to revisit "the Frankenstein idea but told as legitimately as possible" (*17*). Amy Sturgis elucidates in her essay, "In Search of *Fringe*'s Literary Antecedents":

> The difference between the characters of Victor and Walter is one of redemption. In Shelley's story, Victor dies, leaving his cautionary tale as his only positive legacy. Walter, however, has the chance to try to contain some of the damage he's caused. He's released from the mental asylum where he's spent the last seventeen years in order to lend his unique expertise to the science team of the Fringe Division (*18*).

Walter's "creature" was the rupture between worlds caused at Reiden Lake. A benefit of investigating The Pattern is Walter can atone for the damage he caused. After Peter's sacrifice and erasure from reality, he eventually phased back into existence. Walter eventually accepted this version of his dead son despite an initial reticence to do so. Actor John Noble recounts, "When Walter says to Peter, 'I know you're not my son, but you're the closest I'm ever going to get.' It was these people actually accepting that, but that's part of what we do in life, isn't it?" (*8*).

Later in season four, Walter journeyed to the alternate Earth to help uncover a mole, eventually identified as alternate-Broyles. His motivation, like Bishop's earlier, was to save his young son. The two fathers have a poignant conversation:

> Alt-Broyles: If you had to do it all over again, would you make the same choice?
>
> Walter: If you had asked me that question a few months ago, the answer would have been, "No." But now that I have met my adult son, at the moment, I don't know that I'd do anything differently. ("The Consultant")

After Alt-Broyles' arrest, Walter admonished the others, "Don't judge him. No one can be certain what they're capable of... How far they'll go to save the ones they love. I know this more than most" ("The Consultant"). Executive producer Pinkner adds, "Another theme that's always run through the show is the choices you make can actually change who you are. Perception is everything, and the choices you make and the way you perceive the world" (*8*). Pinkner's comments foreshadow *Fringe*'s conclusion and complement remarks September (Michael Cerveris) makes in the final episode. "Destiny can be changed," the former Observer advises Walter. "You have to have the will to change it, even if it requires sacrifice." Both men realize the solution to the Observer occupation is to take the child Michael (also known as Anomaly XB-6783746) 150 years in the future and create a paradox that will reset the timeline. Although September volunteers to perform this act, his untimely demise necessitates Walter stepping in to complete it. His willing sacrifice to separate himself from Peter completes the redemption Sturgis alluded to in her essay and counterbalances the abduction of young Peter, which caused the rift between worlds. Thus, as Jensen suggested in his Writers Guild article, *Fringe* used parallel worlds as a means to explore the age-old debate of destiny v. free will. The title of the series finale, "An Enemy of Fate," and the actions depicted within demonstrate the writers' advocacy of free will.

References

1. Seger, L. *Making a Good Script Great*, 2nd ed.; Samuel French: Hollywood, CA, 1994; p 202.
2. Barsan, R. *Looking at Movies: An Introduction to Film*; Norton: New York, 2004; p 64.
3. McKee, R. *Story*; New York: Regan Books, 1997, p 68.
4. Rice, L. *Entertainment Weekly*, September 28, 2012, pp 26–37.
5. Paul, M. *Yahoo.com*. http://tv.yahoo.com/news/big-bang-theory-draws-highest-ratings-yet-why-184500803.html (accessed November 21, 2012).
6. Grbic, J. *ScriptPhD.com*. http://www.scriptphd.com/comics/2010/07/24/comic-con-day-2/ (accessed July 24, 2010).
7. MacEachern, D. *Television without Pity*. http://www.televisionwithout pity.com/show/fringe/pilot-91-1.php (accessed September 9, 2008).
8. The Culture of Fringe. In *Fringe: The Complete Fourth Season*; Warner Home Video: Burbank, CA, September 4, 2012.
9. Grazier, K. R. In *Fringe Science*; Grazier, K. R., Ed.; BenBella Books: Dallas, TX, 2011; pp x–xi.
10. Levinson, P. In *Fringe Science*; Grazier, K. R., Ed.; BenBella Books: Dallas, TX, 2011; p 42.
11. Wolff, E. *Discover*. http://blogs.discovermagazine.com/sciencenot fiction/2009/07/31/talkin-science and-science-fiction-with-eurekas-jaime-paglia/ (accessed July 31, 2009).
12. Wilson, M. *About.com*. http://scifi.about.com/od/eureka/fr/EURE_ review-401.htm (accessed July 9, 2010).
13. Wilkins, A. *io9.com*. http://io9.com/5842005/eureka-throws-itself-into-a-black-hole-with-awesome-results (accessed September 20, 2011).
14. Jones, S. N.; Maxcy, L. R. *Eureka Unscripted*. http://eureka unscripted.typepad.com/blog/2010/02/blinding-you-with-science-part-one.html (accessed February 3, 2010).
15. Grbic, J. *ScriptPhD.com*. http://www.scriptphd.com/chemistry/2010/04/13/interview-breaking-bad-creatorexecutive-producer-vince-gilligan/ (accessed April 13, 2010).
16. Jensen, T. A. *Written By*; Summer 2011; p 22.
17. Evolution: The Genesis of *Fringe* In *Fringe: The Complete First Season*; Warner Home Video: Burbank, CA, September 8, 2009.
18. Sturgis, A. In *Fringe Science*; Grazier, K. R., Ed.; BenBella Books: Dallas, TX, 2011; p 21.

Chapter 22

Hollywood Science: Good for Hollywood, Bad for Science?

Sidney Perkowitz*

Charles Howard Candler Professor of Physics Emeritus, Emory University, Atlanta, Georgia 30322
***E-mail: physp@emory.edu**

Much of what Hollywood has to say about science and technology appears in science fiction and superhero films. Many of these are money-making blockbusters that influence how millions of people perceive science and scientists. Science onscreen is often inaccurate and can spread misinformation; nevertheless, appropriately used, "Hollywood science" can inspire young scientists-to-be, teach science, and enhance public discussion about science in our society.

Many Hollywood films fit into genres such as action (for example, *The Bourne Legacy*, 2012), crime (*The Girl with the Dragon Tattoo*, 2011), or Western (*True Grit*, 2010), but there is no genre labeled "science" or "technology." Still, these subjects appear in biographical, historical or documentary films about famous scientists like Madame Curie or momentous events like the development of the atomic bomb. Science has its greatest film presence and impact, however, in the science fiction and superhero genres. These highly popular motion pictures are big in Hollywood, earning billions of dollars by appealing to huge U.S. and global audiences. As they do, they carry explicit and implicit messages about science and scientists to millions of people.

That raises a question that I first discussed in my book *Hollywood Science* (*1*) and pursue further here: "Hollywood science" is clearly good for Hollywood, but is it also good for science? Judging by the history and impact of movie science, it can be, if properly used; otherwise, there are risks to science and its credibility.

The Beginnings of Science Fiction and Superhero Films

Science fiction has been embedded in film culture almost since movies began, starting in 1902 with director Georges Méliès' *Le Voyage dans la Lune* (*Voyage to the Moon*), from a story by Jules Verne. Some early science fiction films, such as *Metropolis* (1927) and *Godzilla* (1954), have become classics or at least cult classics; but mostly these motion pictures were low-budget "B" productions with little acclaim until features such as *2001: A Space Odyssey* (1968) and *Star Wars* (1977) brought major commercial and critical success.

Films starring superheroes—imaginary characters whose supernatural or extraordinary abilities are devoted to fighting evil—trace back to the "Golden Age" of comic books, the 1930s to the 1950s. Superman, in many ways the prototype of a superhero, appeared on the comic book scene in 1938, followed by Batman and others. Serialized superhero movies for children came after as did some superhero films in the 1960s; but like science fiction, it took a breakthrough film, *Superman* (1978), to carry the genre to a new level. Now superhero films draw on a whole set of characters such as the Hulk and Spider-Man.

The Reach of Hollywood Science

Since these beginnings, literally thousands of science fiction and superhero features have been released (*2*). The most successful reach enormous audiences as measured by world-wide box office receipts, and comprise over one-third of the 50 all-time top grossing films; for example *Spider-Man 3* (2007, ranked 23rd with a gross of $891 million), *Star Wars: The Phantom Menace* (1999, 10th at $1,027 million), and *Avatar* (2009, at $2,782 million the all time highest grossing film ever). Based on ticket prices, each of these films in the top 50 is estimated to have reached from 100 million to well over 300 million viewers, more than the U.S. population (*3*).

If only because of such huge numbers, these movies are culturally significant. Critics of every cultural and political persuasion routinely analyze what the latest science fiction or superhero film says about contemporary life and society. *2001* and *Star Wars* appear in lists of all-time best movies. Images, ideas, and language from the two genres have become iconic, from Spider-Man casting a web that lets him nonchalantly swing between skyscrapers, to space travel via *Star Trek*'s "warp drive," to "may the Force be with you" and the Hulk's warning "Don't make me angry! You wouldn't like me when I'm angry!"

Much of the meaning of these films derives from the messages they deliver about science and technology, and related societal issues such as climate change.

Science Content and Accuracy

But how much science really shows up in these films and how valid is it? By definition, a science fiction film uses scientific ideas, either as currently understood or by speculative extrapolation. For example, *Voyage to the Moon* and *Destination Moon* (1950, based on a story by Robert Heinlein), used existing science to propose that we could reach the Moon. Now that we have, science fiction like *Star Wars* postulates a next stage still far beyond our reach, travel to the stars. Other ideas, conjectures, or events involving every field of science and technology—alien life; earthquakes, disease outbreaks and other disasters; genetic engineering; artificial intelligence, and more—animate thousands of science fiction films.

Superhero films do not necessarily involve scientific ideas, but these can appear in superhero origin stories and some superheroes begin as scientists. Their ensuing adventures may also have a "sciency" or hi-tech tinge. Physics major Peter Parker is bitten by a genetically modified spider, becomes Spider-Man, and in *Spider-Man 2* (2004), battles a rogue scientist; Superman can fly because the Earth has a lower gravity than the alien planet of his birth; researcher Bruce Banner is exposed to gamma rays and finds that when he's angry, he turns into the incredibly strong and angry Hulk; and industrialist Tony Stark develops a tiny multi-million horsepower "arc reactor" to drive a flying suit that makes him a superhero in *Iron Man, Iron Man 2,* and *The Avengers* (2008, 2010, and 2012).

Whether a given film uses accepted or wildly speculative science, scientists may well have negative views about the science in the film. Some scientists delight in finding the errors, as has been described by astronomer and science blogger Phil Plait. As child and adult, he

> made fun of the science in movies. "That's so fakey!" I would cry out loud when a spaceship roared past…I decided it would be fun to critique the science of movies, and I dove in with both glee and fervor…It was surprisingly easy to deconstruct Hollywood accuracy, or lack thereof. Any mistake was fair game…

(Plait has recently changed his approach) (*4*). Filmmakers also note the lack of scientific accuracy. James Cameron, writer and director of *Avatar* and other science fiction movies, brings an expert's opinion when he says that science fiction films "almost never get their facts right (*5*)."

When the science on screen is wrong, scientists tend to think that filmmakers just do not know or care enough to get it right. Yet writers, directors, and actors have their own reasons to avoid error. They want to pull viewers into their imaginary worlds, but blatant violations of scientific ideas or natural law can ruin the illusion. Often enough, however, since film is a visual medium, "getting the science right" comes down only to "getting the science to look right," which means projecting a seemingly authoritative ambience with sleek lab equipment tended by scientists in lab coats.

The biggest factor playing against scientific exactness, though, is that Hollywood is not in business to produce illustrated lectures about chemistry or astrophysics but to turn out entertaining, money making films. That requires narrative drive, a dramatic arc, and compelling characters. When screenwriters or directors apply their story-telling judgment, they may choose to distort or hype the science in order to tell a better story.

The result is continuing tension between story and science that can be well balanced but can also skew a film to either side. As a pioneering movie about spaceflight, *Destination Moon* benefited from having Robert Heinlein consult about the science and artist Chesley Bonestell design the lunar sets, each at the top of his field and known for his commitment to accuracy. The result is scientifically impeccable, with realistic moonscapes and a clever Woody Woodpecker cartoon segment that illustrates the physics of spaceflight. But the story underplays its inherent drama, producing a correct but unexciting film (Figure 1).

Figure 1. Destination Moon (1950) gets high marks for accuracy in its story of spaceflight to the Moon in an era when that was still science fiction, but lags in drama and excitement. Courtesy of The Kobal Collection at Art Resource, NY with additional credit to George Pal Productions.

In other cases, the science may be distorted for story's sake but is not utterly wrong. *The Day After Tomorrow* (2004) presents basic information about global warming such as how scientists trace the history of global temperatures, the effects of warming on sea level and ocean currents, and the counterintuitive fact that the changed currents can cause an ice age. But to generate dramatic heft and a sense of desperate urgency, the time frame is compressed to days and weeks rather than the actual years and decades. The result is an intense, fast-paced story that uses award-winning special effects to both inform and misinform about global warming (Figure 2).

Figure 2. While getting some of the science right, The Day After Tomorrow (2004) exaggerates and accelerates the effects of global warming, as in this tsunami approaching New York City, rendered by computer generated imagery (CGI). Courtesy of The Kobal Collection at Art Resource, NY with additional credit to 20th Century Fox.

Then there are films that deliberately use completely wrong science. The accepted plot device where spacecraft travel faster than light violates the theory of relativity, but viewers are asked to suspend disbelief so that fictional spaceships can quickly cover cosmic distances. Other films stretch nuclear and genetic science beyond reality to produce impossible but striking mutants like the giant dinosaur Godzilla or the human variant superheroes in *X-Men* (2000) and its sequels. Unfortunately, some films contain unforced scientific errors that could just as easily have been expressed correctly without damaging the story, such as misstatements in *The Core* (2003), a film about what happens when the Earth's core stops spinning.

Scientists should certainly point out wrong science when they see it, but they can also fruitfully work out reasonable balances between story and science with filmmakers, or acquaint them with science that might spark movie ideas. The Science and Entertainment Exchange of the National Academy of Sciences (http://www.scienceandentertainmentexchange.org/) successfully enables such interactions between interested scientists and filmmakers. But as long as science fiction and superhero films generate billions, there is little reason for Hollywood as a whole to change its approach to science.

If Hollywood science cannot be fully trusted, should we conclude that Hollywood's gain is necessarily science's loss? Not at all, in my opinion. Properly used, fictional science can influence real science by helping to educate students and the public, contributing to the general discourse about science, and even inspiring scientists.

Inspiring Careers and Stretching Imaginations

Soon after NASA's Curiosity rover arrived on Mars in August 2012, the space agency named the rover's landing site "Bradbury Landing" after the late Ray Bradbury, author of the science fiction classic *The Martian Chronicles*. Asked why Bradbury was so honored, Michael Meyer, the NASA project scientist for Curiosity, replied: "This was not a difficult choice…Many of us and millions of other readers were inspired in our lives by stories Ray Bradbury wrote to dream of the possibility of life on Mars (*6*)."

Fictional science can inspire dreams and encourage scientific imagination. Bradbury first did so though books, and films can do the same. Ask a group of scientists and many will talk fondly about the science fiction films of their youth. In *The Seven Secrets of How to Think Like a Rocket Scientist,* James Longuski described how space scientists at a "prestigious Government laboratory" (probably NASA's Jet Propulsion Laboratory) gathered regularly to watch science fiction films from the 1950s. They would laugh at the errors, but

> they loved these films. They were like children who want to hear the same fairy tale over and over again. These were the fairy tales of the rocket scientists; their unfettered hearts seeking contact with outer space. Their logic turned off…their dreams turned on. Imagination wasn't silly to them (*7*).

Despite the flaws in Hollywood science, what scientists see on the big screen can motivate them to extend what *is* to what *might be*, using real rather than fictional science.

That inspiration can be particularly important for children and adolescents. Their attitudes toward science and scientific careers are influenced by popular culture, especially as it portrays scientists. In one study, Jocelyn Steinke of Western Michigan University and her colleagues pointed out that "most children do not typically come in contact with actual scientists." Instead,

> …many grow up seeing images of scientists…as depicted by characters and images in books, movies, television programs, magazines, comics, video games, clip art, Web sites, and a variety of other media sources

The study found that over 40% of 300 students 12 to 13 years old said that their images of scientists came from films or television (the study did not distinguish between the two media) (*8*).

Unfortunately, these images often convey stereotypes, not fully realized characters. When James Cameron bemoaned inaccurate science fiction films, he also identified two movie scientist stereotypes as "idiosyncratic nerds or actively the villains." There is also a third less negative stereotype, scientist as hero. Still, a stereotype is a stereotype. All three appear in films and distort the reality of what kinds of people become scientists and why.

For example: mad geneticist Marlon Brando creates unnatural mixtures of human and beast in *The Island of Dr. Moreau* (1996) (Figure 3), and evil geneticist Sean Bean harvests human body parts from clones for profit in *The Island* (2005); climate scientists Dennis Quaid and Ian Holm in *The Day After Tomorrow* heroically risk all (the Ian Holm character dies) to warn the world of coming disaster; Dr. Brackish Okun (Brent Spiner) in *Independence Day* (1996) is unkempt and peculiar; and in an ambivalent, metaphorical turn in *Spider-Man 2*, well-meaning Doctor Octavius becomes destructive Doc Ock (both played by Alfred Molina) through his own science when neural implants he uses to develop clean fusion power take over his mind.

Figure 3. Geneticist Dr. Moreau (Marlon Brando) in The Island of Dr. Moreau (1996, the latest film version of the H. G. Wells story), is a mad scientist who creates a race of half-human monsters. Courtesy of The Kobal Collection at Art Resource, NY with additional credit to New Line.

Such stereotypes may give young people only a confused understanding of what it is to be a scientist. However, some movie characters project more balanced images, such as radio astronomer Ellie Arroway in *Contact* (1997) (Figure 4). Jodie Foster plays the part well and the film convincingly shows Arroway's early interest in science as fostered by her father, the rewards of a commitment to science along with the difficulty of balancing it with a personal life, and Arroway's scientific integrity—all parts of being a real scientist. Anecdotal evidence and analyses show that scientists, non-scientists and media scholars alike find Ellie Arroway an appealing and valid scientist role model (*9–11*).

Figure 4. Radio astronomer Ellie Arroway (Jodie Foster) in Contact (1997) is a relatively realistic and nuanced depiction of a scientist on screen. Courtesy of The Kobal Collection at Art Resource, NY with additional credit to Warner Bros/Southside Amusement Co.

As a bonus, this realistic character is female, not male. The historically low number of women in science and engineering is growing but they remain underrepresented (*12*). As Steinke notes, one approach to increasing their numbers is to give females opportunities to encounter women scientists:

> Interaction with women scientist role models has been singled out as an important factor in fostering positive attitudes toward science and scientific careers in girls and young women…In the absence of real-life role models, images of women scientists in the media may serve as important sources of information (*13*)…

Films can provide this inspiration since women scientists are increasingly appearing on screen. For instance, in 60 movies made from 1929 to 2003, only 18% of the scientists portrayed were female (*14*, *15*). Similarly, in the Internet Movie Data Base IMDB.com I found that in over 800 films with scientist characters released from 1915 to 2012, fewer than 20% included women scientists. But more recent films released from 1991 to 2001 show a big increase to 34% females among scientist and engineer characters (*16*). This correlates with growth from 1993 to 2008 in the female fraction of U.S. workers with science or engineering degrees (21% to 28%) or with careers in those areas (31% to 38%) (*12*). It is unclear, however, how much of this was influenced by movie characters, especially since few scientists are portrayed as well as in *Contact*.

Beyond potentially interesting young people in science careers, Hollywood science can help them learn real science—remarkably, even if the Hollywood science is wrong.

Educating Scientists and Non-Scientists

Essayist Susan Sontag once noted that science fiction films wield a power of "sensuous elaboration…by means of images and sounds…" that gives them an impact far beyond written science fiction (*17*). Today, digitized scenes created by computer—dubbed CGI, computer generated imagery—provide visual power that can give these films a big role in teaching science.

First used in the robot story *Westworld* (1973), CGI can show every kind of advanced technology, distant planet, and alien being that inventive writers can create, and can simulate exotic phenomena like tsunamis and black holes (see Figure 2). In our media-driven age with students attuned to visual representation, these effects can enhance science teaching, especially if combined with absorbing emotion and drama (which may be a problem. Some film critics comment that unimaginative reliance on CGI coupled with a certain corporate style of film making is producing formulaic movies that lack feeling, meaning, and characterization, including many superhero films) (*18*, *19*).

Whatever the cinematic merits, one teaching approach is to enrich regular science courses with lessons from the movies, as Costas Efthimiou at the University of Central Florida has pioneered in "Physics in Films." (Other such courses are described in this volume). For example, he uses *Armageddon* (1998), in which a huge asteroid "the size of Texas" is on a collision course with the Earth. To avert this, NASA sends up a team of oil drillers to plant a thermonuclear bomb deep inside the oncoming rock. When the bomb is set off, it will presumably split the asteroid and push the two pieces sideways to bypass the Earth on either side.

In the movie, this plan saves our planet, but does real science support the happy Hollywood ending? To find out, the students calculate the asteroid's mass, use the laws of mechanics to compute the paths of the two halves, and reach a surprising conclusion: even a powerful hydrogen bomb explosion would separate the two massive chunks of rock by only 400 meters. Both would still hit the Earth, and disaster would not be averted after all. This example engages the students and then channels their interest to exercise their analytical and scientific abilities. The value of the approach is borne out by the fact that students in the course perform better than those in a similar course without films (*20*, *21*).

A course I have co-taught since 2006 at Emory with my colleague Eddie von Mueller of Film Studies exemplifies a different approach (*22*). Rather than work within an existing science course, "Science in Film" was designed to illuminate both science and cinema for science and humanities students, as expressed in the two course texts: my own *Hollywood Science*, about the science in movies; and Vivian Sobchack's *Screening Space*, about the cinematic and cultural meanings of science fiction (*23*).

The course is built around films chosen to cover important scientific ideas and events and their human impact, from climate change to the rise of computers, or that show scientists (Table I). Students hear lectures about the science, given in broad terms rather than quantitative detail, and watch the associated films. Having absorbed all this, the students are led by both instructors to discuss the real and the fictional science and are also required to write film logs and papers.

Table I. Semester topics and filmography for the course "Science in Film" co-taught at Emory by the author and his colleague Eddie von Mueller, Department of Film Studies

Week	*Topic*	*Required films*	*Suggested films*
1	Historical Introduction	*Le Voyage dans la Lune*	*La Hotel Electrico, Destination Moon*
2, 3	Alien Encounters	*War of the Worlds* (1953, 2005) *The Day the Earth Stood Still*	*The Thing From Another World, Invasion of the Body Snatchers* (1956, 1978), *E. T., Close Encounters of the Third Kind, Prometheus*
4	When Worlds Collide	*Armageddon*	*Deep Impact, When Worlds Collide*
5	Worlds Gone Mad	*The Day After Tomorrow*	*Soylent Green, Volcano, Waterworld, 2012*
6,7	Smashing Atoms	*On the Beach* (1959) *Fat Man and Little Boy*	*Godzilla, The Sum of All Fears, The China Syndrome, Chain Reaction*
8	The Third Horseman	*Outbreak, Contagion*	*The Omega Man, The Andromeda Strain, Panic in the Streets*
9	Send in the Clones	*Gattaca*	*Jurassic Park, The Sixth Day, The Boys From Brazil, The Island of Dr. Moreau* (1996), *The Island, Splice*
10, 11	Men and Machines	*Colossus: The Forbin Project Terminator*	*A.I., 2001: A Space Odyssey, Westworld, RoboCop, I, Robot, Prometheus*
12, 13	The Movie Scientist	*Contact Dr. Strangelove*	*Metropolis, Frankenstein, The Boys from Brazil, Spider-Man 2*
14	Reel Scientists	*Kinsey*	*Gorillas in the Mist, Dr. Ehrlich's Magic Bullet, Infinity, A Beautiful Mind*

The course enhances science literacy by introducing big scientific ideas and also through its cinematic perspective. The students are asked to consider how a film packages its science and scientists—what choices are made, what attitudes and agendas underlie them. This encourages critical thought about varied approaches to science in outlets from entertainment and news media to research journals. That makes students better able to thoughtfully weigh scientific claims, for instance in public policy debates, where film itself can also play a role.

Public Discourse

Some science fiction or superhero films influence public discussion of science because they express a particular viewpoint. Others do so as products of their times that reflect existing values and concerns. As they echo and amplify current issues, their wide exposure makes them one more voice that discusses science in our lives.

These influences were apparent in the 1950s and 1960s during the Cold War when the world faced serious nuclear threat. The classics *The Thing from Another World* (1951) and *The Day the Earth Stood Still* (1951) commented on nuclear dangers, *Godzilla* conveyed fears of radiation, and *On the Beach* (1959) and *Dr. Strangelove* (1964) dealt with nuclear war. Later *The China Syndrome* (1979) presented a fictional civilian nuclear accident nearly simultaneously with a real reactor meltdown at Three Mile Island, and *The Sum of All Fears* (2002) was about nuclear terrorism. Now, nearly 70 years after the first nuclear explosion, films still use radiation or nuclear angst as a *deus ex machina* and in their back stories, as in the origins of the Hulk, and of superhero Dr. Manhattan in *Watchmen* (2009).

Many films express new concerns about genetic engineering and cloning. After the structure of DNA was determined in 1953, genetic manipulation started to become a reality. It entered into *The Boys From Brazil* (1978), *Jurassic Park* (1993), *The Island of Dr. Moreau* (1996), *The Sixth Day* (2000), *The Island* (2005), *Splice* (2009), and more. Most of these films are pessimistic about evil geneticists or societies based on genetic discrimination—the theme of *Gattaca* (1997)—and some tackle the morality of genetic engineering.

Tension between science and religion also enters into a recent high-profile film, *Prometheus* (2012). This semi-prequel to *Alien* (1997) has flaws along with high points, but with its story about the origins of life, a scientist character with religious convictions, and a near-human android character, it evokes discussion about how and why life was created (Figure 5).

Figure 5. By speculating about how life began on Earth and including a scientist with religious faith (Noomi Rapace, center) and an artificial human (Michael Fassbinder, right), Prometheus (2012) invites thoughts about how, why, and by whom humanity was created. Courtesy of The Kobal Collection at Art Resource, NY with additional credit to Scott Free Productions/20th Century Fox.

Among films that have strongly affected public consciousness, the dramatically enhanced climate change presented in *The Day After Tomorrow* created much controversy about the film's scientific accuracy and political influence. After it opened in 2004, Anthony Leiserowitz of the University of Oregon (now Director, Yale Project on Climate Change Communication), surveyed 529 adults and found that the film had "significant impact." It led viewers to

> higher levels of concern and worry about global warming [and] encouraged watchers to engage in personal, political, and social action to address climate change and to elevate global warming as a national priority…the movie even appears to have influenced voter preferences.

This influence was widespread. Domestic (U.S. and Canada) filmgoers bought 30 million tickets, and worldwide box office sales of $544 million represent many more millions of viewers (*24*). Also, during the weeks before and after its release, internet traffic to websites about global warming grew significantly, probably aided by the major marketing effort for the film (*25*).

It is illuminating to compare this film to a less sensational documentary released two years later. In *An Inconvenient Truth* (2006), former Vice-President Al Gore made the case for global warming using mostly accepted science without major distortion or dazzling CGI, essentially as an illustrated lecture (Figure 6). The film won a Best Documentary Academy Award and grossed $25 million each domestically and overseas—excellent for a documentary, but representing fewer than 4 million domestic tickets compared to 30 million for *The Day After Tomorrow*.

Figure 6. An Inconvenient Truth (2006), with former Vice-President Al Gore, presented global warming less dramatically and more accurately than The Day After Tomorrow but reached only a fraction of its viewers. Courtesy of The Kobal Collection at Art Resource, NY with additional credit to Lawrence Bender Prods/Lee, Eric.

This disparity shows in no uncertain terms that distorted science reaches more people than real science, which returns us to the original question "Is Hollywood science good for science?" The question needs an answer because science fiction and superhero films will continue to reach big audiences, with all that implies.

The Future of Hollywood Science

Hollywood sees a bright future in fictional science. IMDB.com lists about 200 new science fiction and superhero titles, mostly Hollywood products, due for release in 2013–2016 or in development. Some will likely be major hits, such as continuations of successful franchises like *Iron Man 3*, *RoboCop*, *The Hunger Games: Catching Fire*, and *Star Trek Into Darkness,* in 2013; *Jurassic Park IV* and *X-Men: Days of Future Past* in 2014, and *The Avengers 2* in 2015.

These upcoming films carry assorted possibilities for real science. Most are science fiction rather than superhero stories, where science is often less prominent or may be displayed along with magical or divine powers. In *The Avengers*, for instance, the supernatural abilities of the gods Thor and Loki are confusingly featured on an equal basis with Iron Man's flying suit and other hi-tech elements (Figure 7). But this scenario too can be used to teach science, or of equal value, the distinction between science and magic. Some of the films also cover big issues like climate control (also known as geoengineering), a controversial method under study by scientists and policy makers to reverse climate change, and so could produce an impact like that of *The Day after Tomorrow.*

Figure 7. Some superhero films mix scientific or hi-tech elements with fantastic or supernatural ones. The Avengers (2012) features both Iron Man (top, Robert Downey Jr.) in his advanced flying suit and the god Thor (bottom, Chris Hemsworth) with his magic hammer. Courtesy of The Kobal Collection at Art Resource, NY with additional credit to Marvel Enteprises.

The Day After Tomorrow is eight years old and *Contact* came out in 1997. It is striking that many of the films that convey good science or plausible scientists, or inspire discussion, date back years. These older movies would occasionally pause the action to deliver some exposition, exercising viewer's minds even if the science was fanciful. CGI now tempts filmmakers to replace exposition with non-stop spectacle. If fantasy hugely (sometimes numbingly) amplified by computerized filmmaking is "chasing human temperament and destiny—what we used to call drama—from the movies" as David Denby writes, it is also chasing away ideas (*26*).

Still, it may take spectacles to attract millions of viewers and give a scientific issue wide circulation. Scientists have to face the quandary that though high popularity for a film translates into broad exposure for its science, it probably correlates with low accuracy. The dilemma is well put by Ron Von Burg at Christopher Newport University, who studies how science is communicated (*27*):

> [H]ow can scientists marshal the increased public attention that accompanies a popular film to help communicate important scientific matters to non-scientists without undermining their scientific credibility (*28*)?

There is no single or simple answer, but scientists can use film to their advantage if they remember that movies are not lectures in Science 101. If scientists look for the true scientific nugget behind even outlandish screen science, and express it accurately within the sense of wonder that Hollywood can create, the result can be inspirational power, better teaching, and greater outreach.

Like any moviegoer with popcorn in hand, any scientist can enjoy watching a science fiction classic from the 1950s or Hollywood's latest superhero effort; then he or she can step into a classroom or behind a podium and draw on what the film offers in drama or visual richness or excitement to make meaningful points about, and for, science.

References

1. Perkowitz, S. *Hollywood Science: Movies, Science and the End of the World*; Columbia University Press: New York, 2010.
2. The Internet Movie Data Base IMDB.com lists over 6,000 feature films in the genre "sci-fi" or under the keyword "superhero" ("superhero" is not listed as a stand-alone genre), with some overlap between the two areas and with the related genres of "fantasy" and "horror."
3. Box office and ticket price data from http://boxofficemojo.com. Box office figures not adjusted for inflation.
4. Plait, P. How I Stopped Worrying about Science Accuracy and Learned to Love the Story. *The Science and Entertainment Exchange*, May 24, 2012 [Online]. http://tinyurl.com/83r4dmy (accessed August 13, 2012).
5. Revkin, A. Filmmaker Employs the Arts To Promote the Sciences. *The New York Times*, February 1, 2005, p D2.
6. Vejvoda, J. NASA Names Mars Curiosity Landing Site after Ray Bradbury. *IGN*; August 23, 2012 [Online]. http://tinyurl.com/crj6hf5 (accessed August 30, 2012).

7. Longuski, J. *The Seven Secrets of How To Think Like a Rocket Scientist*; Springer: New York, 2010; pp 5–6.
8. Steinke, J.; Lapinski, M.; Crocker, N.; Zietsman-Thomas, A.; Williams, Y.; Evergreen, S.; Kchibhotla, S. Assessing media influences on middle school-aged children's perceptions of women in science using the draw-a-scientist test. *Sci. Commun.* **2007**, *29*, 35–64.
9. Steinke, J. Women scientist role models on screen: A case study of *Contact*. *Sci. Commun.* **1999**, *21*, 111–136.
10. Perkowitz, S. Hollywood Physics. *Physics World*, July 2006, pp 18−23.
11. Perkowitz, S. *Hollywood Science: Movies, Science and the End of the World*; Columbia University Press: New York, 2010, pp 181−184.
12. National Science Board. *Science and Engineering Indicators 2012*; NSB 12-01; National Science Foundation: Arlington VA, 2012. [Online] http://www.nsf.gov/statistics/seind12/c3/c3h.htm (accessed September 13, 2012).
13. Steinke, J. Women scientist role models on screen: A case study of *Contact*. *Sci. Commun.* **1999**, 111–112.
14. Flicker, E. Between brains and breasts: Women scientists in fiction film. *Public Understanding Sci.* **2003**, *12*, 307–318.
15. Flicker, E. Representation of Women Scientists in Feature Films: 1929 to 2003. *Bridges*, Vol. 5, April 14, 2005. [Online] http://tinyurl.com/9fdrzjw (accessed September 13, 2012).
16. Steinke, J. Cultural representations of gender and science: Portrayals of female scientists and engineers in popular films. *Sci. Commun.* **2005**, *27*, 27–63.
17. Sontag, S. *Against Interpretation and Other Essays*; Picador: New York, 2001; p 212.
18. Dargis, M.; Scott, A. Super-Dreams of an Alternate World Order. *The New York Times*, June 27, 2012. [Online] http://tinyurl.com/d6jj9r8 (accessed September 27, 2012).
19. Denby, D. Has Hollywood Murdered the Movies? *The New Republic*, October 4, 2012, pp 29−40.
20. Efthimiou, C.; Llewellyn, R. Cinema as a Tool for Science Literacy. April 16, 2004 [Online]. http://arxiv.org/PS_cache/physics/pdf/0404/0404078v1.pdf (accessed September 27, 2012).
21. Efthimiou, C.; Llewellyn, R. Avatars of hollywood in physical science. *Phys. Teach.* **2006**, *44*, 28–33.
22. Perkowitz, S.; von Mueller, E. Communicating Real Science through Hollywood Science. In *Taking Science to the People*; Johnson, C., Ed.; University of Nebraska Press: Lincoln, NE, 2010; pp 81−88.
23. Sobchack, V. *Screening Space: The American Science Fiction Film*; Rutgers University Press, New Brunswick, NJ, 1997.
24. Leiserowitz, A. Before and after *The Day After Tomorrow*. *Environment* **2004**, *46*, 22–37.
25. Hart, P.; Leiserowitz, A. Finding the teachable moment: An analysis of information-seeking behavior on global warming related websites during the release of *The Day After Tomorrow*. *Environ. Commun.* **2009**, *3*, 355–366.

26. Denby, D.Has Hollywood Murdered the Movies? *The New Republic*, October 4, 2012, p 33.
27. Von Burg, R. Decades away or the day after tomorrow? Rhetoric, global warming and science fiction film. *Crit. Stud. Media Commun.* **2012**, *29*, 7–26.
28. Von Burg, R. Decades Away or The Day After Tomorrow? *Communication Currents*, February 7, 2012 [Online]. http://www.natcom.org/CommCurrentsArticle.aspx?id=2156 (accessed September 16, 2012).

Chapter 23

Hollywood Aliens

Seth Shostak*

SETI Institute, 189 Bernardo Ave., Mountain View, California 94043
***E-mail: seth@seti.org**

Intelligent beings from other worlds have become a frequent staple of cinema sci-fi, appearing in at least a half-dozen films each year. To some extent this is the consequence of computer animation techniques that allow these imaginary characters a wide range of appearance and actions. However, recent developments in astronomy and space research have also primed the public to believe that such beings might actually exist, increasing their credibility with movie audiences. Although Hollywood aliens are very much like us in appearance and abilities, they are generally bereft of credible incentives for their (frequently malevolent) behavior. We describe the limited view of extraterrestrials given in films, and compare their construction and motives to those that real aliens might have.

Now here's an interesting story idea for a movie. Imagine visiting a large moon, four light-years away, populated by blue-skinned aliens who, despite a degree of social sophistication that might impress your bowling buddies, are still reliant on animal transport. These empathetic and appealing characters are in a battle to avoid having their world turned into a massive, open-pit mine.

No? Well, here's another plot that's sure to boost popcorn sales: It's a long time ago, and some random galaxy far, far away is roiled by the endless intrigues of a massive empire (democracy is apparently not popular in this part of the cosmos). Faceless warriors, hermetically packaged in white plastic, spend years duking it out with young and humanoid knights. The knights wield weapons that became obsolete on Earth when gunpowder was invented.

Nutty? Not very. According to the movie web site Box Office Mojo, "Avatar" is the biggest grossing film ever, raking in ticket sales rivaling the national budget of Paraguay. "Star Wars: Episode 1" is the number ten money-maker, despite the goofy premise that defenders of freedom would trust their lives to a light saber and some mysterious scalar field called "the Force."

What do these pictures have in common, other than market success? They feature *aliens*—extraterrestrial sentients. And so do three of the all-time, top-ten money-earning movies. Aliens are definitely trending. Mind you, in the real world, scientists haven't found good evidence for *any* life beyond Earth—sentient or otherwise. But aliens have been discovered by the entertainment industry, and the discovery has proven lucrative.

Sadly, however, these other-worldly beings are typecast. Celluloid extraterrestrials are subservient creatures who usually act *en masse*. They're mostly unattractive, crippled in their abilities, and generally propelled through the story by trivial motives.

But real aliens are likely to be different.

Why They're Popular

Cinema aliens are all-purpose characters, although the overwhelming majority are slotted to be antagonists. An exception was *E.T., The Extraterrestrial*, a friendly creature who fell to Earth as a botanist and eventually became a mascot adopted by suburban kids. But far more typical is the type of extraterrestrial found in *War of the Worlds* or the *Alien* franchise: pitiless and peckish creatures who have nothing to offer humans, other than a tour of their other-worldly digestive tracts.

While often lacking in breadth of character, aliens have utility, as they can be safely cast as thoroughly evil beings. No organizations will protest that they, as a group, are being unfairly maligned, or that these entities from afar are merely misunderstood. Aliens can also exhibit the most exotic of appearances and behaviors without fear of contradiction, although these possibilities are only spottily exploited. Extraterrestrials are the natural inhabitants of distant locales and the otherwise-sterile acreage of deep space, and they provide both a reason to go into the cosmos, and a danger in doing so.

The popularity of extraterrestrials with movie makers bears witness to these plus points. Aliens have appeared in at least 300 theatrical films since the Second World War. Part of this success can undoubtedly be ascribed to technical factors, such as the development of computer animation techniques that have given these beings a wider range of behavior and appearance. Gone are the days when they were stiffly brought to life either with puppets or actors in rubber suits. In addition, developments in astronomy—including the discovery of planets around other stars and the robotic exploration of the nearby solar system—have undoubtedly raised public awareness of the possibility of life in space, making alien characters both plausible and current. There's also a practical advantage in that these synthetic actors don't demand residuals from the film production companies.

While these factors may account for the durability of aliens in the movies, it's noteworthy that these stars from the stars have become especially alluring in the last two dozen years. There are undoubtedly multiple causes for this, including the rapid improvement in computer animation noted above. In addition, by the beginning of the 21st century, astronomers were finding planets by the bucketful, suggesting that worlds amenable to life could be commonplace.

However, another plausible inducement for making aliens part of the casting call was the collapse of the Soviet Union. Aliens comprise a handy pool of bad guys to replace the sinister characters from the USSR that had been the steadfast movie heavies of the Cold War. From 1949 to the demise of the Soviet Union in 1991, the number of films featuring extraterrestrials averaged 3.4 annually. Since then, the average has more than doubled, to 8.0 per year. Aliens are everywhere.

In addition, there's the widespread appeal of these beings to young people. Evolution has ensured that we are hard-wired to be interested in predators and potential competitors. Like the dinosaurs—which are intriguing to nearly every child—extraterrestrials are seen as fascinating in their latent danger, and yet at a safe remove from everyday life. They're scary without being threatening.

Physiology

Hollywood aliens mostly come in two flavors: child-like and friendly, and relentlessly nasty. The latter category is more common, given that aggressive aliens produce instant conflict and a straightforward story line. Nonetheless, sympathetic aliens, like *E.T.*, *Starman*, *Close Encounters of the Third Kind*, and *Paul* have been popular, and in the case of the Spielberg films, *extremely* popular.

Usually, friendly aliens are built like children, with small bodies, big eyes and—in the case of *E.T.*—a wrinkly face. The noxious aliens are often supersized versions of arthropods (*Independence Day*), insects (*Starship Troopers*) or squids (*War of the Words*, *Monsters*). This is no more than the sci-fi equivalent of the general cinema conceit in which evil people are physically unattractive.

When it comes to alien appearance, one morphology is considerably more fashionable than others: The short, smooth-skinned gray with large eyes and no smile. This type of being is not only frequently seen in movies and on television, but is regularly described by the many thousands of people who claim to have encountered extraterrestrials on Earth.

The grays are popular with the public, which makes them popular with film makers. After all, they comprise an easy visual shorthand for visitors from another world, and that reduces the need for time-wasting exposition. Films then reinforce this campy icon, completing the circle.

The grays are familiar and terrestrial, with their bilateral symmetry, upright stance, four appendages, and clearly demarcated heads. This is undoubtedly a consequence of a simple assumption: Any beings with the ability to come to Earth (as they often do in films) will be technically more advanced than us. Consequently, we naively assume that they will also be more advanced in an evolutionary sense. Extrapolating a popular view of our own Darwinian trajectory, such creatures are expected to have larger heads and smaller bodies—better adapted to the cerebral lifestyle we assume the future will hold. The grays are our distant descendants.

One could object to the easy belief that evolution will inevitably favor bigger brains and smoother complexions. While these are facile ideas, they're hardly realistic. Evolution has no goal, and shifts strategies on the fly. In addition, there's the highly suspect, self-centered nature of this forecast. Imagine if trilobites were asked to describe putative life forms visiting Earth. These ancient arthropods might earnestly argue that extraterrestrial beings will resemble trilobites, differing only in a few plausible refinements. Our popular perception of aliens is strongly anthropocentric.

But having recognized this bias, is there anything we could say about how the real E.T. might appear? A casual perusal of Earth's fauna will demonstrate that there are many body plans that are functional, and only a few resemble us. On the other hand, there are some features of our physiology that seem particularly important for an intellectual and technically competent species: for example appendages able to manipulate tools, stereo vision, and a capability for speech. Being a land-dwelling animal is also undoubtedly helpful, as this has allowed us to develop metallurgy, radio, and astronomy.

But while these simple considerations demonstrate that humans have an appropriate design for a technologically adept species, it hardly ensures that ours is the *only* design. Many things could be different. Why not six appendages (like insects) rather than four? The latter number is simply an accident of evolution, and hardly prescribed by any plausible requirements. Why not add an eye that scans behind us, or a better sensitivity to smells? Perhaps a workable ability to gauge magnetic fields?

These matters are of interest to biologists, undoubtedly, but not so much to Hollywood. Media mavens are rightly concerned with the requirements of story telling, and those often favor alien construction that is similar to ours.

Again, this is a matter of narrative economy. Consider gestures—such as expressions of anger, fear, or friendship. Snarling is surely species specific, but when we see a gray alien snarl in some sci-fi offering, we are automatically on alert—a shortcut that aids the film maker. Steven Spielberg crafted E.T. with big eyes and a big head—not because these are especially likely characteristics of someone from the Andromeda galaxy, but because they gave him a child-like appearance that provided audiences immediate insight into his character.

The benefit of other-worldly beings that are similar to us is obvious. And not infrequently, the idea is taken to its logical conclusion: cinematic extraterrestrials that are physically *identical* to humans. From *Invasion of the Body Snatchers* to *The Day the Earth Stood Still*, invading aliens will frequently make the decision to masquerade as *Homo sapiens*, a tactic that will presumably give them greater knowledge of our societal workings, or allow them to infiltrate and eventually overwhelm our planet with less effort. Sometimes a small concession is made to their transcendental origins, just so that you can be sure that they're really alien. Spock's ears, or Klaatu's flat personality are examples.

But while aliens in human form have obvious practical advantages for actors as well as the audience, they don't make much sense as science. It's difficult to believe that when Charles Darwin visited the Galapagos, he would have considered it beneficial to adopt the appearance of an iguana or a finch in order to better study these creatures, or dominate them.

Aside from appearance, there's the matter of the chemical arrangement of these beings from other worlds. Few films bother to delve into the biochemistry of extraterrene characters, although there are hints to be gleaned from their vulnerabilities. The malevolent extraterrestrials in *Signs* would dissolve into harmless mush if splashed with ordinary water (could they be related to the Wicked Witch of the West?) Sea water was also a fatal fluid for invaders from afar in *The Day of the Triffids*. Given that most astrobiologists consider liquid water to be the common denominator for life, this particular Achilles heel seems unlikely, but in any case mandates a substantially different construction than terrestrial biota.

Other suspect, but largely uninformative, vulnerabilities include aliens who could be stopped by microwaves in *Darkest Hour* (a film whose ratings were lower than Death Valley) and of course the Martian invaders in *The War of the Worlds*, who were felled by terrestrial microbes.

These too bespeak a body makeup that's either very unlike our own, or—in the case of susceptibility to our germs—too much like our own. After all, infectious diseases are the result of specific biochemical interactions that have co-evolved between agent and host over the four billion years of life's presence on Earth. It's stunningly unlikely that our bacteria could sicken the extraterrestrials.

What about DNA? Would the aliens have it? While all terrestrial life uses DNA for information storage, it's hardly clear that it's the only molecular structure that would serve this function. But whether movie aliens have DNA, XNA, QNA or some compound that's not even a nucleic acid is almost never made clear, probably because audiences aren't terribly interested in disquisitions on organic chemistry. The exception would be those films (like *Prometheus*) in which we are explicitly related to the extraterrestrials. According to *Star Trek* lore, Mr. Spock was spawn of a Vulcan dad and a human mom, which says that the two species share not only DNA, but virtually total reproductive compatibility: again, an unlikely but seductive suggestion. Even less believable, if that's possible, are scenarios in which the aliens come to breed with us, such as in *Decoys*, *Species 2*, and countless TV shows. The interactions violate both biology and the standards of polite behavior.

Issues of DNA and reproductive compatibility aside, even the question of whether the aliens sport carbon-based chemistry is seldom established in movies (with the exception of *Star Trek*). Silicon-based biology has a long pedigree in written sci-fi, and to chemistry cognoscenti seems reasonable because silicon sits just below carbon in the periodic table. Consequently, it has many similar properties. But this interesting biochemistry alternative has been sparsely used in film and television (although the extraterrestrials in the *Alien* franchise are said by some to be silicon-based creatures). Again, these technical details would only be worth the explanatory trouble if they reinforced a particular ability or susceptibility. Apparently, this isn't often the case.

Finally, and aside from the general uncertainty concerning their biochemistry, it seems to be *de rigueur* for otherworldly visitors to have truly remarkable metabolism. They often grow quickly, reaching full size in a matter of hours or days (think of the creature in the original *Alien*). They generally do this despite the lack of any apparent natal food supply, although some aliens will chow down on anything that moves. (They seldom exhibit much interest in plants.) These indiscriminant carnivores apparently come from a world on which virtually every animal inhabitant is suitable meat-on-the-hoof. And that, presumably, implies an impressive supply of herbs or herbivores to feed the hoofed meat.

Earth, by comparison, might look like a culinary desert to such ravenous creatures. But then again, since their biochemistry is likely to differ in many respects from ours, there's little reason to expect that any of the life forms on this planet would provide them with nourishment no matter how many animals (including humans) they consumed.

Abilities

Since movie extraterrestrials come from another world, sporting their own evolutionary history and possibly their own biochemistry, many Hollywood writers feel free to bless them with extraordinary powers. After all, who knows what might develop on another planet? Ignorance becomes power.

Shape-shifting is an occasional talent, and the mysterious extraterrestrials in *Contact* were able to appear as Ellie Arroway's long-lost dad. Some can manage remarkable medical stunts. In *Paul*, the alien could heal with touch, and the robot in *The Day the Earth Stood Still* brought Klaatu back from the dead, at least for a short time.

Flying is another alien ability, albeit infrequent. E.T. could sail through the air without flapping his appendages or using any other obvious method for generating thrust, and do so efficiently enough to bring along a friend and a bicycle. While this violates physics as we know it, most audiences—conditioned by years of watching Superman's aerial acrobatics—are unfazed. (Note that technically Superman is also an alien.)

When it comes to conversation, Tinseltown sentients often use telepathy. This is less for novelty than for utility. It obviates the usual pretext that the extraterrestrials have mastered colloquial, American English. It also allows them to communicate with one another without spouting the gobbledy-gook of an invented alien tongue. Their telepathic powers are invariably coupled with some sort of universal translator, so that their thoughts appear in the minds of humans in the correct language and, one presumes, vice versa. Failure to speak also gives the extraterrestrials an inscrutable look.

Visitors from afar are credited with what seems to be superior intelligence, although this is sometimes hard to judge, perhaps because of the limitations of our own mental abilities. When the animal-like aliens of Spielberg's *War of the Worlds* paw through the basement of a suburban home, there's the sense that they are an advanced species bemused by the paraphernalia of a more primitive one (ours). And certainly, the very fact that movie aliens have managed to construct the transport hardware that can bring them to our planet bespeaks technology that's far beyond our own. So it's easy and somewhat natural to assume that beings who are our technological superiors are also our intellectual superiors.

However, despite their better cognition, it's not essential that—as individuals—they have better brains than ours. Some manifest their A-list intellect with a hive mind. They're an alien swarm with a collective consciousness, usually masterminded by a single queen. Examples appear in *Star Trek: First Contact*, *Avatar,* and both the *Alien* and *Terminator* franchises. In terrestrial settings, collective intelligence is familiar in bees and ants, and has the advantage of being less vulnerable (the killing of many individuals might not neutralize an aggressive species). In addition, building intelligence from simpler, soulless, and highly redundant components eliminates the necessity of a long period of education, as the individuals can be "born smart," given their limited individual complexity.

But despite the indirect evidence that many Hollywood aliens are our intellectual superiors, the most remarkable fact of cinema aliens is that their SAT scores are apparently not *much* higher than ours. We can outwit them in *Independence Day*, we can interact sympathetically with them in *E.T. The Extraterrestrial* and *Avatar*, or we can hire them as crew members for our own space missions, as in the *Star Trek* and *Star Wars* series. This is perhaps the most glaring error in the Hollywood depiction of extraterrestrials, simply because it is *a priori* so unlikely.

Consider: *Homo sapiens* has a 200 thousand-year history. While experts debate whether our mental capabilities are still evolving, the mainstream view is that our species has had approximately the same I.Q. for the last 30 thousand years or so. It's possible that we might keep this level of cognitive capability for another few tens of thousands of years, but it seems far more likely that we will soon be augmenting it either by implanting circuitry into our bodies or simply tinkering with our DNA. That would mean that our current mental ability existed on our planet for only 0.001 percent of the time that life has been here, and is not likely to last more than a few centuries more. It would be a considerable coincidence if another species just happened to be at our present intellectual level.

Consequently, the idea that we could meaningfully interact with aliens from another world—whether as their friends or foes—is far-fetched. It is, instead, a dramatic necessity. If visiting aliens were—as they might be—hundreds of millennia beyond our level, then not only would we have a hard time comprehending their intentions, we'd be incapable of interfering with their plans, assuming they had any. And I.Q. aside, visiting aliens will, as noted, have a considerable technical edge. The idea that Earthlings could successfully repulse an invasion by creatures able to traverse light-years of distance is akin to the thought that howler monkeys could effectively take on the Air Force.

Motives

Encounters between Earthlings and aliens—the fundamental premise of the type of film described here—are almost always deliberate. Either the extraterrestrials are paying a house call because our planet offers something they want (e.g., *The Arrival*, *Battle Los Angeles*, and many more), or they lure us to their world because we're necessary to their survival (*Alien*). So while there are occasional plot lines in which terrestrials and extraterrestrials meet by accident (e.g., *District 9*), in most cases the film makers need to suggest a reason for an interstellar rendezvous.

They don't always bother. In *Battleship*, the aliens are here for no discernible purpose other than to take part in a naval exercise, and in many other invasion pictures the extraterrestrials' motives seem to extend no farther than "take over the planet," preferably using as much ordnance as possible. The reasons for destroying Earth are only occasionally given.

However, that's not the case for *War of the Worlds*—whose story line dates back more than a century. The Martians have come to Earth because their own planet is drying up. They envy, and apparently need, our navy-crowded seas. The idea that water is the honey that will lure extraterrestrials to Earth is a frequent sci-fi trope, but it doesn't make much sense. Hydrogen is an extremely abundant element, constituting three-quarters of the cosmos, by weight. Oxygen is the third most-abundant. Water, therefore, is a common constituent of the universe—and no aliens would have to travel far to find it. In addition, it's dense, and therefore expensive to lift off Earth and transport.

But if water isn't what our cosmic confreres are after, then what is? In the original version of *The Day the Earth Stood Still*, the extraterrestrial comes to warn us against our build-up of nuclear weaponry. In the remake, he comes out of concern for our environmental depredations (much as in *Star Trek IV*). However, neither of these motives passes the baloney test, because it's hard to understand how any extraterrestrials could know of such behavior (let alone have much interest in it). News broadcasts that would inform them of such things are only tens of light-years out into space, and consequently it's not likely that any extraterrestrials have yet learned about our poor conduct, let alone had time to come to our planet to seek redress. The credibility of stories in which aliens attempt to correct our bad deportment is on shaky ground.

There is at least one motive for alien visits that is predicated on actual history: proselytizing. Carl Sagan once pointed out the possibility that aliens might come here as missionaries, hoping to convert us to some unspecified belief system. It's a premise that Hollywood has yet to embrace.

But aside from spreading their religion, it's difficult to understand why anyone would spend the enormous amount of energy and/or time necessary to visit Earth. There are no physical resources present in our Solar System that couldn't be found in *their* solar system. Earthly science, while impressive to us, is not likely to be much of a draw since—if they have the ability to come here—they know all of that already. Perhaps they just want to be friendly, and welcome us into the Galactic Club, as in *Star Trek: First Contact* or *Contact*. But as a practical matter, the chances are slim that a neighboring society is even within a few thousand years of our level of development. It doesn't seem reasonable that such disparate neighbors would make for chummy club companions.

Indeed, about the only thing that might be *special* about Earth and therefore enticing to visiting aliens are its life and its human culture. Alien biologists might wish to come simply as part of a research project. Or they might be interested in our music, our art, or our political systems. Few films seem to adopt these motives however, with the exception of *E.T. The Extraterrestrial*, in which the aliens came to collect plants, and *Indiana Jones and the Kingdom of the Crystal Skull*, in which visitors from afar amass some archeologically important *objets d'art*.

They're often here, but it's seldom clear why.

The Real Aliens

In reviewing the history of extraterrestrials in film, one notes that these popular characters are most frequently ciphers for human bad guys. Occasionally, they're child-like strangers who come here as friends or to save us from ourselves. Their morphology, if not their biochemistry, is frequently similar to our own, and their intellectual level is comparable to ours. At a fundamental level, they are most often modeled on us: a social species of intelligent beings, acting in concert to accomplish a goal more defined by its tactics than by its strategy or motives.

The aliens in movies are exotic *Homo sapiens*. But would actual aliens be so similar?

That's doubtful. Even aside from the expected variety in alien appearance and construction produced by each world's own evolutionary history, there's another possible development path that implies that real aliens could be quite different from the usual cinema versions. In particular, because most sci-fi stories involve physical encounters here on Earth or on some planet that's home to neither them nor us (e.g., the *Alien* films), this implies an extraterrestrial society at least centuries more technically advanced than our own. They must be capable of interstellar travel.

An important consequence is the following. Consider the time-line of technical development on Earth. Practical radio was developed at the beginning of the twentieth century. Five decades later, the first computers appeared. While not a certainty, many technologists predict the development of strong artificial intelligence—machines with the cognitive capability of humans—sometime in this century (and quite likely long before the appearance of practical interstellar travel). In other words, within a few centuries following the development of radio and rocketry, our species will invent our intellectual successors.

If this chronology applies even approximately to other sentient beings, then it's overwhelmingly probable that any aliens capable of traveling between the stars have long ago moved from biological intelligence to synthetic intelligence. The principal advantage of the latter is that it can then design its own successor (much as today's microprocessors benefit from the design capabilities of previous versions of themselves). This Lamarckian evolution can enormously outstrip the speed, and ultimately the abilities, of Darwinian evolution.

So this is the real disconnect between Hollywood aliens and the type of intelligence that might actually dominate the cosmos. While film makers envision hordes of little gray guys or arthropod monsters cursed with ugly bodies and a surfeit of mucous, the real aliens are unlikely to be protoplasmic at all. If we ever meet someone from beyond the Solar System, we should expect them to be clever, of course. But, unlike us, they will be deliberately engineered. And they won't be alive.

Chapter 24

Beyond Teaching and Learning: Bringing Together Science and Society with and through Movies

Erik Stengler*

Science Communication Unit, University of the West of England, Coldharbour Lane, Bristol BS16 1QY, United Kingdom
***E-mail: Erik.Stengler@uwe.ac.uk**

Science and movies seem to be getting along very well lately. Old complaints about unrealistic and derogatory depictions of science and scientists have given way to an apparent honeymoon between filmmakers and scientists which includes more and better science on the big screen and on TV. So much so that the main goals and motivations for science communication can be mapped to the ways science is depicted in fictional film. This scenario may not always be the consequence of a targeted effort by the scientific community, but is certainly welcomed in any case. What scientists can undoubtedly do now is to contribute to making it last by helping filmmakers achieve their ultimate goal of entertaining.

Introduction

When we think of science education one of the ideas that first comes to mind is inspiration. If we are looking back at our own experience of classroom science we invariably remember those teachers who were able to enthuse us with their passion for the topics they taught and who made memorable and life-changing events out of otherwise routine-laden lessons. We might or might not recall any specific piece of knowledge of those classes, but they certainly inspired us to look at science with different eyes, and in some cases set us off on the path of a career in science. If we look forward along the current trend and need to bring much needed new students into scientific disciplines, we all agree that the way to go is to inspire young students in that same way.

But the world has changed substantially, and there are now many more and different ways the message can reach these young students. Effective teaching in the classroom certainly is still a major factor, but present-day children, youths and adults are exposed to science in various different contexts and education is acknowledged to occur not only at early ages but throughout life. Let us thus understand here the term education in its broadest sense, which encompasses both formal and informal education, for people of all backgrounds, ages and professions. Under this perspective let me first review different aims and motivations behind the pursuit of science education of students and the public.

- One major aim is, of course, education in the strict sense of **teaching and learning**. This takes place in the classroom and is complemented by a wide range of informal education initiatives, from open day visits to labs to purpose built interactive science centers, planetariums, aquariums, and zoological and botanical gardens.
- As mentioned above, inspiring teaching and learning has always helped to **motivate students towards scientific careers**. But the current alarming decrease in the numbers of students of fundamental scientific disciplines turns this into an explicit and urgent need, in order to ensure the continuity of basic research in several important research fields.
- Scientific institutions also feel the need to inform the public about the research they undertake as a duty towards the taxpayer as its funder. This results in an approach to science communication as a **showcase of current and anticipated research**, presenting it as important for society, and highlighting those aspects relevant to citizens.
- There is also the concern to promote **scientific literacy** in order to build a society of informed citizens and voters. These will need to have educated views on the various areas where science intersects with democracy, from deciding whether a line of research is worth funding from a merely economic point of view, to morally and ethically controversial issues. This scientific literacy approach also includes a concern about the increasingly worrying lack of critical thinking in certain population segments and their defenseless exposure to the pseudosciences.

Communication and Emotions

Science in movies can and does contribute to each of these general aims of science education and communication. At the heart of its potential to do so is the importance of the emotional aspect of any attempt to communicate. If we remember those inspirational teachers of our childhood it is certainly not because of the particular interest or relevance of the contents of their lessons at a given time. It is rather related to their ability to spur in their pupils strong emotions that irreversibly reinforce a positive attitude towards the subject, regardless of what contents were covered during a certain class. The pleasurable emotions of discovery by being able to ask the right questions; of understanding thanks to a well prepared and explained lesson; of joining the adventure of humankind's quest to know thanks to knowledge being presented in a relevant wider context; these are all examples of emotions that make for a lasting impact that goes a long way towards achieving the aims of science education considered above.

Documentary filmmaking has gradually realized the need to appeal to viewers' emotions to reach a broad audience. A good example is the BBC's recent series *The Great British Countryside* (*1*). This show is not even presented as a science documentary. It appeals to viewers' emotional connections to certain landscapes through the relationship between their geology and their population´s everyday life, from economy sustaining activities to pure leisure, encompassing trade, hiking, sports, traditions, gastronomy and much more. Science is embedded in a relatable manner. Another emerging trend that responds to the need to stir viewers' emotions are "live" documentaries like Planet Earth Live (*2*) and Volcano Live (*3*), where viewers are drawn in to share the presenters' live experience on site, in jungles and near active volcanic sites, respectively. Nevertheless this kind of factual show demands that viewers place themselves in a learning attitude in one way or another. The Oxford English Dictionary reflects this perception of documentary features defining "documentary" as "*Factual, realistic; applied esp. to a film or literary work, etc., based on real events or circumstances, and intended primarily for instruction or record purposes*" (*4*). This is often hardly compatible with a context of leisure time in which audiences primarily seek entertainment. Viewers seek entertaining shows irrespective of whether their subject matter is interesting to them or not.

It does therefore not come as a surprise that, in addition to the immense wealth of high quality documentaries being available and increasingly striving to appeal to viewers' emotional engagement, a trend is emerging and consolidating, of scientists and science communicators concerning themselves with the educational value of fiction movies, where emotions are a central way of communicating with the audience. Emotional involvement scaffolds engagement and thus provides unique educational opportunities. Hopefully, this trend will be underpinned by research on its effectiveness. Such research is still scarce and fragmented as excellently reviewed by Orthia *et al* (*5*), and opens a whole new field for science educators and communicators.

Movies and Science Education and Communication

Science in Movies as an Educational Resource

The use of science in movies as an educational resource considers the way scientific principles are taken into account (or ignored) in this or that sequence of a film, such as Special Relativity in *The Planet of the Apes* (*6*) or Evolution in *The Time Machine* (*7*), or the didactic value of historically accurate accounts of the lives and feats of scientists, from Marie Curie (*8*) to Charles Darwin (*9*). These are important ways to utilize movies as teaching tools or illustrations in popular talks, which I frequently undertake and enjoy. This approach has become the subject matter of educational resources in the form of websites such as Teachwithmovies.com (*10*) or Film Education (*11*) and books (*12–16*). Also among works aimed at the general public there are numerous titles that directly address the science of TV shows and movies with the title or subtitle *The Science of...* (*17–23*). A specific way that has proven to be particularly effective along these lines is to structure a session entirely around five or six brief video clips from different movies about a certain topic, ordered in a way that they help move the lesson forward, while at the same time "recover" those students lost due to their short attention span. Even the most reluctant ones end up engaging with the lesson if they can relate to the clip shown, because they went to see that movie and therefore can speak out knowingly in front of their peers. Luckily, there are sufficient movie clips to choose from so that the science they address fits different stages of the development of a particular subject.

But there is a deeper and farther reaching way in which movies map to the other goals, motivations and messages that the scientific community tries to address through science education and communication.

Science in Movies for the Promotion of Scientific Careers

As mentioned above, one major and common aim of educating young students in science is to promote scientific disciplines as professional options, especially in recent years, when there is an alarmingly decline in the uptake of scientific careers, and less qualified professionals are becoming available to replace retiring investigators/researchers. As there is a strong emphasis on motivation in this approach to science education, it usually focuses on the benefits, both personal and professional, such a career can provide in life. Scientists tend to be presented as explorers in a quest, as solution providers for many important issues of modern life, as citizens who are recognized for their achievements…

This endeavor has certainly been both mirrored and supported by the way science and scientists have been portrayed in movies. We have had mad, selfish and sometimes even evil scientists in movies as recent as *Spiderman 2* (*24*), but gradually this distorted view has given way to more neutral representations, in which the villain is not the scientist but a 'regular' wrongdoer who lures and forces the scientist into his service for a misuse of his scientific discovery, as seen recently in *The Dark Knight Rises* (*25*), or where there are scientists on both sides, as in *Coma* (1978) (*26*).

It is increasingly frequent, too, that the scientists play a pivotal role to counteract or fight the villain as seen in the three strands of *CSI* (*27–29*), *Bones* (*30*) or *Numb3rs* (*31*), or even become the heroes themselves, often unwillingly, as it is expected from the corresponding archetype (*32*), as seen for example in *The Andromeda Strain* (*33*) or *Sphere* (*34*). A typical case of this is the ubiquitous natural disaster movie, in which the tension between the scientist-hero and the politician/military-antagonist is revisited again and again as the latter usually does not want to lose decision-taking power nor risk losing authority by raising the alarm too early. The methodological and even philosophical question of scientific certainty is then often addressed as the wider debate in which the argument is contextualized, as for example in *Dante's Peak* (*35*).

Science in Movies and Scientific Literacy

Issues like the latter, as well as ethical and legal implications of scientific research, need to be made accessible to the wider public in the effort to keep society informed and even involve the public in decision-making. This falls within the area of the promotion of scientific literacy, whose aim is to foster well-informed voters and users of technology. In this regard, movies are again a powerful channel to raise awareness of these far reaching questions. Numerous films have addressed such issues in more or less depth. Some deal with them directly, as in *Godsend* (*36*), others just as a backdrop of the actual story, like *Jurassic Park* (*37*).

Noteworthy in this context is the evolution of the scientific questions attached to superheroes: Ang Lee´s vision of *The Incredible Hulk* (*38*) includes nanotechnology; both Batman in Christoper Nolan´s version (*25*) and *Spiderman 2* (*24*) are involved in the development of unlimited clean energy through fusion, and Spiderman´s powers have their origin in genetically modified spiders, both in Sam Raimi´s version (*39*) and in the latest reboot by Mark Webb (*40*). All of these movies reflect the controversial aspects of leading edge research of the moment and contribute to the public awareness about them. A further notch in this direction is offered by science fiction dystopias, in which the controversial science is addressed through its extrapolation to an imagined future in which its effects are taken to extremes. This is the way in which the issues related to artificial intelligence have been addressed most often, with cinematographic landmarks such as *Blade Runner* (*41*), *A.I.* (*42*) or *The Matrix* (*43*), and so many more.

Science in Movies as a Showcase of Research

But current research does not need to be controversial. The scientific community is eager to make known to the public any research that is being carried out with taxpayers' funds. This showcases science communication as a duty towards the public (which provides funds through taxes) and has a clear reflection and support in filmmaking. *Contact* (*44*), *Mission to Mars* (*45*), and *Red Planet* (*46*) depict imaginary scenarios that lay within the horizon of research carried out at the time the movies were made (although real life science went into different directions thereafter). We could also include in this category accounts of fairly recent science history like *The Dish* (*47*) or *Apollo 13* (*48*), or certain disaster movies as they show how science helps to face such disasters with ever more refined and effective warning techniques, such as *Twister* (*49*), defense mechanisms, as in *Deep Impact* (*50*), or solutions, as in *The Andromeda Strain* (*33*).

Storytelling and the Scientific Message(s)

However, it may be considered by some to be an unduly interference with filmmaking to proactively try to influence how science is portrayed in movies with any of the abovementioned purposes in mind, as if it were a kind of subliminal advertising or education. In this regard, let me argue that fictional film is a modern way of storytelling, a deeply human activity of emotional communication that goes back to the beginning of mankind. From cave paintings and the stories told around a fire, to theatre plays, movies and computer generated audiovisuals projected onto the domes of digital theatres, humans have devised so many different ways to carry out the one and same activity of storytelling. Storytelling is so embedded in human nature that it has been found in serious psychology to have therapeutic value in interventions with emotionally damaged children (*51*, *52*) and even with patients with Alzheimer´s disease (*53*).

Movies would not even be the most modern form of storytelling any more. Computer games could be viewed as an even more sophisticated and immersive way to tell a story, and it is becoming more and more frequent that a game version of a film is released while it is still being shown in cinemas. The computer game of *The Matrix* (*54*) should not be viewed as just an entertaining merchandise of the fictional world, but rather yet another way of immersing the audience in an experience that has numerous and deep philosophical layers (*55*). More recently the fictional world of *Defiance* (*56*) is expanded into a game (*57*) simultaneously to the TV series to provide a uniquely immersive, and therefore engaging opportunity.

And storytelling has always had an educational angle, sometimes even in a very explicit manner, as was the case in Haesop´s fables or Jesus´s parables. In its cinematographic version, nobody will be surprised at, let alone object to, movies being a channel to spread values like generosity, self-sacrifice, patriotism, the joy of living, the power of love, etc… Like anything in life, this can certainly be taken too far. When the educational intention is simply juxtaposed to a story and it hinders or even spoils the entertainment value, a movie can certainly come across as patronizing or worse. Movies can even be used as sheer propaganda and even a tool for indoctrination. But within a reasonable and universally acceptable parameter space, stories always use their entertaining value to convey a message through emotions, which is the most effective way to make an impact in people´s minds and hearts. A recent clear example is the message of environmental concern of *The Lorax* (*58*). If this is considered acceptable and even positive in other subjects, why should science be excluded from the educational potential of storytelling through movies?

The Scientific Community and the Movie industry

In fact, in view of the apparently successful marriage between science and film one might be tempted to conclude that this is due to a deliberate collaboration with all those purposes of science communication in mind. But reality tells us otherwise: there are still relatively few instances in which a proactive approach by scientists to filmmakers has had an effect on a successful movie. The trend towards a perhaps more realistic and fairer depiction of science and scientists in movies has resulted primarily because filmmakers have considered in recent years that it might add some value in their works. Features like plausibility, realism, relatability come immediately to mind.

Plausible and coherent scientific explanations increase the credibility of a story, particularly among educated audiences. Realistic depiction of scientists makes the audiences' identification with them as story characters easier. The inclusion of real life hot topics, an increasing number of which are scientific and technological, makes a movie more relatable to the viewers, who will more easily be drawn into the argument. The fact that this way of using science in film is also convenient from an educative point of view is only a collateral effect which is not sought directly by filmmakers. It should not be forgotten that if it would hinder, rather than help, their primordial aim of entertaining, they would as easily revert to a more caricaturesque, inaccurate and derogatory portrayal of science and technology.

What the scientific community needs to do now, is to make the most of the current trend, and contribute to make it last as long as possible. The only way to go is to reinforce the added value this kind of scientific garnish provides to movies, and to make an effort to come up with new ways it can keep doing that.

As extensively argued by David Kirby (*59*), a successful and lasting relationship between a scientific consultant and the director of a movie has to be based on the former being convinced that the latter's only goal is to make a movie that will attract and entertain big audiences. The filmmakers have the power to ignore and to disregard any suggestion by the scientific consultant or even to get rid of him altogether. The scientific consultant, if hired at all, is very low in the hierarchy of professionals that contribute to the success of a movie. No director will even think of not having a soundtrack of some sort, or allow the actors to act without make-up, or work without a lighting expert. But a scientific consultant will be easily let go in case of financial constraints, or even in view of severe disagreement with the senior members of the creative team.

It is not realistic to dream of a future scenario where things are much different in this regard. The scientific community needs to be realistically humble in the approach to science in the movies and just play the game to their own interest while trying to make it last. But within these terms and boundaries, the prospect is very promising. Science and technology are providing topics and developments at increasing rates, and there is always more to pick from by filmmakers. There is a growing awareness of potentially controversial issues with science and technology in society, and filmmakers will naturally draw from them to build the essential ingredient of conflict into their stories.

So, let the conflict stay in the realm of fiction and let us procure a harmonious collaboration between science and fictional film for the benefit and enjoyment of everyone.

References

1. Summerhill, M. (Producer) *The Great British Countryside*, BBC, February 2012.
2. Webb, R. (Producer) *Planet Earth Live*, BBC, May 2012.
3. Holland, A. (Producer) *Volcano Live*, BBC, July 2012.
4. The Oxford English Dictionary. http://www.oed.com/view/Entry/56332?redirectedFrom=documentary#eid (accessed January 2, 2013).
5. Orthia, L. A.; Dobos, A. R.; Guy, T; Kan, S. Z.; Keys, S. E.; Nekvapil, S.; Ngu, D. H. Y. *Int. J. Sci. Educ. Part B* **2012**, *2*, 149–174.

6. Schaffner, F. J. *The Planet of the Apes*, 1968, Twentieth Century Fox.
7. Wells, S. *The Time Machine*, 2002, Warner Bros.
8. LeRoy, M. *Madame Curie*, 1943, Metro-Goldwyn-Mayer.
9. Amiel, J. *Creation*, 2009, Recorded Picture Company.
10. Teachwithmovies.com. http://www.teachwithmovies.com (accessed September 1, 2012).
11. Film Education. http://www.filmeducation.org (accessed September 1, 2012).
12. Perkowitz, S. *Hollywood Science: Movies, Science and the End of the World*; Columbia University Press: New York, 2010
13. Weiner, A., *Don't Try This at Home: The Physics of Hollywood Movies*; Kaplan Publishing: NewYork, 2007.
14. Rogers, T. *Insultingly Stupid Movie Physics: Hollywood's Best Mistakes, Goofs and Flat-Out Destructions of the Basic Laws of the Universe*; Sourcebooks Hysteria: Naperville, IL, 2007
15. Lambourne, R. J., Shallis, M. J., Shortland, M. *Close Encounters?: Science and Science Fiction*; Taylor & Francis: Oxon, U.K., 1990.
16. Cavanaough, T. W.; Cavanough, C. *Teach Science with Science Fiction Films: A Guide for Teachers and Library Media Specialists*; Linworth Publishing: Worthington OH, 2004
17. Cavelos, J. *The Science of the X-Files*; Berkeley Boulevard Books: New York, 1998.
18. Simon, A. *The Real Science Behind the X-files: Microbes, Meteorites, and Mutants*; Simon & Schuster: New York, 1999.
19. White, M. *The Science of the X-Files*; Legend Books: London, 1996.
20. Cavelos, J. *The Science of Star Wars*; St. Martin's Griffin: New York, 2000.
21. Grazier, K. R., Ed. *The Science of Michael Crichton*; Benbella Books: Dallas, TX, 2008.
22. De Salle, R.; Lindley, D. *The Science of Jurassic Park and the Lost World*; Harper Collins: London, 1997.
23. Gresh, L. H.; Weinberg, R. *Why Did It Have To Be Snakes?: From Science to the Supernatural, The Many Mysteries of Indiana Jones*; John Wiley & Sons: Hoboken, NJ, 2008.
24. Reimi, S. *Spider-Man 2*, 2004, Columbia Pictures.
25. Nolan, C. *The Dark Knight Rises*, 2012, Warner Bros.
26. Crichton, M. *Coma*, 1978, Metro-Goldwyn-Mayer.
27. Donahue, A.; Zuiker, A. E. *CSI: Crime Scene Investigation*, 2000, Jerry Brickheimer Television.
28. Donahue, A.; Mendelsohn, C.; Zuiker, A. E. *CSI: Miami*, 2002, Jerry Bruckheimer Television.
29. Donahue, A.; Mendelsohn, C.; Zuiker, A. E. *CSI: NY*, 2004, Jerry Bruckheimer Television.
30. Hanson, H. *Bones*, 2005, Josephson Entertainment.
31. Falacci, N.; Heuton, C. *Numb3ers*, CBS Television Studios.
32. Vogler, C. *The Writer's Journey. Mythic Structure for Writers*; Michael Wiese Productions: Studio City, CA, 1998.
33. Wise, R. *The Andromeda Strain*, 1971, Universal Pictures.

34. Levinson, B. *Sphere*, 1998, Warner Bros.
35. Donaldson, R. *Dante's Peak*, 1997, Universal Pictures.
36. Hamm, N. *Godsend*, 2004, 2929 Productions.
37. Spielberg, S. *Jurassic Park*, 1993, Universal Pictures.
38. Lee, A. *Hulk*, 2003, Universal Pictures.
39. Reimi, S. *Spider-Man*, 2002, Columbia Pictures.
40. Webb, M. *The Amazing Spider-Man*, 2012, Marvel Studios.
41. Scott, R. *Blade Runner*, 1982, Warner Bros.
42. Spielberg, S. *Artificial Intelligence*, 2001, Amblin Entertainment.
43. Wachowski, A. Wachowski, L. *The Matrix*, 1999, Silver Pictures.
44. Zemeckis, R. *Contact*, 1997, Warner Bros.
45. De Palma, B. *Mission to Mars*, 2000, The Jacobson Company.
46. Hoffman, A. *Red Planet*, 2000, Mars Production Pty. Ltd.
47. Stich, R. *The Dish,* 2000, Dish Film Ltd.
48. Howard, R. *Apollo 13*, 1995, Imagine Entertainment.
49. De Bont, J. *Twister*, 1996, Universal Pictures.
50. Leder, M. *Deep Impact*, 1998, Zanuck/Brown Productions.
51. Therapeutic Storytelling. http://older-child.adoption.com/parenting/therapeutic-storytelling.html (accessed September 8, 2012).
52. Older Children Adoption Support. http://www.olderchildadoptionsupport.com/therapeutic-stories/ (accessed September 8, 2012)
53. Silberner, J. In SHOTS, NPR's Health Blog. http://www.npr.org/blogs/health/2012/05/14/152442084/alzheimers-patients-turn-to-stories-instead-of-memories (accessed September 8, 2012).
54. Ohkata *Enter the Matrix*, 2003, Warner Entertainment Japan.
55. Oreck J. *Return to Source: Philosophy & 'The Matrix'*, 2004, Warner Bros. Entertainment.
56. Alexander, B. A.; Nankin, M.; Bullock, J. P., III *Defiance*, 2013, Universal Cable Productions.
57. Trion Worlds, 2012. http://defiance.com/en/game/ (accessed January 2, 2013).
58. Renaud, B.; Balda, K. *The Lorax*, 2012, Universal Pictures.
59. Kirby, D. *Lab Coats in Hollywood: Science, Scientists and Cinema*; The MIT Press: Cambridge MA, 2011.

Chapter 25

How Hollywood Inspires the Exploration of Space

Randii R. Wessen*

Jet Propulsion Laboratory, California Institute of Technology, National Aeronautics and Space Administration, Pasadena, California 91109
***E-mail: randii.r.wessen@jpl.nasa.gov**

Science fiction takes us to the future and to worlds yet to be discovered, but science fiction is much more than just entertainment. It is a communication device that gives us a glimpse of what could be. The master of this communication technique is Hollywood. Their scope ranges from documentaries, to non-fiction, to the very essence of imagination, fiction. These fictional stories, if done correctly, guide scientists and engineers to consider realities that currently don't exist…but could. All one has to do is scan the science fiction stories of the 1950s to see that much of their imagined capabilities are in existence today. Communication satellites in geostationary orbit, humans landing on the Moon, space stations, and interplanetary travel all started as fiction. Science fiction is not a luxury but a societal necessity. Science cannot move forward without some idea of what future state must be proved or disproved. Fundamental research has to make a hypothesis, which is its own kind of science fiction, which can be tested through the scientific method. Realities of today have many technologies first articulated in fiction. Even the names of these technologies are sometimes taken from the science fiction as a kind of tribute. Hollywood entertains all of us, but its inspiration is what helps us develop tomorrow.

Introduction

Marvins, Lukes, Hals, and, of course, Kirks were everywhere. When one walks down the halls of the operations buildings at NASA's Jet Propulsion Laboratory (JPL) and peers into offices, one is more likely to find Marvin the Martian dolls, Star War's *Millennium Falcon* models, *2001: A Space Odyssey* movie posters, or *Star Trek* memorabilia then you are to find NASA logos, Caltech coffee mugs or MIT notebooks. The reason for this is, that, as kids, we all watched science fiction on television, in the theaters, or both. These shows were more than just mindless images flickering at 24 frames per second, but rather adventures that took us to new worlds. These experiences had very profound effects on our interests, our education, and, ultimately, our careers.

Today at NASA, all of us were, and still are, deeply moved by technical engineering endeavors like the Mercury Program to launch humans into space, the Gemini Program to develop space faring technologies, the Apollo Program to land humans on the Moon and the Space Shuttle Program to attempt to make space travel routine. We were also moved by 20th Century Fox's *Lost in Space*, Stanley Kubrick's *2001: A Space Odyssey*, Steven Spielberg's *E.T.: The Extra-Terrestrial*, and who doesn't remember Robby the Robot from the classic science fiction film *Forbidden Planet* (Figure 1)? One particular Science Planning Manager for the Cassini Saturn Mission had so many *Star Trek* action figures that his office looked more like the apartment of Sheldon Cooper and Leonard Hofstadter on *The Big Bang Theory* than an office for a NASA flagship mission (Figure 2). That manager, Brian Paczkowski, shares, "I grew up during the Gemini and Apollo era of manned spacecraft, which was very inspirational for me and helped direct my career goals - I wanted to be involved, in any way, with the space program! At the same time, I got to experience the awe and wonder of what the future could be like with every new episode of the original *Star Trek*. This combination motivated me to pursue a science and engineering degree in college and, ultimately, a career with NASA."

The Importance of Science Fiction

Science fiction is the logical extrapolation of what is known today. Good science fiction does not include things that are impossible or at least tries to keep the number of impossible things to a minimum. "Impossible" means something that "violates the laws of nature." As an example of what is or isn't possible, it is impossible to travel faster than the speed of light. On the other hand, can humans live to be 400 years old? The answer is yes. It doesn't violate a law of nature, we just don't know how to achieve it. Sometimes, some aspects of a science fiction plot must violate "what is impossible" in order to create a story. Science fiction adventures sometimes use faster-than-light travel because the distances between stars are so vast. A space thriller at another world in another Solar System must assume that we can get there in a reasonable amount of time before the story can be told.

Figure 1. Robby the Robot, from the 1956 science fiction film Forbidden Planet, poses at the 2006 San Diego Comic Con in honor of the 50th anniversary of his appearance in that movie. Courtesy Wikimedia Commons.

Unfortunately, the general public does not realize just how far away other stars are from our Sun. To illustrate this, we'll use objects built by humans that are the most distant things humanity ever built, namely the *Voyager* spacecraft (Figure 3). These two ambassadors from Earth are traveling at about 1,000,000 miles every day. Traveling at those speeds, *Voyager* would take approximately 78,000 years to get to the closest star to our Sun…and neither of the *Voyagers* are even traveling in that direction! And the stars themselves travel through space faster than any spacecraft people have ever built! However, if storytellers cannot write about "warp drive," traveling faster than the speed of light, then many stories can't even be written.

Figure 2. Evidence that television series like Star Trek and Doctor Who, as well as movies like Star Wars, influence space exploration is ubiquitous in the office of personnel on the Cassini/Huygens Mission at NASA's Jet Propulsion Laboratory. Courtesy Kevin Grazier.

Science fiction, whether in books, on television, or in movies give us a brief glimpse of what tomorrow can be. It teaches us how to imagine. Albert Einstein once said, "Imagination is more important than knowledge. For knowledge is limited to all we now know and understand, while imagination embraces the entire world, and all there ever will be to know and understand." Though we really need to be able to acquire knowledge, there can be no increase in knowledge without imagination. Without imagination, there is no creativity. Our mind's eye must be able to consider a future possibility before we can attempt to make it a reality, and that's where scientists, engineers, artists, futurists, and the leader of creativity, Hollywood, comes in. Hollywood is all about imagination. Filmmakers make "real" that which no longer exists, that which could've existed, or that which could possibly exist sometime in the future.

Figure 3. The Voyager spacecraft with its science boom and radioisotope thermoelectric power supply deployed. Courtesy NASA/JPL/Caltech.

When Hollywood and Science Join Forces

One of the more successful engineering and artistic collaborations was between Wernher von Braun, the chief architect of the Apollo Program to the Moon, and Walt Disney, the father of Mickey Mouse and the creator of the Walt Disney Corporation. It is no secret that the Disney Corporation was, and still is, in the business of imagination for the purpose of entertaining and educating people. Both von Braun and Walt Disney (Figure 4) realized that media (TV and film) were powerful tools to entertain and inform.

In the 1950s von Braun and Disney teamed up to produce three space-related TV shows. The Disney Corporation produced and visualized the shows while von Braun provided the scientific content. On March 9, 1955, the first show entitled *Man in Space* aired on television (*1*). The main thrust of the show was the possibility that a passenger rocket could be designed, built and operated within 10 years.

Although it took more than fifty years, the show correctly predicted the advent of commercial spaceflight. Today humans have paid for trips into space and many companies are working on rockets to launch people into low Earth orbit on a regular basis. The individual that has the honor of being the first commercial astronaut belongs to Dennis Tito (Figure 5), an ex-engineer at NASA's Jet Propulsion Laboratory (*2*). In the 1960s, Mr. Tito, using his mathematical skills in orbital mechanics, applied them to investing. His financial skills grew so great that he eventually left the Laboratory to start his own investment company. Using his own funds, Mr. Tito was launched to the International Space Station on April 28, 2001 aboard a Russian rocket.

Figure 4. Wernher von Braun (right) posing with Walt Disney (left) in front of a model of a German V-2 rocket. Courtesy NASA.

In addition, the first commercial rocket company, Virgin Galactic established by Richard Branson was founded in 2004, will start routine commercial suborbital missions sometime before 2015 (*3*). Though these flights will not make it to orbit, it does bring von Braun's dream of commercial spaceflight that much closer to reality. In addition, many other commercial companies are developing human-rated spacecraft to take paying customers to low Earth orbit. It's just a matter of time when the general public will be able to have their vacations in space, and then sometime later, vacations on the Moon.

The second von Braun/Disney space episode was aired on December 28, 1955 and was entitled *Man in the Moon* and highlighted a space station around the Earth with human missions to our natural satellite (*1*). Today, that space station is a reality and is called the International Space Station (ISS) (Figure 6). The ISS is now the residence for six astronauts and cosmonauts and orbits the Earth once every 90 minutes. It has had a permanent human presence in space since October 2000 and is now a major international research laboratory (*4*).

Figure 5. The Soyuz TM-32 crew. From left to right Dennis Tito, Commander Talgat Musabayev, and Flight Engineer Yuri Baturin. Mr. Tito spent 7 days, 22 hours and 4 minutes in space (3). Courtesy NASA.

Figure 6. An image of the International Space Station as seen from the Space Shuttle Endeavor (Mission STS-134) as it began its journey back to Earth. Courtesy NASA.

As for sending humans to the Moon, that was accomplished with the Apollo 11 mission which had Neil Armstrong and Buzz Aldrin successfully land the Eagle lunar lander on its surface (Figure 7). Five other 2-person crews followed but the economics for sending humans to the Moon were just not there. The Apollo Program ended in 1972 with the completion of the Apollo 17 mission. However, this is not the end of human exploration of the Moon. Humans will eventually return when the technology becomes cheaper, the engineering becomes safer and the economic reasons for working on the Moon become more profitable. In the year 2013 these conditions have not been meet, but it's getting closer. I tell my children that they will be the last generation of human beings to say, "I remember looking up at the Moon when there were no city lights on it."

Von Braun was right about the space station and human trips to the Moon, however, his estimate of when this would be accomplished was off by more than a half century. That error does not reduce the value of dreaming or reduce the need for imagination. Rather, imagination is the first step needed to socialize humanity with a new idea. After all, many people felt that landing a human on the Moon was impossible (not the proper use of the word). After the success of the Apollo 11 mission, landing humans on the Moon's surface was not only possible but by the end of the Apollo 17 mission, had became routine. We used our imagination to dream about landing humans on the Moon. Because of the Apollo Program, we no longer have to dream about it, we've already done it. Now we only have to wait for the economic conditions to "turn positive."

Figure 7. Neil Armstrong on the surface of the Moon during the Apollo 11 mission. Courtesy NASA.

The final von Braun/Disney episode aired on December 4, 1957 and featured the prospect of a nuclear powered spacecraft (*1*). Nuclear-powered spacecraft have been dreamt about for decades. The first technical treatment of such a craft was called Project Orion and was studied by engineers in the 1950s at Los Alamos Laboratory (*5*). It is clear that von Braun was not the first to think of this type of rocket, but he was the first to propose one for the general public with the help of Walt Disney. Nuclear power offers humanity a way to far exceed the current capability of liquid-fueled chemical rockets.

More recently *Star Wars*, created by George Lucas, used TIE Fighter spacecraft to fight for the evil Empire. The acronym "TIE" was not an imaginary 3-letter acronym but rather an acronym for a very plausible "Twin Ion Engine" concept (*6*). Ion engines, sometimes referred to as electric propulsion, use charged particles electrically accelerated out of its engine rather than the chemical explosion of a fuel and an oxidizer. They have been used for decades for station-keeping on Earth-orbiting satellites. An ion engine was finally space-qualified (Figure 8) as a main propulsion system 27 years after the first *Star Wars* movie with the launch of *Deep Space 1* (*DS1*) on October 28, 1998 (*7*). *DS1* validated a xenon solar electric propulsion engine. Since then the Japanese have used "ion-drive" for their *Hayabusa* spacecraft to return samples from asteroid Itokawa (*8*). Today the United States *Dawn* spacecraft is using ion-drive to achieve a space exploration "double header." *Dawn* will orbit two of the major asteroids in the Solar System. The craft has already completed its Vesta portion of the mission and is now heading for a 2015 rendezvous with Ceres, the largest asteroid in the Asteroid Belt (*9*).

Figure 8. Testing of Deep Space 1's NSTAR ion thruster. This "hot fire" test at NASA's Jet Propulsion Laboratory was used to validate the function of this thruster. Courtesy NASA/JPL/Caltech.

Science Meets Science Fiction

Though scientists and engineers use nature and physics to guide their imagination, many ideas are first born out of fiction. It doesn't matter if its written, animated or filmed. The result is the same. It directs the imagination to consider places that do not exist…but could. Sometimes the idea is so far ahead of its time that decades must pass before the technology of the day can catch up to the fiction. Sometimes the science fiction is so technically correct that when the capabilities of the day do catch up, the new technology is given the name used in the fictional story because it was "done right."

For an example of how reality can mirror fiction, one only has to read Jules Verne's *De la Terra a la Lune* (French for "From Earth to the Moon") written more than a hundred years before the Apollo Program (*10*). In Verne's timeless story, members of the Gun Club decide to launch individuals to the Moon with a huge cannon. The cannon was built at Stone's Hill Florida, exceedingly close to what is today the Kennedy Space Center. As if this wasn't enough of a coincidence, both the cannon's human-carrying projectile and the Apollo 11 spacecraft were both called "Columbia"; they both had a crew of three; and both ended their mission with a splashdown in the Pacific.

Still, Verne's story is just one of many examples of these 'coincidences.' The United States Constellation Program, which began in 2004 by the second Bush administration, was composed of an *Ares 1* launch vehicle to send a crew of four astronauts to low Earth orbit in an *Orion* capsule (*11*). Once in orbit, the *Orion* capsule would rendezvous with the transfer orbit injection stage and lunar lander launched by the much larger *Ares V* launch vehicle. If one reads Arthur C. Clarke's *2001: A Space Odyssey*, one would notice a very interesting fact. In his *Space Odyssey* story, as Dr. Heywood Floyd was approaching Space Station One, he notices other spacecraft in the area. There was the *Titov-V*, which is an acknowledgment to Gherman Titov, the second man to orbit the Earth, but there were two other vehicles in the story. The two craft were named the *Ares-1B* lunar carrier and the *Orion III* (*12*). Do you really believe the names of the vehicles in the Constellation Program, and the names of the craft in *2001: A Space Odyssey*, was just a coincidence?

Making something "real" by using the fictional names and applying it to real hardware or program is not as uncommon as one might think. In Arthur C. Clarke's 1973 story called *Rendezvous with Rama*, Clarke writes about an early warning system to protect the Earth from future impacts by large cosmological objects (e.g., asteroids and comets) (*13*). In Clarke's story, on September 11, 2077, a meteor exploded in the sky above northern Italy. Six-hundred-thousand individuals died in this fictional event. The inhabitants of Earth decided never to be surprised again by such a colossal impact so they built an early warning system called Project SPACEGUARD.

A comet impact with the Earth would produce large-scale devastation. As an example of the extent of the destruction, in 1994 the Comet Shoemaker-Levy 9 was torn into 23 pieces by Jupiter's intense gravitational field when it came too close to the planet. The fragments were labeled A, B, C, D, E, F, G, H,…all the way to W, and all impacted the planet (Figure 9). The largest of these fragments was the

G-fragment, about a mile across, and when the G-fragment collided with Jupiter, the collision released 6 million megatons of TNT (*14*). That amount of energy would be equal to the total energy released from one small atomic bomb dropped every second…for eleven years! Obviously, one small "cosmic event" can ruin your whole day. An informal joke at JPL is, "the reason why the dinosaurs are not alive today is because they didn't have a space program."

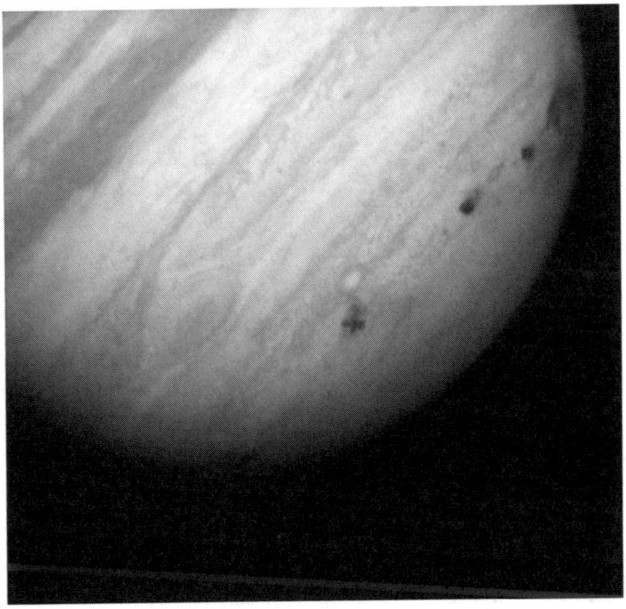

Figure 9. An image of Jupiter after the fragments of Comet Shoemaker-Levy 9 slammed into its atmosphere. The brown spots are the scars left behind from this July 1994 event. Courtesy NASA/Hubble Space Telescope.

Fast-forward 17 years. The US House of Representatives included the following text in their 1990 Authorization Act for NASA:

> The Committee believes that it is imperative that the detection rate of Earth-orbit-crossing asteroids must be increased substantially, and that the means to destroy or alter the orbits of asteroids when they threaten collision should be defined and agreed upon internationally (*15*).

The committee directed NASA to begin this process with two workshops. The first would be used to design a program that would dramatically increase the detection rate of Earth-orbit-crossing asteroids. The second would define systems for altering the trajectory of or destroying those asteroids that pose a danger to life on Earth. The name of this NASA program to increase the detection rate of asteroids is called SPACEGUARD (*15*).

There are many examples of the intimate relationship between science fiction and future scientific endeavors. In the late 1970s when the Space Shuttle was becoming a reality, a name was needed for the first shuttle. It was only an engineering model and would never actually be launched into space. This craft was needed to perform flight tests in Earth's atmosphere. The name chosen for this shuttle was *Enterprise*. The US Navy has had 9 ships previously named *Enterprise* (*16*). One belonged to the US Continental Navy in 1776, another was a *Yorktown*-class aircraft carrier that was the most decorated vessel in the US Navy and one was the first US nuclear powered aircraft carrier. The question is, was this shuttle named after one of these very famous Naval ships or was it named after James T. Kirk's famous starship (Figure 10)?

Figure 10. A photograph of the original production model of the USS Enterprise at the National Air and Space Museum in Washington, D.C. This "Constitution-class" Enterprise starship "boldly went were no man had gone before.". Courtesy Wikimedia Commons.

All of these examples have a common theme: imagination becoming reality. Imagination is needed for entertainment and it's needed for increasing our understanding of the Universe. Hollywood is in the business of entertainment through imagination. They manufacture dreams. These dreams can be comedies, horrors, mysteries, or dramas. Any one of these themes can use space as a backdrop. Those "space movies" we call science fiction and can give us brief glimpses of tomorrow. The closer these stories are to a believable future (or a future we want), the more we celebrate them with dolls, models and posters. At NASA we dream of sending robots to other worlds, finding life on other surfaces and one day, visiting them ourselves. Walking the halls of our laboratory for deep space robotic exploration, one comes into contact with those dreams as they become reality.

...the dream of yesterday is the hope of today and the reality of tomorrow
-- Dr. Robert H. Goddard

Acknowledgments

This work was done as a private venture and not in the author's capacity as an employee of the Jet Propulsion Laboratory, California Institute of Technology.

References

1. Wright, M. Article on Von Braun and Walt Disney: The Disney–Von Braun Collaboration and Its Influence on Space Exploration, 1993. http://history.msfc.nasa.gov/vonbraun/disney_article.html.
2. Biographies of U.S. Astronauts. http://www.spacefacts.de/bios/astronauts/english/tito_dennis.htm.
3. Virgin Galactic. http://www.virgingalactic.com/.
4. International Space Station. Wikipedia. http://en.wikipedia.org/wiki/International_Space_Station.
5. Project Orion (nuclear propulsion). Wikipedia. http://en.wikipedia.org/wiki/Project_Orion_(nuclear_propulsion).
6. Star Wars. http://www.starwars.com/explore/encyclopedia/technology/tie/
7. Deep Space 1. http://nmp.jpl.nasa.gov/ds1/.
8. Hayabusa. Japan Aerospace Exploration Agency. http://www.isas.jaxa.jp/e/enterp/missions/hayabusa/index.shtml.
9. Dawn. NASA. http://www.nasa.gov/mission_pages/dawn/main/index.html.
10. Verne, J. *De la Terra a la Lune*; Pierre-Jules Hetzel: Strasbourg, France, 1865.
11. http://www.nasa.gov/mission_pages/constellation/main/index2.html
12. Clarke, A. C. *2001: A Space Odyssey*; New American Library: New York, 1968.
13. Clarke, A. C. *Rendezvous with Rama*; Harcourt Brace Jovanovich: Orlando, FL, 1973.
14. One Universe. http://www.nap.edu/jhp/oneuniverse/linked_motion_54-55.html.
15. Asteroid and Comet Impact Hazards. http://impact.arc.nasa.gov/downloads/spacesurvey.pdf.
16. USS *Enterprise*. Wikipedia. http://en.wikipedia.org/wiki/USS_Enterprise.

Editors' Biographies

Donna J. Nelson

Donna J. Nelson, Ph.D., is a professor of chemistry at the University of Oklahoma. Nelson researches and teaches organic chemistry and has also conducted research into ethnic and gender diversity among highly ranked science departments of research universities. She has many activities to foster conveying correct science and images of scientists to the public, such as being a science advisor for the television show *Breaking Bad.* She is a proponent of scientific accuracy in television and film. She has received several honors, including being named American Chemical Society (ACS) Fellow, American Association for the Advancement of Science Fellow, Guggenheim Fellowship, and National Science Foundation Special Creativity Extension. Nelson has written more than 100 peer-reviewed publications and books, and she has given hundreds of invited presentations to national meetings of professional societies and organizations, universities, and radio and TV. Additional information is available at http://en.wikipedia.org/wiki/Donna_Nelson#Television.

Kevin R. Grazier

Kevin R. Grazier, Ph.D., a recovering rocket scientist, spent fifteen years on the Cassini/Huygens mission at JPL. His research areas include numerical method development, and large-scale, long-term simulations of Solar System dynamics, evolution, and chaos. Grazier is the science advisor for TNT's *Falling Skies*, Syfy's *Defiance*, and the film *Gravity*, having previously served the same role on *Eureka*, the Peabody-award-winning *Battlestar Galactica*, and several other television series and movies. Grazier also co-wrote *The Science of Battlestar Galactica*, and was an editor and contributing author for *The Science of Dune*, and *Fringe Science: Parallel Universes, White Tulips, and Mad Scientists*.

Jaime Paglia

Jaime Paglia is a film and television writer/producer/director who has developed projects with New Line Cinema, MTV Films, The Canton Company, Akiva Goldsman's Weed Road Pictures, Universal Cable Productions, ABC Studios, TNT, and MGM Television, among others. He is also the co-creator of the record-setting Syfy Channel television series, *Eureka*. His mom enjoyed the show very much.

Sidney Perkowitz

Sidney Perkowitz, Ph.D., Candler Professor of Physics Emeritus, Emory University, has written or edited over 100 research papers, six research monographs, and six books of popular science including *Empire of Light*, *Universal Foam*, and *Hollywood Science*. He writes and speaks nationally and internationally about science in entertainment, science and art, the science of food, and other popular topics, with media appearances that include CNN, NPR, and the BBC. He is on the Board of Directors and blogs for the Science and Entertainment Exchange, National Academy of Sciences. He is a Fellow of the American Association for the Advancement of Science. Website: http://sidneyperkowitz.net/.

Indexes

Author Index

Subject Index

E

F

H

M

N

P

R

S

T

U

W